AGRICULTURAL RESEARCH POLICY

COMPANION VOLUME:

ISNAR Agricultural Research Indicator Series: A Global Database on National Agricultural Research Systems. Philip G. Pardey and Johannes Roseboom. Published by Cambridge University Press. 1989. ISBN 0 521 37368 9

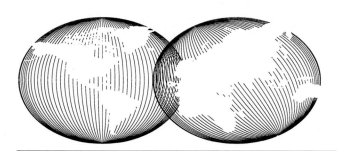

AGRICULTURAL RESEARCH POLICY
INTERNATIONAL QUANTITATIVE PERSPECTIVES

PHILIP G. PARDEY

JOHANNES ROSEBOOM

JOCK R. ANDERSON

Editors

Published for the International Service for National Agricultural Research

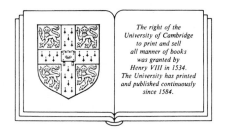

The right of the
University of Cambridge
to print and sell
all manner of books
was granted by
Henry VIII in 1534.
The University has printed
and published continuously
since 1584.

CAMBRIDGE UNIVERSITY PRESS

Cambridge

New York   Port Chester

Melbourne   Sydney

Published by the Press Syndicate of the University of Cambridge
The Pitt Building, Trumpington Street, Cambridge CB2 1RP
40 West 20th Street, New York, NY 10011–4011, USA
10 Stamford Road, Oakleigh, Melbourne 3166, Australia

First published 1991

Printed in Great Britain at the University Press, Cambridge

A catalogue record of this book is available from the British Library

Library of Congress cataloguing in publication data available

ISBN 0 521 40009 0 hardback

Contents

12 Challenges to Agricultural Research in the 21st Century 399
    Vernon W. Ruttan

    12.1 Technology, Institutions, and the Environment 399
    12.2 Biological and Technical Constraints 402
          12.2.1 Advances in Conventional Technology as the Primary
                  Source of Growth 402
          12.2.2 Advances in Conventional Technology Will Be Inadequate 403
          12.2.3 Issues to Be Met over the Next Two Decades 403
    12.3 Resource and Environmental Constraints 406
          12.3.1 An Overview 406
          12.3.2 Emerging Research Implications 408
    12.4 Concluding Perspective 411

    Appendix   National Agricultural Research Expenditure and Personnel
                Estimates, 1961-85 413

    References 423
    Author Index 447
    Subject Index 451

List of Tables

List of Illustrations

Foreword

After the publication of the *ISNAR Agricultural Research Indicator Series* (Cambridge University Press, 1989), the obvious next step was an in-depth analysis of the Indicator Series data as they relate to agricultural research policy. This volume provides such analysis and more.

The editors and the authors of the various chapters have grappled with some of the most important facets of what is an extremely complex subject, and many issues have been raised. Naturally, not all the issues of agricultural research policy have been addressed here, but those that are included have been placed in a quantitative framework to the extent possible. The results of the analyses and the views expressed by the authors are informative and thought-provoking, providing a more solid basis for policy-making than has been available in the past.

This book is aimed at agricultural research policymakers and analysts as well as those who influence present and future policies, be they in government, at universities, at research institutes, or in the donor communities. It will also be of special interest to policymakers within national agricultural research systems who want comparative data of a regional or international nature.

We at ISNAR are confident that this book will become an essential input to research policy development. The potential beneficiaries of agricultural research, of both the present and the new century, urgently require far-reaching decisions and action.

Christian Bonte-Friedheim

Acknowledgements

The completion of this book owes much to the patience and encouragement of ISNAR's past and present management. The seeds of the endeavor were sown by Bill Gamble, were nurtured by Alexander von der Osten, and have come to fruition under Christian Bonte-Friedheim, while Howard Elliott provided a continuity of support throughout the project. Meanwhile, there were changes in the lives of the team that worked on this book, including at least a successful gestation and delivery, a hip replacement, a marriage, and a job change, to mention just a little of the larger history of the project.

The work reported here involved the collection, cleaning, compilation, and analysis of a good deal of data. In this regard we have been ably and cheerfully assisted by Wilhelmina Eveleens, Sandra Kang, and Bonnie Folger. Our persistent demands for information have been well served by ISNAR's library staff, Peter Ballantyne and Sandra Gardner.

Many colleagues at ISNAR, the Universities of Minnesota and New England, the World Bank, sister CGIAR institutes, the TAC and CGIAR Secretariats, and national agricultural research systems offered constructive criticism and suggestions on this material as it devel^ped. We would particularly like to single out Christian Bonte-Friedheim, John Dillon, and Wilhelmina Eveleens who each read and commented on major sections of the manuscript.

This work would not have reached this stage but for the sustained and highly creative endeavors of Fionnuala Hawes on the word processing, layout, and related technological fronts, and Kathleen Sheridan in the editorial and publishing department. We are most grateful for their faith and endurance in the enterprise. Richard Claase's design flair is also well represented.

The on-going financial support of the Italian government for the data base and research policy work at ISNAR is especially acknowledged.

The authors who worked with us through several drafts in order to produce what we feel is a coherent body of work, are also deserving of a final thank-you. We are grateful to ISNAR for providing this forum; naturally, final responsibility for the opinions expressed herein rests with the respective authors.

Philip Pardey, Johannes Roseboom, and Jock Anderson

List of Contributors

Jock R. Anderson	World Bank and University of New England
Barbara J. Craig	Oberlin College
Ruben E. Echeverría	ISNAR
Shenggen Fan	ISNAR and University of Minnesota
Theodore Graham-Tomasi	University of Minnesota
Guido Gryseels	TAC Secretariat - CGIAR, FAO
George W. Norton	Virginia Tech and ISNAR
Philip G. Pardey	ISNAR and University of Minnesota
Carl E. Pray	Rutgers University
Terry L. Roe	University of Minnesota
Johannes Roseboom	ISNAR
Vernon W. Ruttan	University of Minnesota
G. Edward Schuh	Hubert H. Humphrey Institute, University of Minnesota

Introduction

Policy cauldrons have a way of fermenting vigorously even in the absence of applied heat, presumably because public investment in policy analysis obliges activity, even when there is little at stake other than the livelihood of the analysts themselves. Heat is, however, applied from time to time and then the analysts must get really busy.

In the agricultural research policy arena, the close of the century is definitely a period of noteworthy ferment in almost every conceivable dimension of policy discussion. Geopolitical developments are changing the political map of the world and are unlikely to stabilize by the imminent end of the millennium. The old cliches of first, second, and third worlds may soon lose much of their descriptive value as new alliances and priorities emerge.

One thing seems certain. The flow of official development assistance to what have generally been recognized as less-developed countries seems destined to be seriously compromised. Agricultural research assistance has been a small but economically significant part of official development assistance, especially in terms of fostering agricultural productivity and growth. Depending on donors' perceptions and priorities, commitments to this minor but crucial component of official development assistance are thus in question as competing demands on always-limited resources are made.

It was T. W. Schultz (1964, p. vii) who likened the sophistication of those involved in the agricultural policy process to farmers who planted crops according to the phases of the moon. Be that as it may, it is our presumption, perhaps (but hopefully not) naive, that such policy decision making can only be aided by access to better information. Accordingly, the intention in this volume is to move the policy dialogue beyond merely qualitative impressions and toward a process that is underpinned with data — data that are new, cogent, and informative. These data derive from a long-standing ISNAR-based endeavor to describe just what has been happening in the world of agricultural research — and it is this world rather than the related worlds of agricultural extension and technology that we address in this volume.

Many of the data were reported in an antecedent volume (Pardey and Roseboom 1989) in a rather undigested form. What is attempted in this new volume is analysis and interpretation of these data from several contrasting perspectives. Some of these perspectives are fairly predictable; for instance, no serious observer of the policy forces that shape the nature and level of public investment in agricultural research should be surprised to see chapter 1 address the public-good dimension of agricultural research from a political economy perspective.

This public-policy worldview is maintained in chapter 2, which addresses more international dimensions of agricultural research policy and focuses on the "interconnectedness" of the human species through trade and the profound benefits that can, and indeed should, be derived from it. Agricultural research, through its influences in changing resource productivity in a world of diverse resource endowments, plays a vital role in international competitiveness and trade, and thus also in international patterns of growth and development.

If only life could be so simple. Some of the concerns for the environment raised in chapter 2 are taken up in the review of the major contemporary concerns for "sustainability" in chapter 3. The semantics here are anything but settled, but what is certain is that this issue will add to the challenge facing policymakers for decades to come. Concerns over the environment, broadly defined, are variously popular, pressing, and imperative, but what makes them particularly fascinating and challenging is the lack of certain resolution for most of them.

Agricultural research, of course, is no stranger to the lack of certainty in its accomplishments. Analysts typically, and perhaps often quite defensibly, seem to act as if uncertainty did not pervade both the agricultural sector and the research endeavors within it. The purpose in chapter 4 is to indicate when this approach might be appropriate or (for the somewhat rare cases) when something more interventionist may be justified.

The broad context of agricultural research as an ingredient in economic development is taken up in part II. Taking at face value the idea of crawling before we walk, chapter 5 grapples with measurement issues that cannot be dodged in dealing with data from different countries and different data bases at different times under different economic regimes. Not all the answers are entirely happy but procedures that seem "most reasonable" are identified and thus provide the basis for much of the quantitative material that follows. In chapter 6, the preferred procedures (in short, deflate-first and then convert for best international comparability) are used to describe patterns of growth and development for major regions.

This theme is explored in much greater depth in part III, which begins in chapter 7 with detailed regional quantitative descriptions of recently available data. The regions used for this purpose are the much-troubled sub-Saharan Africa, the large and rather successful Asia & Pacific, as well as China separately, debt-ridden Latin America & Caribbean, and the agroecologically challenging West Asia & North Africa. A further "region" is introduced for comparative insight, namely the more-developed countries of the first world, which, in their historical development, surely have many lessons for their later-developing counterparts. The next two chapters in this part take up issues less regional in orientation. Chapter 8 brings new data to bear on several broad policy issues facing national systems, such as how many, what sort, why, when, on what, whom, and so on. The very significant and complementary international initiatives that bear on many of these same questions are addressed in chapter 9, where the evolutionary issues of the CGIAR are especially addressed.

In this day and age, when the issue of private-public balance has virtually joined the rhetoric of the street, no consideration of agricultural research policy would pretend to be adequately complete without due consideration of the roles of the private sector. This is taken up in part IV.

The book is completed with a final chapter that places the subject matter of this volume in the context of the challenges that face agricultural research policymakers as we move forward to the 21st century.

Philip Pardey, Johannes Roseboom, and Jock Anderson

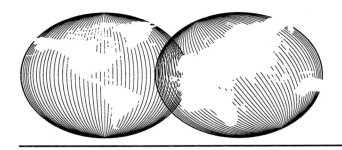

PART I

AGRICULTURAL RESEARCH
IN A POLICY CONTEXT

Chapter 1

Economic Policy and Investment in Rural Public Goods: A Political Economy Perspective

Terry L. Roe and Philip G. Pardey

This chapter focuses on the interdependence between countries' agricultural and foreign trade policy and their ability and commitment to increase the productivity of resources in agriculture. Governments in less-developed countries typically play a pervasive role in their economies. This role may be warranted in agriculture for those cases where markets fail to provide socially optimal levels of agricultural technology, rural infrastructure, education, information, and other services that lower market transaction costs[1] and increase the productivity of land and labor. However, the amelioration of market failure is hampered in less-developed countries for many of the same reasons that cause markets to fail (Stiglitz 1989); these include the lack of human capital and inadequate public infrastructure required to identify and assess the opportunity costs of market failures, and to perform the fiscal and allocative functions required to address them. Under such circumstances "government failure" may be more limiting than market failure (Krueger 1990).

Bates (1983) and numerous others (e.g., Srinivasan 1985; de Janvry and Sadoulet 1989) have pointed out that attempts to address market failure are often exacerbated by the collective action of special interest groups. These groups tend to place pressure on governments to seek their own differential advantage with the unintended effect of taxing others and directing resources away from productive and into unproductive profit-seeking activities (Bhagwati 1982). In the case of less-developed countries, these pressures often result in economic policy having an urban bias (Braverman and Kanbur 1987). For reasons discussed later, governments often respond to these groups by pursuing market-oriented policies as a means of redirecting income flows to those with the strongest influence.

Policies with regard to the trade sector are particularly important. The familiar

---

[1] Nugent (1986) includes in this category the costs of information, negotiation, monitoring, supervision, coordination, and enforcement.

argument is that foreign trade policy determines the degree to which international markets for final goods and services, information, and technology can interact to yield a growth path along which patterns of production, investment, and capacity creation are determined. Numerous policy instruments, including import and export taxes, quotas, and licences, nontariff barriers to restrict the quantity exported or imported, currency exchange rates, and controls of foreign exchange, are used to impact directly upon these markets. These instruments alter domestic resource flows, filter the transmittal of foreign market shocks into the domestic economy, and control the allocation of foreign capital.

Interventions in agriculture include subsidized inputs, farm-gate price ceilings or floors, quota allotments, parastatal enterprises to control the spatial and temporal allocation of commodities, wholesale and retail price ceilings, middleman subsidies for processed staple foods, food stamps, and ration schemes. In other cases, agencies or quasi state enterprises simply transact with wholesalers and/or retailers at prices which are not sufficient to cover their marketing margins. Operating deficits are then usually covered by profits earned from monopoly rights in other markets, transfers from the treasury, or from debt creation.

These interventions have a myriad of impacts on the rural sector of the economy. Frequently overlooked is their impact on (a) a country's fiscal capacity to invest in rural public goods such as agricultural research, infrastructure, and education, (b) a tendency to bias the public allocation of resources toward the modern farm-household subsector relative to the traditional and typically labor-surplus subsector of the rural economy, and (c) an urban bias in the provision of public goods.

The chapter is divided into three main sections. We focus on the direct and then the indirect effects of intervention in agriculture in the first section. The direct effects include the distortions in agricultural prices from interventions within the domestic sector and in the sector's foreign trade markets. The indirect effects come about from interventions that protect the industrial sector and distort the value of a country's currency. The impact of these interventions on fiscal deficits, exchange rates, inflation and real interest rates, and on a country's susceptibility to economic shocks are discussed. We also consider how these interventions tend to limit a country's capacity to invest in public goods such as agricultural research.

The second section focuses on the political economy of economic policy. This section addresses the question: why have countries continued their pursuit of interventions that result in an inefficient allocation of resources and exacerbate adjustments to economic shocks? We draw upon recent literature that Colander (1984) has termed neoclassical political economy. We suggest that, in part, policy is the result of domestic interest groups seeking to achieve outcomes that while socially wasteful, provide them with a differential advantage relative to other groups in society. It is also suggested that insights into this process are important in clarifying the possibilities for realignment of economic policies, from which those dependent on agriculture can reap considerable benefits. We show that, in principle, capital accumulation and technological change in agriculture can play an important role in helping to realign these policies and, in so doing, contribute to agriculture's role as an engine of economic growth.

In the third section we turn our attention to the nature of public investment in agricultural research. We present quantitative evidence on the structure of public support for agricultural research and place that evidence in the political economy context developed in the prior sections of the chapter.

1.1 POLICIES THAT TRANSFER RESOURCES FROM AGRICULTURE

The studies by Balassa (1986) and Mitra (1986) on country adjustments to world economic shocks suggest that countries following policies to maintain internal market distortions in spite of changes in world market conditions experienced slower rates of growth than countries that followed more outward-oriented policies.[2] The countries attempting to maintain internal market distortions can be characterized as having pursued policies of import-substituting industrialization while attempting to maintain abundant supplies of low-cost staple foods to urban centers. Agriculture is an integral part of these policies for reasons discussed later.

Policies to protect the domestic industrial sector typically draw upon many of the same policy instruments that discriminate against agriculture. These include quotas, tariffs, and import licenses. Many of the countries that have high rates of protection for manufacturers also allow imports of raw materials intended for export production to enter duty free (IMF 1985, p. 74).

These policies typically result in a transfer of resources from agriculture through implicit taxation. Capital and currency markets are also used for this purpose. Another source of implicit taxation of the agricultural sector is the tendency to underinvest in the provision of rural public goods such as education, rural infrastructure, agricultural research and extension programs, and so on. The outcome of these policies in the presence of the world market shocks experienced during the mid and late 1970s often became even more punitive as countries experienced trade imbalances, fiscal deficits, and an appreciation of their real exchange rates.

1.1.1 Policy Interventions that Distort Incentives in Agriculture: Evidence on Selected Countries

Policies with Direct Impacts

The extent of agricultural price distortions from interventions in foreign trade and domestic markets is shown for selected countries in tables 1.1 and 1.2 for export and import crops,

---

[2] Of the selected countries studied, Balassa (1986) listed Egypt, Morocco, Philippines, Jamaica, Peru, Tanzania, Indonesia, and Nigeria as pursuing inward-oriented policies.

respectively. These tables are taken from a study at the World Bank in which the senior author participated. The study focused on 18 countries for the period from 1966 to 1984 (Krueger, Schiff, and Valdés 1988).

Table 1.1 shows that the average direct nominal protection rate on the producers of selected agricultural exports was 11% during both the 1975-79 and the 1980-84 periods. Negative rates can be viewed as an implicit output tax in the sense that producers are receiving a lower price for their commodities, and hence lower returns to factors of production, than they would receive in the absence of intervention. Similarly, positive rates

Table 1.1: *Direct, Indirect, and Total Nominal Protection Rates for Exported Products*

Country	Product	Nominal protection rate (1975-79)			Nominal protection rate (1980-84)		
		Direct	Indirect	Total	Direct	Indirect	Total
		%	%	%	%	%	%
Argentina	Wheat	−25	−16	−41	−13	−37	−50
Brazil	Soybeans	−8	−32	−40	−19	−14	−33
Chile	Grapes	1	22	23	0	−7	−7
Colombia	Coffee	−7	−25	−32	−5	−34	−39
Cote d'Ivoire	Cocoa	−31	−33	−64	−21	−26	−47
Dominican Rep.	Coffee	−15	−18	−33	−32	−19	−51
Egypt	Cotton	−36	−18	−54	−22	−14	−36
Ghana	Cocoa	26	−66	−40	34	−89	−55
Malaysia	Rubber	−25	−4	−29	−18	−10	−28
Pakistan	Cotton	−12	−48	−60	−7	−35	−42
Philippines	Copra	−11	−27	−38	−26	−28	−54
Portugal	Tomatoes	17	−5	12	17	−13	4
Sri Lanka	Rubber	−29	−35	−64	−31	−31	−62
Thailand	Rice	−28	−15	−43	−15	−19	−34
Turkey	Tobacco	2	−40	−38	−28	−35	−63
Zambia	Tobacco	1	−42	−41	7	−57	−50
Average		−11	−25	−36	−11	−29	−40

Source: Krueger, Schiff, and Valdés (1988, p. 262).

Note: The direct nominal protection rate $= \left(\dfrac{P_i - e \cdot P_{wi}}{e \cdot P_{wi}} \right) \cdot 100$, while the total nominal protection rate

$= \left(\dfrac{P_i / P_{NA} - e^* \cdot P_{wi} / P_{NA}^*}{e^* \cdot P_{wi} / P_{NA}} \right) \cdot 100$ where P_i is the domestic producer price of a tradable agricultural

commodity i; P_{wi} is the border or world price for the commodity i; e is the official nominal exchange rate; e^* is the equilibrium rate of exchange, i.e., the rate that would equilibrate the current account in the absence of distortions; P_{NA} is the price index of the nonagricultural sector; and P_{NA}^* is the price index of the nonagricultural sector that would prevail in the absence of distortions. See Krueger, Schiff, and Valdés (1988, Appendix) for more details.

can be viewed as an implicit output subsidy. Of the 16 countries, these estimates indicated that only one country subsidized the producers of exports by more than 2% during the first period, while three countries subsidized producers by more than 2% in the second period. The range in the direct tax was from 26% (Ghana) to 36% (Egypt) during the first period and from 34% (Ghana) to 32% (Dominican Republic) during the second period.

These effects are referred to as direct because they exclude exchange rate and price distortions in other sectors of the economy. They were implemented by using a variety of policy instruments. In the foreign trade markets, instruments include tariffs, quotas, export taxes, and subsidies, and in domestic markets, they include farm-gate price ceilings or floors and quota allotments where a proportion of total output must be sold to the government or a parastatal at a given price, plus various marketing and middleman subsidies that serve to alter farm-level prices. Distortions in the prices of agricultural inputs are not included in these measures.

Note the contrast between these results and the level of direct, nominal rates of protection experienced by producers of the import-competing crops reported in table 1.2.

Table 1.2: *Direct, Indirect, and Total Nominal Protection Rates for Imported Products*

Country	Product	Nominal protection rate (1975-79)			Nominal protection rate (1980-84)		
		Direct	Indirect	Total	Direct	Indirect	Total
		%	%	%	%	%	%
Brazil	Wheat	35	−32	3	−7	−14	−21
Chile	Wheat	11	22	33	9	−7	2
Colombia	Wheat	5	−25	−20	9	−34	−25
Cote d'Ivoire	Rice	8	−33	−25	−16	−26	−10
Dominican Rep.	Rice	20	−18	22	6	−19	7
Egypt	Wheat	−19	−18	−37	−21	−14	−35
Ghana	Rice	79	−66	13	118	−89	29
Korea	Rice	91	−18	73	86	−12	74
Malaysia	Rice	38	−43	46	8	−10	58
Morocco	Wheat	−7	−12	−19	0	−8	−8
Pakistan	Wheat	−13	−48	−61	−21	−35	−56
Philippines	Corn	18	−27	−9	26	−28	−2
Portugal	Wheat	15	−5	10	26	−13	13
Sri Lanka	Rice	18	−35	−17	11	−31	−20
Turkey	Wheat	28	−40	−12	−3	−35	−38
Zambia	Corn	−13	−42	−55	−9	−57	−66
Average		20	−25	−5	21	27	−6

Source: Krueger, Schiff, and Valdés (1988, p. 263).

Note: See table 1.1.

The direct subsidy ranged from a 1975-79 average of 20% to 21% for 1980-84. Of the sixteen countries, only four (Egypt, Morocco, Pakistan, and Zambia) were found to directly tax the import-competing crops studied during 1975-79, while six taxed these crops during the 1980-84 period. The remarkable result is the degree to which the domestic prices of importable crops are raised or protected relative to exportable crops.

Policies with Indirect Impacts

The allocation of resources to either agriculture or the urban-industrial sector (e.g., labor, new and selected old capital, credit) depends on the relative rates of return they are expected to earn in these sectors. The returns to the sector-specific factors in agriculture (such as land, land improvements, fixed structures, agricultural technology) and the wealth embodied in them are also influenced by these relative rates of return. The importance of wealth embodied in these factors is frequently overlooked. The value of sector-specific factors affects farmers' incentives to invest in their maintenance (e.g., land improvements). Their value also largely determines the capacity of the sector to obtain credit. Hence, distortions that undervalue these factors also tend to decrease the level of private investment in the sector.

Returns to resources in the urban-industrial sector are influenced by interventions to protect the sector from foreign competition (e.g., import quotas, export subsidies, and tariffs) and by interventions within the sector itself (e.g., subsidized credit, public utilities, and licensing). Currency exchange-rate policy — essentially, the regime used to allocate foreign exchange and to control the flow of foreign capital — can also influence the relative rates of return between the sectors. An overvalued currency can serve to tax the producers of export- and import-competing goods, to subsidize the consumption of imported goods, and to push more resources into the production of home goods. Estimates of the overvaluation of selected country currencies appear in table 1.3.

Trade distortions in the nonagricultural sector were measured by estimating the price index of nonagricultural goods and services that would be expected to prevail in the absence of distortions. The expected value of a country's currency that would prevail in the absence of trade interventions was estimated using an elasticity approach to the supply and demand for foreign exchange. The cumulative effect of these two sources of distortion on the relative nominal rate of protection appears in the column labeled *indirect effects* in tables 1.1 and 1.2. The total or cumulative effects are also reported in these tables.

The glaring outcome of this analysis is that the indirect effects are negative, and for the most part, they tend to dominate the direct effects. The indirect tax on the producers of agricultural exports in the eighteen countries averaged 25% during 1975-79 and between 27% and 29% (depending on the countries included) during 1980-84. With the exception of a single country, the range was from 66% (Ghana) to 4% (Malaysia) in the first period and from 89% (Ghana) to 7% (Chile) in the second period. The result was an average total tax on the producers of selected agricultural exports that averaged 36% in the first period and 40% in the second period.

Table 1.3: *Fiscal Deficit as a Percentage of GNP and Official Currency Exchange Rates as a Percentage of Computed Rates, Selected Countries*

	Argentina		Egypt		Morocco		Philippines		Zambia	
Year	Fiscal deficit relative to GNP [a]	Official currency exchange rate to computed rate [b]	Fiscal deficit relative to GNP	Official currency exchange rate to computed rate	Fiscal deficit relative to GNP	Official currency exchange rate to computed rate	Fiscal deficit relative to GNP	Official currency exchange rate to computed rate	Fiscal deficit relative to GNP	Official currency exchange rate to computed rate
	%	%	%	%	%	%	%	%	%	%
1966-69	-1.9	-20.6	-10.0	-17.1	-4.3	-14.9	-1.2	-1.5	NA	14.3
1970-74	-4.4	-0.1	-7.9	-15.5	-3.0	-14.2	0.2	0.6	NA	11.9
1975-79	-6.9	-5.3	-22.5	-17.7	-9.0	-22.6	-0.5	-4.9	-12.9	-2.1
1980-84	-6.9	-32.1	-21.1	-13.3	-9.0[c]	-20.5	-2.6	-5.8	-11.9	-8.6

Source: Computed from the working papers of the World Bank comparative study of the political economy of agricultural pricing policies, see Krueger, Schiff, and Valdés (1988). Country authors are A. Sturzenegger and W. Otrera (Argentina); J-J. Dethier (Egypt); H. Tuluy and L. Salinger (Morocco); P. Intal and J. Power (Philippines); D. Jansen (Zambia).

[a] A negative sign denotes a fiscal deficit.
[b] The computed currency exchange rate, P_e, is the rate that is expected to prevail in the absence of trade distortions. See Krueger, Schiff, and Valdés (1988) for the general method used to compute these values and Green and Roe (1989) for the procedure used in the case of the Dominican Republic. The percentages represented here measure $((P_o - P_e)/P_e) \cdot 100$ where P_o is the official market rate. A negative sign thus denotes an overvalued exchange rate.
[c] Average for the years 1980-82.

In the case of selected import-competing crops, the total tax on producers was far lower, averaging 5% and 6% during 1975-79 and 1980-1984, respectively. These small negative rates for selected agricultural import crops stand in contrast to the protection rates for import-competing industrial goods. Results from an IMF (1985, table 64) study of 35 less-developed countries found that the rates of protection of manufacturing were often higher than in most more-developed countries. The average effective rate of protection was 50% during 1966-72 and 60% in the late 1970s.

1.1.2   Direct Effects of Distorted Agricultural Incentives

First, in the absence of other distortions and resource transfers to agriculture,[3] the implications of these results for the resources employed in agriculture are clear: (a) the returns to resources in agriculture were, for most countries, more adversely affected by macroeconomic policies and policies pursued to benefit the urban-industrial sector than they were by policies within the sector alone, (b) policy has served to decrease the returns to resources employed in the production of export crops, i.e., in crops where returns to resources are relatively high, compared with the resources employed in the production of import-competing crops, and (c) policy has served to decrease the returns to resources employed in agriculture relative to the urban-industrial sector of the economy.

It is likely that nontraded commodities too are affected by the distortions in traded commodities. For reasons discussed below, the ratio of the price of nontraded to traded agricultural commodities will tend to rise in the presence of an overvalued currency, although the prices of nontraded commodities may fall in absolute terms. Included in this category for many countries are highly perishable commodities such as cassava and, in some cases, livestock products. Consequently, the implicit taxes imposed on traded commodities can be expected to "push" more resources into the production of nontraded commodities.

Second, protection of the industrial sector tends to induce a structure that is capital intensive, with small, relatively high-cost plants that are not able to compete in world markets.[4] Scale economies are limited to the domestic market. As the industrial structure

---

3   Care should be exercised so as to not overstate these results. Since political pressures and the lack of infrastructure often make it difficult for countries to use first-best policy instruments to fund expenditures on public goods, it may be argued that interventions of the form discussed here are the only means available to meet these needs. Indeed, results for the Dominican Republic (Greene and Roe 1989) suggest that government transfers back to agriculture exceed the sum of the effects of direct and indirect transfers out of agriculture in some years. However, while some of the transfers back to agriculture were in support of infrastructure, the largest proportion of transfers supported parastatal marketing firms, food subsidies, and land-reform programs. Hence, it is questionable whether the social profitability of the transfers back to agriculture equaled or exceeded the social opportunity cost of the resources transferred from the sector. See Lipton (1977) and Braverman and Kanbur (1987) for further discussion of government expenditures.

4   See Krueger (1978) and Bhagwati, Brecher, and Srinivasan (1984) for an insightful discussion of

becomes more concentrated and less competitive, agriculture tends to suffer another source of taxation. The intermediate industrial goods it obtains from protected industries (fertilizers, machinery) tend to be of inferior quality relative to those available in world markets and they tend to rise in price, while the price of agriculturally sourced goods sold to domestically protected processing industries (e.g., cotton) tends to fall. Also, technological advances embodied in imported capital and intermediate goods tend to become less available to agriculture as the domestic industrial sector attempts to supply these needs.

In the case of Brazil, for example, Brandao and Carvalho (1989) report that the farm-gate prices of soybeans were lowered by export taxes placed on soybeans to encourage the domestic milling of oil, while Intal and Power (1989) report that the Philippines banned the export of copra to encourage the domestic processing of oil. If inputs to agriculture are subsidized, then part of the burden is passed to the government, although poor quality and problems of timely delivery can be viewed as an increase in the real price of inputs to producers.

Third, prospects of relatively high real wages in urban areas tend to induce a rural-to-urban migration. Off-farm migration may be exacerbated as a consequence of these policies, which tend to draw more resources into the production of nontraded goods produced in urban areas. In spite of the migration into urban areas, the absorptive capacity of urban labor markets is limited because of the capital-intensive industrial structure that import-substitution policies tend to induce. Labor, which for numerous reasons finds it difficult to migrate, tends to get "locked" into agriculture. In the presence of high population growth rates, the absence of technological change, and increased capital inputs, land-labor ratios can decline leading to a decline in the real wage in agriculture (Hayami and Ruttan 1985, table 13-1). These outcomes often create the illusion of economic problems in agriculture when the actual problem lies with the industrial sector of the economy.

Fourth, the narrowing of the marketing margins that intervention in agricultural input and output marketing systems commonly implies often leads to an exodus of the private sector from these activities.[5] Effectively, the public sector assumes many of the functions of resource allocation over time (storage), space (transportation), and form (processing). While these interventions tend to lower temporal variation in prices (Krueger, Schiff, and Valdés 1988, table 3), the result is inefficiency in both public and private resource allocations and the emergence of parastatal fiscal deficits that are eventually funded through domestic resource transfers, money creation, or foreign borrowing.

Fifth, since protection makes the industrial sector appear profitable relative to agriculture, agriculture is forced to compete for resources that are artificially made more expensive. This includes peak seasonal demand for labor and credit. Agriculture must also compete for public investments. If the analysis of the net social value of public investments

---

import-substitution industrialization policies.

[5] For specific examples, see von Braun and de Haen (1983) and Greene and Roe (1989) for the case of Egypt and the Dominican Republic, respectively.

by authorities does not adequately take into consideration the artificially induced profitability of returns to investments in the protected sectors, then public investments in the rural economy, and in agricultural technology in particular, are likely to be less than they would be in the absence of protection.

1.1.3   More Complex and Indirect Effects of Economic Policy

Many additional indirect impacts on agriculture come about through fiscal deficits and other macroeconomic imbalances that seem to accompany the policies discussed above.

Fiscal Deficits

Large fiscal deficits are often associated with both external and internal macroeconomic imbalances that have adverse impacts on agriculture. The results reported in table 1.3, which show fiscal deficits associated with estimates of currency overvaluation for five countries, and those in table 1.4, which show a positive correlation between fiscal deficits and price distortions for the same countries, tend to be typical of less-developed countries pursuing import-substitution industrialization policies.[6]

These policies tend to decrease a country's participation in foreign trade because they lead to a decrease in exports while excess demand for imports is restrained through tariffs, quotas, and other mechanisms to protect the domestic industrial sector and to save foreign exchange.[7] Tax revenues decline because foreign trade taxes tend to be the single most important source of revenue for many less-developed countries.[8] As mentioned, public expenditures are usually required to implement and maintain agricultural price distortions, the maintenance of low real prices to consumers for staple foods, and protection of the industrial sector. With governments often reluctant to alter interventions in the presence of declining revenues, fiscal deficits are the inevitable result.

Fiscal deficits can be financed in several ways; by money creation, domestic and foreign borrowing, and by drawing on foreign exchange reserves, although the latter two forms are commonly used to finance trade imbalances. In the presence of fixed nominal rates, the real value of a country's currency can appreciate as the financing of fiscal deficits (particularly through monetization) generates income flows that in turn increase aggregate

---

6   *World Development Report 1988* shows that public-sector deficits averaged over 23 less-developed countries reached a peak of 8% to 11% of GDP in 1981-82. Deficits, while particularly high for the indebted countries, then declined to a range of about 4% to 6% in 1985.

7   Greene and Roe (1989, p. 290) estimate that the total effects from the removal of price distortions on rice, sugar, and coffee alone would have increased foreign exchange earnings by an average of about 21% between 1974-84.

8   For 86 countries, Tanzi (1987) found that foreign trade taxes accounted for an average of over 30% of total tax revenue. For several countries, tax revenues from imports alone exceeded 50% of total government revenue.

Table 1.4: *Correlation between the Fiscal Deficit as a Percentage of GNP and the Nominal Rate of Protection, Selected Countries*

	Argentina	Egypt	Morocco	Philippines	Zambia
Constant	−2.935	−12.235	−3.624	−0.419	−5.878
	(2.680)[a]	(5.801)	(4.206)	(1.570)	(4.454)
Nominal rate of protection	0.093[b]	0.233[b]	0.146[b]	0.040[c]	0.306[d]
	(0.031)	(0.049)	(0.053)	(0.022)	(0.212)
R^2	0.28	0.51	0.32	0.14	0.21
Degrees of freedom	24	23	16	21	8

Source: See Table 2.3.

[a] Standard error in brackets.
[b] The nominal rate of protection for wheat.
[c] The nominal rate of protection for copra.
[d] The nominal rate of protection for corn.

demand. In the case of Egypt, for example, Scobie (1983) found that a 10% rise in government expenditures decreased the stock of net foreign assets by 1.7%, increased inflation by about 5.3%, and thus led to an implicit appreciation of the Egyptian pound.

Efforts to attract domestic savings in order to finance deficits often decreases the pool of savings available to the private sector. This crowding-out effect is made even more detrimental when earnings on savings deposits are held artificially low to minimize the cost of servicing the public debt. The pool of savings tends to decline while parallel markets for credit emerge that convey higher rates of interest than would likely prevail in a liberalized credit market.

Exchange Rates

Monetary and fiscal policy, foreign trade policy, and direct sectoral interventions of the type discussed above are among the determinants of real exchange rates. But, the form of exchange regime is also important. Most less-developed countries employ some form of controls on capital flows. And, in the case of exchange-control regimes, the nominal exchange rate takes on a more important role as a policy instrument because the domestic price of tradable commodities is cut off from the world price. We consider some of the effects that have a major impact on agriculture.

Take first the possible impact of an overvalued currency on consumer demand for goods and services.[9] In models of the current account (e.g., Krueger 1985, ch. 3), house-

---

[9] For our purposes, an overvalued currency is typified by a situation where the excess demand for foreign exchange is positive, the trade in goods and nonfactor services is negative, and a country must either ration foreign exchange, draw upon international reserves, borrow from world capital markets, or utilize some other means to maintain the trade imbalance that, in any case, may not be sustainable in the longer run. The

holds respond to the increased income streams by increasing consumption expenditures. If increased consumption has no impact on world market prices, i.e., a country cannot alter its terms of trade, the prices of traded goods in the economy remain unchanged. The effect of increased expenditure on nontraded goods, however, gives rise to an increase in their price, i.e., inflation. Consequently, a larger portion of the traded goods produced in the economy is consumed in domestic markets and their production declines as producers respond to rising prices of home goods relative to traded goods.

If exchange controls are maintained, the result is an imbalance in the country's external account and the need to ration foreign exchange through import quotas, licensing, or other means that can induce additional distortions in the economy. While total consumption expenditure is expected to increase, the income of households whose major income is derived from traded commodities tends to decline. The decline in income may be partially offset depending on the difficulty of allocating resources from production of traded to nontraded goods. Since a relatively large share of agricultural output in less-developed countries is traded, incomes of agricultural households invariably decline. Real wages may rise or fall depending on the capital intensity of the traded goods relative to the nontraded sector of the economy.[10] However, rural wages generally decline as it becomes more difficult for labor to find employment in the urban-industrial sector of the economy.

Consider next some expected impacts on the demand for agricultural inputs. An overvalued currency can lower the domestic price of imported capital (such as agricultural machinery and chemicals) relative to the price of domestically produced inputs in much the same way that overvaluation led to the indirect effects reported in tables 1.1 and 1.2; except in this case overvaluation amounts to an input subsidy. These lower relative prices can, in turn, encourage the substitution of imported inputs for domestically supplied inputs. Whether an overvalued currency has actually led to the substitution of capital for labor in agriculture, as has occurred in the industrial sector of many countries, has not, to our knowledge, been documented. Instead, one of the following two outcomes would seem to be more common.

First, in the presence of import-substitution industrialization, it seems likely that the existence of trade barriers, when combined with the need to ration foreign exchange, would limit the importation of capital. Thus, in spite of the artificially lower prices of imported goods induced by overvalued exchange rates, capital is unlikely to be available in quantities that would replace agricultural labor on a large scale. Nevertheless, those with special

---

real value of a currency under conditions of no excess demand (i.e., floating rates) may still be overvalued in the sense that the opportunity cost of resources employed in home goods production is low relative to the actual opportunity cost that would prevail in the absence of interventions that distort exchange markets. In part, the problem lies in determining the value that would prevail in the absence of distortions.

10 Models of the capital account suggest that overvaluation can lead to a decrease in foreign investment, low real interest rates, and capital flight as households prefer to place more savings in foreign assets. See Krueger (1985, ch. 5) for a discussion of exchange-rate regimes and the incentives provided to the holders of assets.

access to foreign markets through licenses or such may overemploy imported inputs and thus give rise to modern and traditional farms within the sector.

Second, as indicated, the rationing of foreign exchange requires some type of import licensing or control regime. Firms obtaining these rights often earn substantial rents from reselling the imported goods in domestic markets. In this setting, the prices of both the imported and domestically supplied inputs will have a tendency to rise, thus increasing production costs of purchased inputs. It is likely that exchange rate and import controls of the form discussed here will not enhance the capital deepening in agriculture relative to the deepening that would be expected to prevail in more outward-oriented economies.[11]

Inflation and Real Interest Rates

If the nominal exchange rate were permitted to adjust to the increase in demand for foreign exchange, the currency would depreciate and aggregate real income would tend to decline as the domestic price of traded goods increased relative to home goods prices. However, if this adjustment were permitted to occur, the import-substitution industrialization policies would be at least partially undone as the negative protection to producers of traded agricultural commodities and positive protection to consumers of these commodities relative to the industrial sector (i.e., the indirect effects) would decline.

Hence, governments often respond by some form of exchange controls that serve to fix or peg the nominal value of the currency. In this situation the rise in the price of nontraded goods leads to a further appreciation in the real value of the currency. This rise in value contributes further to the adverse indirect effects on agriculture and the need to strengthen measures to protect the industrial sector from foreign competition.

Since the increase in the rate of inflation depends on a number of factors, the experience among countries varies considerably. A major difference among less-developed countries in Asia and in Latin America is that Asian countries tend to limit fiscal deficits to their ability to finance government in a noninflationary manner. In their study of this contrast, Dornbusch and Reynoso (1989) found that high inflation in Latin America was related to the indexation arrangements that link current to past inflation. The other difference was that a significant amount of the budget deficit was financed by money creation. Households tended to respond by protecting their assets from inflation and possible future devaluation through capital flight and dollarization. These efforts largely removed this source of savings from investment in the domestic economy. Firms were found to react by forestalling investments and holding paper assets rather than investing in real resources. Dornbusch and Reynoso (1989, p. 209) concluded that "... the scope for deficit finance as an engine of economic growth is extremely limited and extraordinarily hazardous."

---

[11] Households' protection of money assets from macroeconomic imbalances tends to make these assets unavailable to industry that might otherwise borrow and use them to enhance the technology embodied in inputs supplied to agriculture.

Controls on domestic capital markets are also coincident with import-substitution industrialization policies. Interventions in domestic capital markets have given rise to negative real interest rates in part because nominal rates remain fixed during periods of high inflation. At various times during the 1970s, negative real interest rates were particularly large in Brazil, Ghana, Jamaica, Nigeria, Peru, and Turkey (*World Development Report 1985*). The experiences documented by Balassa (1984) for Brazil and Turkey and by Corbo, de Melo, and Tybout (1986) for the Latin American countries of the Southern Cone support the view that the effects of below-equilibrium real interest rates tend to lower the efficiency of investment by discriminating among capital users, to favor the application of capital intensive techniques, to discourage domestic savings, and to encourage capital outflows.

Susceptibility to Economic Shocks

Countries pursuing economic policies of this nature tend, as the external debt crises of many countries illustrate, to be more susceptible to shocks to world markets, i.e., they are unable to sustain their policy-induced distortions. Their response to shocks has typically been money creation, an increase in arrears, and still further increases in the level of distortions until illiquidity forces an abrupt policy liberalization effort.

The Dominican Republic's response to world market shocks tends to be typical of countries pursuing import-substitution industrialization policies. The county's response to the 1973/74 shock was to forestall adjustments because the demand on foreign exchange earnings associated with food grain and petroleum imports was partially offset by the rise in earnings from sugar exports (Greene and Roe 1989). Fiscal and trade deficits were met by money creation and external debt, which further contributed to inflation, overvaluation in the real value of the country's currency, and many of the other distortions mentioned. The 1979/80 shock to world markets precluded the continuance of these policies as debt restructuring and rather abrupt policy liberalization were required to maintain foreign exchange liquidity.

Since it is politically difficult to reduce current spending in the short run, the adjustment pressure is often shifted to capital spending (Tanzi 1986). This would be appropriate if only unproductive investment projects were eliminated. Instead, fiscal austerity often results in budget cuts to education, infrastructure, and the numerous other areas where markets fail to optimally allocate resources. The rapid changes in policy that were required when the distortions precipitated a liquidity crisis have almost always given rise to declining real incomes, increased unemployment, a decline in the quality of diets, and a deterioration in the quality of health and increased infant mortality (Pinstrup-Andersen 1988a, b).

The IMF (1989a, p. 25) reports that those countries that continue to face persistent external financing constraints tend to respond by curtailing public investment and resorting to at least partial monetization of the deficit. The result has been an acceleration of inflation and a deterioration in growth prospects. Countries that have fared reasonably well tend to be characterized as those that have pursued efforts to improve economic efficiency through trade liberalization and to promote growth, while at the same time undertaking fiscal reforms to broaden their revenue base and target public expenditures to areas that promote market efficiency. In the case of agriculture, this includes education, rural electrification, roads, agricultural research and extension programs, and so on. Fiscal reforms include broadening a country's tax base while at the same time increasing local capacity to generate revenues that can be allocated to the maintenance of rural infrastructure.

Additional Impacts of Macroeconomic Imbalances on the Rural Economy

The adverse impacts of the macroeconomic imbalances discussed above also tend to compound the impact of distortionary sectoral policies mentioned in the previous section. To these lists can be added the following.

First, it bears reemphasis that fiscal deficits give rise to a tendency to underinvest in areas where markets fail. Underinvestment in these areas is particularly deleterious to agriculture since, as is well known, the efficiency with which labor, purchased inputs, and output markets function in rural areas is particularly dependent on access to educational opportunities, market, and technological information, production technology, capital markets, and the level of spatial costs. Furthermore, Elias (1985) and Binswanger et al. (1987) suggest that public-sector investments in these areas induce private-sector investments as well, so that supply becomes more elastic to output price changes and less elastic with respect to changes in the price of an input. In other words, the brunt of adjustment tends to be spread over more inputs.[12] Herein lies an important source of economic growth for agriculture.

Second, through a combination of price distortions and macroeconomic imbalances, both the demand for and supply of agricultural technology can be altered. Not unlike the industrial sector, the agricultural sector can be launched on a growth path that cannot be sustained when policies are liberalized, and it is the sector most unlikely to attain its potential level of economic efficiency so that it can be competitive in world markets.

In the case of a single commodity, producers have an incentive to adopt a cost-reducing technology even though its price is distorted downward relative to industrial-sector prices. In the case of multiple commodities, the producer has the incentive to adopt the technology that maximizes expected net profits. As mentioned, distortions have a tendency to raise the price of nontraded and import-competing commodities relative to exportable commodities. All else being constant, producers will have a tendency to adopt technologies in the production of nontraded and import-competing commodities instead of exportable commodities. This will tend to occur when the expected gains in net profits from the effects

---

[12] This is the point of Mundlak (1985) that agricultural supply response to price occurs through capital accumulation in the rural sector and that technological change is central to that process. Since, as we maintain here, price distortions, macroeconomic imbalances, and fiscal deficits are symptoms of the same policy, the debate between getting prices "right" as opposed to investments in rural education and infrastructure (e.g., Delgado and Mellor's [1987] reply to Schiff [1987]) seems somewhat misdirected.

of their rising relative prices, plus technological cost savings, exceed those of the exportable commodity for which policies are inducing a decline in its relative price. Hence, in a distorted economy, there is a tendency for the adoption of technology in the production of commodities that the country may not have a comparative advantage in producing.

The induced-innovation hypothesis suggests that producers have an incentive to adopt those technologies that save on the relatively scarce factors of production. Sectoral and macroeconomic policies can induce the adoption of nonoptimal technologies to the extent that these policies have an impact on relative input prices so as to disguise the factors that are actually in relatively scarce supply.

Many of the factors that misdirect the demand for technology can also misdirect its supply. To quote de Janvry, Sadoulet, and Fafchamps (1987, p. 14), "If ... the state is equally responsive to market signals in the delivery of public goods as are private agents, the technology induced in public research institutions for one particular product will be uniquely determined by relative factor prices, the size of the research budget, and the state of scientific knowledge." If such is the case and output and input prices are distorted, then the state can be led to the production of technology that, in the absence of distortions, is nonoptimal for the same reasons that producers make nonoptimal choices. But, there are additional factors.

If there are declining marginal productivities in the allocation of research budgets to the discovery of technological advances, de Janvry, Sadoulet, and Fafchamps (1987) find that, as the budget increases, technological advances tend toward neutrality. However, since the policies mentioned are invariably associated with fiscal deficits, the opportunity cost of an additional unit of public revenue occurs at a higher cost as the deficit increases.[13] Hence, governments in these environments are likely to underinvest in agricultural research and, for that matter, rural infrastructure. De Janvry, Sadoulet, and Fafchamps (1987) suggest that, when research budgets are such that they do not capture sizable economies in research discovery, the supply of technology produced is more likely to be factor biased. Now, in the presence of market failures, in particular credit constraints, technological change will be biased and will tend toward mechanical innovations where average farm size and/or inequality in land distribution is greater. If, in addition, the economies of collective action[14] favor those with access to more resources, technology may be even more skewed to the saving of factors of production that are scarce to this special interest group, notably labor-saving land-using technologies.

Together, these policies can launch the sector along a growth path that cannot be

---

[13] This occurs because, in the presence of second-best tax instruments, the rate of deadweight losses increases as distortions increase. Or, from another perspective, if the policies observed are consistent with the government's optimization of its view of society's social welfare function, then an additional unit of tax revenue can only serve to lower the value of its function. The larger the deficit, the larger will be the shadow price of a unit of additional revenue allocated to agricultural research.

[14] The concepts of collective action are discussed in more detail in the next section.

sustained when policies are liberalized. When liberalization occurs, the sector will tend to face larger adjustment costs as the rewards to learning, experimentation, adaptation, and other activities associated with technological adoption during the period of economic distortions are largely lost. The greater the extent to which the growth path diverges from the path associated with liberalized policies, the greater the adjustment costs are likely to be when policies are liberalized. Also forgone are the technological advances that could have occurred in the production of exportable commodities and the development of the institutions supporting them. The forgoing of these advances can serve to lessen a country's comparative advantage in those exportable commodities where it was formerly competitive.

Third, as Srinivasan (1985), Bates (1983), and others have pointed out, the presence of distortions implies that selected producers and consumers earn rents that will disappear if an economy is liberalized. Essentially, the pursuit of these policies tends to redirect income flows and to filter the effects of adverse world-market conditions from the special-interest groups that hold, relative to others, more political influence. Consequently, it is natural and rational for these groups to resist policy liberalization and structural adjustment. Thus, policy adjustments become "sticky" and can easily be undone, as the experience in the Southern Cone of Latin America illustrates.

In the next section we address several questions, including the central one: why have countries persisted in their pursuit of interventions that yield an inefficient allocation of resources and exacerbate adjustments to external shocks? Another related question is: are the mentioned interventions the result of policy mistakes? But, if this were the case, why have countries failed to learn from these mistakes? Insights into these questions are important in order to realign economic policies (as opposed to realignment coming about through a liquidity crisis) and to induce efficient economic growth in agriculture.

1.2   SOME POLITICAL ECONOMY DIMENSIONS OF ECONOMIC POLICY

An overview of the contemporary political economy literature is provided in the introduction to Colander (1984) and by Srinivasan (1985). A component of the literature that we draw upon in this section falls in the category of neoclassical political economy. The key strands of this literature are distinguished by those from political science, typified by Bates (1983) on the behavior of governments in East Africa and by Olson (1982) on distributional coalitions and the free-rider problem. Contributions have evolved from the public choice school (e.g., Buchanan 1980) and the field of trade and development where emphasis is placed on rent seeking (Krueger 1974) or, as Bhagwati (1982) has suggested, on directly unproductive profit-seeking (DUP) activities. The various approaches have focused on questions of tariffs versus quotas as rent-seeking instruments (Bhagwati and Srinivasan 1980) and on rent-seeking and rent avoidance (Applebaum and Katz 1986). Extensions of these approaches have amounted to a broadening of the channels through which agents can influence economic policy. Examples include the presence of regulators of policy instruments that can induce rent seeking (Applebaum and Katz 1987) and the presence of both voting and lobbying behavior (Young and Magee 1986).[15]

An important contribution of this literature is the recognition that it is rational for individuals to allocate resources (e.g., to lobby, protest, vote, or engage in other forms of collective action) in ways that promote their self-interest in the policy process. Hence, public policy can, in part, be viewed as the outcome of various political pressures exerted by members of the domestic economy seeking their own self-interests. This behavior should not necessarily be viewed as undesirable. It often adds to the social good as in the case of local, state, and national governance. Collective action is important when it is focused to resolve problems where markets otherwise fail to allocate society's resources optimally. However, this process frequently fails too when individuals and coalitions, seeking their own self-interest, lobby public authorities for purposes of implementing policy to alter income streams in their favor but to the disadvantage of others.[16] The empirical evidence and policies discussed previously clearly fall into this category.

Among the key insights of this literature are the following: (a) while it is in an individual's own self-interest to engage in collective action for purposes of influencing policy outcomes, this action can be socially wasteful of resources, and (b) the ability to influence policy decisions by some groups in society is greater than others for reasons that relate to the cost of coalition formation, their willingness to pay more to influence policy outcomes, and, of course, the fact that institutions can give unequal access to political authority.

A general overview of some of the forces motivating policy is presented next. Since these forces depend on pressures exerted by special-interest groups, we focus more narrowly in the second section on the factors that motivate these groups to expend resources to influence policy.

1.2.1   An Overview of Some Forces Motivating Economic Policy

When viewing policy choices from an *ex post* point of view, the motivation for the policies discussed in the previous section might be sketched as follows. Since food is a wage good in many countries,[17] policies to lower food prices amount to an increase in real wages and, hence, are an important benefit to (food-deficit) households that do not produce food in excess of their consumption. The interests of urban consumers thus tend to coincide with those of domestic industrialists who view low-priced food as serving to decrease the pressure on nominal wages.

In the case of less-developed countries, Bates (1983, p. 169) argues that urban

---

[15] In the spirit of the latter contributions, Roe and Yeldan (1988) developed a general equilibrium model in which coalition formation, rent seeking, and the government's choice of price policy are endogenous.

[16] See Roe and von Witzke (1989) for a more in-depth discussion of these issues.

[17] That is, food expenditures are a proportionally large component of the consumer price index in low-income countries.

consumers are potent pressure groups demanding low-priced food. They have political influence because of their geographical concentration and strategic location. They can quickly organize and they are largely employed in providing public services, so they can, with relative ease, impose deprivation on others. Bates (1983) notes that urban unrest forms a significant prelude to changes of governments in Africa, as indeed it has elsewhere. Industrialists are also effective in obtaining protection from imports because of the notion that the key to development lies in industrialization. Furthermore, since industrial goods account for a small share of most such households' budgets, import protection of industrial goods will not have a large direct impact on the expenditures of most households. The common interests of these groups suggest that, in the short run, they can benefit from policies that support both import substitution and low-cost food, albeit at the costs discussed in the previous section.[18]

The outcome of policies that discriminate against agriculture in favor of import-substitution industrialization tends to make agriculture poor relative to the urban industrial sector and to decrease a country's foreign exchange earnings. Consequently, rural opposition to these policies tends to increase over time relative to urban industrial support for them. Often, the response is an attempt to save foreign exchange by import substitution in agriculture (recall the results in table 1.2), to subsidize agricultural inputs, to raise farm-level prices by subsidizing the marketing margin for food staples, and a number of other measures. However, the pressures exerted by urban groups are still present. Hence, these measures are often pursued while maintaining both implicit subsidies to consumers and the import-substitution industrialization policies mentioned above.

Corresponding arguments apply to more-developed economies but with opposite consequences.[19] In advanced stages of development, the food share of the consumer's budget declines so that consumers become less sensitive to increases in food prices. Agriculture becomes a smaller component of the total economy and farmers tend to be more specialized. Within their area of specialization, they are better able to organize than are urban groups. Moreover, with food a small share of consumer's expenditures, protection demands in agriculture can be met at lower political cost, with the result that the agricultural sector receives more protection relative to the industrial sector.

These arguments provide insights into the motivation for interventions, but why do governments seem to prefer implicit ways to transfer income and to intervene in markets that perform relatively well if left alone when they could accomplish the same objectives

---

[18] Olson (1982, pp. 203-205) suggests that narrowly based coalitions tend to be more interested in the distribution of society's income to members of the coalition and tend to externalize the cost of this action. Pryor (1983 and 1984) attempted to obtain empirical support for the overall implication of Olson's theory, namely that economies characterized by broadly based coalitions should outperform economies characterized by narrowly based coalitions. Pryor (1984, p. 174) concluded that "... Olson's theory is formulated in a manner still too general to prove successful in the empirical tests ..."

[19] See Anderson and Hayami (1986) for a discussion of the political economy of agricultural protection in more-developed economies.

by the use of more direct policy instruments such as land taxes?

Aside from the institutional inability to carry out the implementation and management of first-best policy instruments, Bates (1983) argues that market interventions facilitate the allocation of political rents. Market interventions permit governments to target the allocation of subsidies through control of marketing functions while, at the same time, transferring resources to supporters (civil servants) engaged in carrying out these interventions. For example, the construction of a bridge or the provision of public education yields a service available to all when the same public resources may be redirected to yield a larger benefit to a few. In Bates' (1983) terminology, market interventions facilitate the "organization of the rural constituency," which supports the government, and "disorganize the rural opposition."

Schuh (1983, p. 296) suggests that governments prefer implicit subsidies and taxes (such as those provided by import licenses and overvalued exchange rates) because they tend to be less observable to the body politic except, of course, to those who tend to receive direct benefits. In support of this notion, Greene and Roe (1989) found that the Dominican rice producers' association was well organized and effective in lobbying the ministry of agriculture to obtain farm-gate prices in excess of border prices. However, the countervailing efforts of lobby groups that tended to be urban-based were successful in lobbying for subsidized energy, industrial trade protection, wheat imports at overvalued official exchange rates, and so on, with the end result that the indirect effects reported in table 1.2 dominated the direct effects of protection. It did not appear that rural lobby groups were aware of the implicit taxes being imposed upon them.

In the next section, we focus more closely on the economic factors that motivate individuals to influence economic policy.

1.2.2   The Economics of Collective Action

It is useful to structure this discussion by drawing on a model[20] of rent seeking and coalition formation in an open two-sector economy that produces an export and an import good.[21] The conceptual framework underlying the analytical result reported in table 1.5 contains two parts. The first is a model of a small open economy composed of rural and urban households. Rural and urban households choose levels of food (or rural goods) and nonfood (or urban goods) to consume.[22] Rural households also choose the amount of labor allocated

---

[20] For an earlier attempt to cast the support for agricultural research in a political economy framework see Guttman (1978) and, more recently, Gardner (1989), and de Gorter and Zilberman (1990).

[21] Since interventions that benefit one sector of an economy can implicitly tax other sectors, as the indirect effects reported in tables 1.1 and 1.2 show, the political economy of economic policy is more insightful when viewed from such a general-equilibrium perspective. This is the approach followed by Roe and Yeldan (1988). The version of the model that is the basis for this discussion is summarized in the appendix to this chapter.

[22] This model takes food goods to be synonymous with rural goods and non-food goods to be synonymous

to the production of food, the amount of land to rent in or out, and the amount of labor to hire or the amount of time to work off the farm, given their levels of labor and land endowments. Urban households choose the amount of labor to allocate to the production of nonfood, the amount of plant and equipment to rent in or out to other urban households, and the amount of labor to hire or to work outside the sector, given their levels of labor, plant, and equipment endowments. Market failure is captured by the presence of a rural and an urban public good that are both supplied by the public sector. These goods may be roads, electrification, and other activities such as research and extension which increase the efficiency of production. To this point, the model of the economy is in the neoclassical tradition.

A departure from the neoclassical tradition comes about when we accept the premise that, if policy impacts on the welfare of households, then it is rational for households to allocate resources to influence policy in much the same way they allocate resources to produce income or to buy goods and services. Hence, the second component of the framework is to allow for the formation of coalitions of households that have similar interests — for instance, farmers as one coalition and urban labor as another. This construction entails what Becker (1983) has referred to as the production of political pressure which, at the level of national policy decision making, gives rise to influence. Political pressure is produced by households allocating resources (in our case, labor for lobbying activities) to organize local groups with similar interests which then place pressure on their local representatives or decision makers in, for example, a ministry such as agriculture. Since this pressure may give rise to higher food prices, urban households can countervail these efforts by allocating resources (labor) to place pressure on authorities for lower food prices.[23]

The next step is to assume that governments act as though they form (possibly differential) preferences over the utility levels or well-being of the rural and urban households in the economy. However, these preferences are not exogenous, they depend on the political pressures generated by the rural and urban lobby groups. Hence, at the national level, these pressures are amalgamated to produce an influence that can change, at the margin, the preferences governments hold for rural relative to urban households. To stay in power, the government is assumed to choose policy (in our case, the price of food relative to the urban good, and the level of rural and urban public goods) that makes those the happiest that have the greatest political influence, subject to the structure of the economy (as described by the economic model above).

Clearly, this structure is a fairly gross simplification of reality. We feel, nevertheless, that it captures the stylized facts of political economy and provides rich insights into how,

---

with urban goods. Either good may be exported or imported.

[23] Thus, in this model lobbying is broadly defined to include those activities that serve to influence public authorities. Another approach is to specify a sector of the economy that produces lobbying services from labor and capital for a fee that is paid by special interests. This structure tends to complicate the framework without providing materially different insights.

Table 1.5: *The Rural Household's Decision Rule for Determining Its Willingness to Pay to Influence Economic Policy in Its Favor*

Term 1	Term 2	Term 3	Term 4	Term 5
$(q_{pr} - q_{cr}) [\partial p / \partial l_r]$	$+ (\partial \pi_r / \partial G_r) (\partial G_r / \partial l_{gr}) [\partial l_{gr} / \partial l_r]$	$-w$	$+ (\bar{L}_r - l_r - l_{qr}) [\partial w / \partial l_r]$	$+ (\bar{x}_r - x_r) [\partial c_r / \partial l_r]$
The product of household production (q_{pr}) less consumption (q_{cr}), i.e., marketable surplus; and the change in price (p) from lobbying (l_r).	The product of the household's shadow price of the public good (G_r); the marginal product from allocating labor (l_{gr}) to produce the good; and the change in the amount of labor allocated to produce the good in response to a change in lobbying effort (l_r).	The wage rate, an endogenous variable.	The product of household labor endowment ($\bar{L}_r$) less the labor allocated to lobbying (l_r) and to production (l_{pr}); and the indirect effect of the change in wages (w) to lobbying level (l_r).	The product of the household's endowment of the sector-specific factor (e.g., land, $\bar{x}_r$) less the amount of the sector-specific factor allocated to the production of output (q_{pr}); and the change in the market price (c_r) of the sector-specific factor to lobbying level (l_r).

Note: The corresponding rule for urban households differs in the first term and, of course, the subscripts. These results are based on a two-sector, open-economy model of a small economy with public goods augmenting production (see appendix A1.1).

for example, technical change within the agricultural sector can increase the willingness of rural households to allocate more resources to the political process so that their interests are better represented in national policy.

The household decision rule which is derived from our model appears in table 1.5. The terms in brackets [·] reflect the household's view of the influence that an additional level of lobbying, on the part of the coalition to which it belongs, will have on the government's choice of the two policy instruments, the domestic price (p) of the export good relative to the price of the import good, and the amount of labor (l_{gr}) allocated to the production of the rural public good (G_r). The household is assumed to treat the lobbying level of the urban-based coalition as given and to ignore the effect of its lobbying level on any taxes that might need to be paid to prevent the government from running fiscal and external imbalances.[24] Essentially, the rule is one of equating the marginal returns from lobbying to marginal costs.

Term 1 is the difference between the household's production of the rural good, e.g., food (q_{pr}), and its consumption of food (q_{cr}). This difference is the household's marketable surplus. If the household produces in excess of consumption, $q_{pr} - q_{cr}$ is positive, and its lobbying efforts result in an increase in the price of q_r relative to the urban good (i.e., $\partial \pi / \partial l_r$ positive), then the household realizes a gain from lobbying. If the household is in food deficit, i.e., $q_{pr} - q_{cr}$ is negative, then it would experience a loss, all else being constant, when additional lobbying results in an increase in relative prices, p.

This result has several implications. First, it suggests that the more specialized the household is, the more willing it is to allocate resources to influence policy or, equivalently, to counteract the lobbying of others. Households with access to more resources than others (q_{pr} large relative to q_{cr}), all else being constant, are more willing to influence economic policy. Second, the availability of a cost-reducing technology also tends to increase the household's willingness to influence policy. Effectively, a cost-reducing technology tends to increase the household's market surplus and, thus, the returns to a marginal increase in the resources allocated to lobbying activity.

When food is an important component of household expenditures, the marginal cost of a price increase, given by the product $q_{cr} \, \partial p \, / \partial l_r$, amounts to a relatively large increase in expenditures on food. Then, there is a tendency to either lobby for a decrease in price or to allocate fewer lobbying resources to increase price. Hence, the result is also consistent with the observation that in countries where food accounts for a relatively large share of disposable income, political pressures tend to favor cheap food policies. Typically, in the process of development, marketable surpluses ($q_{pr} - q_{cr}$) increase, while at the same time the proportion of income spent on food decreases and the proportion of income spent on

---

[24] Hence, this is a Cournot-Nash game where behavior that is rational to an individual may result in an outcome where all households end up worse off (i.e., a prisoner's dilemma result). If households are aware of the impact of lobbying on their taxes, the results are largely unchanged except that lobbying tends to be reduced by the marginal change in taxes.

industrial goods increases. Then, rural households are more willing to influence a policy that favors the rural good while urban households tend to be less willing to influence food policy since less of their income will be affected by the lobby resources allocated for this purpose. Hence, this result is consistent with the observation made above that in more-developed countries, where food is a small component of expenditures and q_r is large, pressures tend to favor policies that subsidize food production.

The household's lobbying efforts can also influence the government's choice of the level of the public good (G_r) to supply to the rural sector. The $\partial \pi_r / \partial G_r$ component of term 2 is the marginal product, or shadow price, of the public good (e.g., agricultural research and extension, rural education, infrastructure, etc.) to the rural household; it is expected to be positive, as is the second component $(\partial G_r / \partial l_r)$. This component is the marginal physical product from the government's allocation of labor (l_{gr}) to the production of the rural public good (e.g., agricultural scientists, extension specialists, and so on). Hence, more lobbying, all else being constant, can increase the production of this public good.

This is the social-good side of the lobbying process. The larger these two components of term 2 are, the more willing the household is to influence government to produce the public good. Put another way, the more efficient the government is in producing the public good and the more important the public good is to increasing the production of private rural goods (q_r) then the more willing rural households will be to lobby for its supply to the rural sector. Consequently, all of society can benefit from this effort; lobbying can be a social good in this case.

The efficiency by which publicly sponsored research efforts generate new knowledge and new technologies that enhance agricultural growth and productivity is in turn influenced by a host of factors. The design, operation, and management of efficient agricultural research institutions is necessary, but far from sufficient, to ensure that scientists face a structure of incentives that promotes the cost-effective development of new, highly demanded, technologies. It requires maintaining an optimal balance among various dimensions of a research program including its commodity, site, and technology emphasis. This involves allocating between long- versus short-run research, site- versus nonsite-specific research, and the like. More broadly, attention must also be given to designing mechanisms that efficiently match the (technical) ability of a research system to supply new technologies with the (economic) forces that shape the demand side of the research-technology transfer equation.

To see further how the lobbying process might benefit the rural sector when policies discriminate against agriculture, recall the point made earlier that the provision of public goods tends to induce additional private-sector investments. An example of this interdependence appears in the results obtained by Antle (1983). He found that the density of roads alone has elasticities of 0.12 and 0.20 for aggregate output and a strong effect on the demand for fertilizer and tractors. These investments increase the long-run elasticities of aggregate agricultural supply. In the short run, the direct price elasticity of aggregate supply is inelastic, ranging from 0.05 to 0.25 (Binswanger 1989, p. 233). Peterson (1979) estimates that long-run elasticities of aggregate supply range from 1.27 to 1.66. These higher elasticities in part reflect the process of capital deepening discussed by Hayami and Ruttan (1985, ch. 6) and is a key element to the economic growth of the agricultural sector. Our model suggests that, since the market cannot be expected to provide socially optimal levels of public goods, political pressure by rural sector special interest groups can play an important role in their provision. Moreover, rural political pressure that succeeds in persuading government to invest in rural sector public goods can, in turn, induce private rural sector investment.

The model suggests that public investments further increase the rural sector's wealth and hence its willingness to pressure government on its behalf. The urban sector can also benefit. But, its benefits from the provision of rural public goods tends to be more indirect, especially for traded agricultural goods. In a small open economy these indirect benefits come about from an increase in foreign exchange earnings that can be used to finance intermediate capital goods, a large share of which is demanded by the urban industrial sector and, perhaps more directly, from the opportunity to process, store, and transport a larger volume of agricultural goods.

Of course, the provision of public goods does not yield instantaneous increases in production as the model assumes. In the case of Argentina, Cavallo's (1988) results suggest that ten years may be required to increase the aggregate elasticity of supply from its short-run level of 0.07 to 0.71. The risk faced is that in response to political pressures, a felt need to generate a quick response may result in market interventions (e.g., subsidized input and output prices) instead of investment in public goods, with the deleterious effects mentioned previously.

Not depicted in table 1.5 is the free-rider problem that can increase the cost of lobbying for the public good. Since the provision of the public good yields benefits to all, including those who do not lobby, there is a tendency for some to free ride, thus spreading the lobby cost over fewer households.[25] Higher costs to the remaining households are expected to decrease the level of resources they are willing to commit to lobbying, with the possible result that little effort will be made to lobby for the public good. In this environment, the household is faced with choosing the alternative that yields the higher return to its lobbying resources. Fewer resources may be required to organize a lobby group that shares common interests in, for example, rice production, than to organize a group to lobby for a public good whose benefits are spread more broadly. Rice producers may find that their lobby resources yield higher returns when they lobby for an increase in the price of rice than if they were to allocate the same resources to lobby for an increase in the provision of a public good. Hence, a situation could arise where political influence yields a distortion in markets and an underinvestment in a public good that would benefit all.

For many less-developed countries, import-substitution industrialization policies

---

[25] Becker (1983) depicts the free-rider problem as the cost of organizing local groups to place pressure on political authority. As the number of households increase, the cost of free riding also increases.

appear to have led to an overinvestment in public goods in urban centers relative to the rural sector of the economy. Braverman and Kanbur (1987, p. 1180) argue that the urban bias in many countries is reflected in the pattern of government expenditures. They note that government expenditures are not typically directed to rural infrastructure but, instead, to nontraded services targeted for the urban sector and often supplied by state or quasi-state enterprises. This tendency is partially reversed in more-developed economies.[26] Thus, the evidence, though sketchy, seems to suggest that those who have been successful in lobbying for market distortions that benefit them, have also been successful in lobbying for public goods.

Term 3 is the opportunity cost (wages, w, in this case) of the resources allocated to lobbying. An increase in output price will tend to increase wages, depending on which sector of the economy is more labor intensive and on the amount of this resource withdrawn from production and allocated to lobbying activities. The allocation of resources to influence policy, and away from the production of goods and services, can decrease a country's production possibilities and add to the social cost of any existing distortion.[27] Further, these results suggest that factor-market imperfections that lower w will tend to increase lobbying activities, all else being constant.

Term 4 is the rural household's net labor position; $\overline{L}$ is its labor endowment, while l_r is the amount of labor allocated to lobbying and l_{qr} is the amount allocated to production of the rural good qr. If $\overline{L} - l_r - l_{qr}$ is negative, then the household is hiring labor from other sectors of the economy. In this case, the household would prefer to pursue cheap wage policies. Households with a small endowment of land would be likely to be labor-surplus, where $\overline{L} - l_r - l_{qr}$ would be positive, and hence, they, along with urban households, would tend to prefer policies that increase real wages. An example of policies that can be expected to stimulate an increase in wages are those that increase the provision of public goods (G_r). Investments in roads, infrastructure, technology, and so on tend to increase the productivity of labor, and hence, real wages should increase.[28]

The first part of the fifth term depicts the amount of the sector-specific factor that the

---

[26] Pardey, Kang, and Elliott (1989, p. 271) conclude that "with rising per capita incomes there appears to b a substantially enhanced incentive for rural 'distributional coalitions' to secure disbursements of public expenditures in their favor."

[27] An implication is that applied welfare measures of economic distortions that do not account for the resources allocated to influence policy underestimate the total losses to welfare. See Srinivasan (1986a) for further discussion.

[28] Of course, the sign of $\partial w / \partial l_r$ depends on a number of factors and is not unambiguously positive, even though this is the most likely result for the rural sector of a less-developed economy. The Stopler-Samuelson condition suggests that if the relative price of a sector's output increases and the sector employs labor intensively relative to other factors of production, then wages can be expected to increase. The converse follows for a price decrease. Other factors influencing the change in wages are the magnitude of the increase in the work force for various skill categories, the rate of labor migration between sectors of the economy, and the extent to which public goods induce a substitution for capital and labor.

household is employing in production. For rural households, $\bar{x}_r$ can be viewed as farmland, and as plant and equipment in the case of urban households. The factor is sector-specific in the sense that, while it can be traded within a sector, e.g., renting land, it cannot be traded between sectors. If $\bar{x}_r - x_r$ is positive, the household is renting the factor out to other households in the sector. Conversely, if this term is negative, the household is renting the factor from other households in the sector. The second component is the change in the rental value in response to a change in lobbying level. Sector-specific factors are felt to influence the household's willingness to pay to influence policy for a number of reasons.

First, if a sector's endowment is held by a small share of the sector's households (such as occurs with large land holdings), then economic policy that raises the value of the endowment will tend to benefit a small share of the sector's households. Also, the type of public goods (G_r) that benefit these households may not benefit other households in the sector. For example, households that control large tracts of land may benefit from technological packages that allow economies of scale from mechanization, whereas smallholders and renters may benefit from technological packages that favor labor-intensive technologies. Controllers of large tracts of land can benefit from investments in major transportation networks, whereas smallholders require more extensive investments in feeder roads and local infrastructure. This divergence in the effects of policy on income streams can give rise to rural-based coalitions that reflect the narrow interests of only a small fraction of rural households.

Second, policy that has an adverse impact on the value of sector-specific endowments can also have an adverse impact on the sector's capacity to obtain collateral to support land and capital improvements in the sector. The inability to make these improvements not only decreases future income streams, but the lowering of the sector's wealth relative to the urban sector tends to decrease the rural sector's willingness to pay to influence economic policy.

Third, policy that has been in place for an extended period can induce structural changes in an economy. If the rental value (c_r) of sector-specific assets is forced upward by policy, then households are more likely to invest in maintenance, upgrading, and expansion. Land improvements and expansion of plant and equipment are examples. This type of asset can also include human capital that is trained to perform tasks and learn skills that are specific to a sector. In the case of agriculture, this may include the skills and agronomic practices acquired to produce crops that are not profitable when a policy change requires that they be produced at world market prices. More likely candidates are, of course, the skills required in manufacturing processes that tend to be unique in production and fabrication; skills that are not easily transferable to another production process.

Hence, policies that alter the rental values of these assets influence a country's capital stock. If increases in capital stock occur in industries that cannot compete in world markets, then in the presence of policy liberalization, the task of realigning a country's capital stock to sectors that can compete in world markets is made much more arduous. As policy liberalization leads to a decline in these industries, labor of this type can face the loss of seniority rights, unemployment, and the need to undergo retraining to obtain equivalent wage levels in other activities. Some of the displaced workers may enter the surplus labor pools of the lower skilled, thereby placing downward pressure on wages in these markets too. The end result can be lower earnings to unskilled labor so that lower-income households bear a disproportionate loss in income compared with households of higher-skilled, though perhaps displaced, workers.

Effectively, the value of protection gets built in to the value of sector-specific assets so that in the short run, policy that results in a decline in its rental value can have large negative wealth effects. Since the factor is sector-specific, the household has no opportunity to transfer it to another sector of the economy that may benefit from a change in policy. In this environment, households that previously may not have been willing to influence economic policy now become ardent supporters of the status quo; they become reluctant to alter policy because of the loss in real wealth that policy liberalization may cause, even though the longer-run prospects for economic growth may be extremely promising.

In the context of table 1.5 and the type of model structure used to guide our discussion in this section, there is a corresponding decision rule for the urban sector of the economy. An interdependence exists between the rural and urban rules in the sense that lobbying to increase rural output price may benefit the rural sector and harm the urban sector. Thus, there is a tendency for one group of households to countervail the lobbying efforts of the other. The outcome of these efforts can lead to a situation known as the *prisoner's dilemma* where all households are made worse off because of their lobbying efforts. However, in the presence of economic growth, this result may not hold as the relative powers of the various groups to influence policy change.[29]

Numerical simulations based on an empirical model of an archetypal economy that embodies the type of structure depicted in table 1.5 confirmed many of the implications discussed above. In addition, these simulations suggested that (a) a concentrated industrial structure tends to induce the urban industrial sector to expend more resources to lobby for policies that benefit this sector and (b) changes in a country's terms of trade that benefit a sector also induce the sector to increase its willingness to influence policy.

1.3   PUBLIC INVESTMENT IN AGRICULTURAL RESEARCH

Placing public sponsorship of agricultural research within a broader political economy framework reinforces the notion that research expenditures are but one of a multitude of competing claims on the public purse. These competing claims impose what at times may be severe (political) constraints on the public resources that can realistically be allocated to research, particularly in many less-developed countries where practical considerations limit governments ability to even generate public funds (Goode 1984). As a consequence, securing and maintaining domestic political support for the public-sector component of

---

[29] K. Anderson (1986) notes in his study of the growth of agricultural protection in East Asia that countries tend to switch from taxing to subsidizing agriculture in the course of economic development.

NARSs, and translating that into financial support for agricultural research is a fundamental issue confronting all national research policy makers.

Traditional Perspectives

Agricultural research intensity (ARI) ratios that express expenditures on public-sector agricultural research as a proportion of agricultural product (AgGDP) are commonly cited measures of the support afforded NARSs. Data for 110 countries grouped on the basis of simple and weighted averages by region and per capita income are given in table 1.6. They show an approximate doubling of intensity ratios for both more- and less-developed countries alike over the 1961 to 1985 period. The data also confirm the positive correlation between income levels and ARI ratios noted by earlier observers, with ARI ratios for high-income countries, (when expressed as simple averages) approximately double those of low- and middle-income countries. Weighted average ratios are often half the corresponding simple average ratios for the less-developed countries but there is relatively little difference across corresponding averages for the more-developed countries. This is due to the tendency for research expenditures to increase less than proportionately with the absolute size of the agricultural sector among less-developed countries whereas this pattern is far less pronounced in the case of more-developed countries.[30]

Naturally the averages presented in table 1.6 mask quite a deal of cross-country and temporal variation in ARI ratios. All but 18 countries spent more on agricultural research relative to AgGDP in 1981-85 than they did in 1961-65. But, over the more recent 1976-80 to 1981-85 period, 37% of the less-developed countries in our sample had declining ARI ratios with approximately half of these countries (i.e., 16 in all) located in sub-Saharan Africa. By contrast only 3 (17%) of the more-developed countries experienced declines in their ARI ratios over the corresponding period.

The traditional view is that governments, or political processes in their wider sense, have done a poor job in securing socially optimal levels of public support for agricultural research. Empirical evidence that historical rates of return to agricultural research — often in excess of 35% — are high relative to other (public or private) investment opportunities is frequently cited evidence of a general tendency to underinvest in agricultural research (Ruttan 1980).

Various explanations for this apparent underinvestment have been offered. One notion is that research managers have been unusually successful in selecting "efficient" research portfolios, but this leaves unanswered the question of why additional funds are not forthcoming. Others (e.g., Ruttan 1980) argue that the inability of governments, be they

---

[30] A double log regression of agricultural research expenditures on AgGDP, involving a pooled 1961-85 sample, yields highly significant elasticity estimates (which measure the percentage change in agricultural research expenditures for a given percentage change in AgGDP) of 0.74 in the case of the less-developed countries and 0.93 in the case of the more-developed countries.

Table 1.6: *Agricultural Research Intensity Ratios, Simple and Weighted Averages*

Region/income group[b]	Simple average					Weighted average[a]				
	1961-65	1966-70	1971-75	1976-80	1981-85	1961-65	1966-70	1971-75	1976-80	1981-85
	%	%	%	%	%	%	%	%	%	%
Nigeria	0.11	0.21	0.29	0.48	0.35	0.11	0.21	0.29	0.48	0.35
Western Africa (15)	0.42	0.50	0.56	0.80	0.91	0.40	0.51	0.55	0.66	0.79
Central Africa (6)	0.51	0.61	0.51	0.55	0.77	0.32	0.34	0.39	0.29	0.28
Southern Africa (8)	0.71	1.09	1.00	1.08	2.04	0.47	0.57	0.64	0.82	1.02
Eastern Africa (7)	0.40	0.57	0.50	0.51	0.63	0.27	0.47	0.42	0.40	0.38
Sub-Saharan Africa (37)	*0.49*	*0.65*	*0.63*	*0.75*	*1.06*	*0.26*	*0.39*	*0.42*	*0.51*	*0.49*
China	*0.41*	*0.31*	*0.39*	*0.47*	*0.39*	*0.41*	*0.31*	*0.39*	*0.47*	*0.39*
South Asia (6)	0.13	0.15	0.19	0.28	0.29	0.11	0.13	0.17	0.27	0.28
Southeast Asia (7)	0.49	0.81	0.56	0.53	0.68	0.21	0.31	0.32	0.31	0.38
Pacific (2)	0.47	0.81	0.86	1.07	1.36	0.43	0.50	0.72	1.16	1.30
Asia & Pacific, ex. China (15)	*0.34*	*0.55*	*0.45*	*0.50*	*0.62*	*0.14*	*0.18*	*0.22*	*0.29*	*0.32*
Caribbean (8)	0.71	1.02	0.93	1.12	1.34	0.25	0.35	0.38	0.44	0.41
Central America (7)	0.25	0.23	0.26	0.32	0.45	0.12	0.11	0.19	0.30	0.43
South America (11)	0.32	0.51	0.64	0.62	0.68	0.38	0.59	0.59	0.67	0.65
Latin America & Caribbean (26)	*0.42*	*0.59*	*0.63*	*0.69*	*0.82*	*0.30*	*0.44*	*0.46*	*0.56*	*0.58*
North Africa (5)	0.73	0.88	1.19	1.05	1.14	0.47	0.59	0.61	0.61	0.69
West Asia (8)	0.53	0.60	0.76	1.04	1.35	0.22	0.45	0.47	0.44	0.47
West Asia & North Africa (13)	*0.60*	*0.71*	*0.93*	*1.05*	*1.27*	*0.28*	*0.49*	*0.50*	*0.48*	*0.52*
Less-Developed Countries (92)	*0.46*	*0.62*	*0.64*	*0.73*	*0.94*	*0.24*	*0.29*	*0.34*	*0.41*	*0.41*

Table 1.6: *Agricultural Research Intensity Ratios, Simple and Weighted Averages (Contd.)*

Region/income group[b]	Simple average					Weighted average[a]				
	1961-65	1966-70	1971-75	1976-80	1981-85	1961-65	1966-70	1971-75	1976-80	1981-85
	%	%	%	%	%	%	%	%	%	%
Japan	1.29	1.48	1.96	2.22	2.89	1.29	1.48	1.96	2.22	2.89
Australia	1.97	2.71	3.49	2.91	4.02	1.97	2.71	3.49	2.91	4.02
Northern Europe (3)	0.76	1.14	1.42	1.97	2.01	0.73	1.09	1.18	1.59	1.76
Western Europe (7)	0.83	1.32	1.61	1.93	2.06	0.72	1.13	1.51	1.76	1.99
Southern Europe (4)	0.22	0.23	0.31	0.40	0.59	0.17	0.19	0.26	0.35	0.65
North America (2)	1.82	2.80	2.21	2.45	3.27	1.60	2.11	1.67	1.92	2.42
More-Developed Countries (18)	*0.88*	*1.30*	*1.48*	*1.72*	*2.02*	*0.96*	*1.29*	*1.41*	*1.60*	*2.03*
Total (110)	*0.53*	*0.73*	*0.78*	*0.90*	*1.12*	*0.48*	*0.60*	*0.65*	*0.72*	*0.76*
Low (30)	0.30	0.37	0.40	0.53	0.65	0.22	0.21	0.27	0.36	0.35
Lower-middle (28)	0.49	0.70	0.69	0.74	1.00	0.24	0.33	0.35	0.39	0.40
Middle (18)	0.47	1.06	0.58	0.58	0.84	0.25	0.44	0.46	0.49	0.57
Higher-middle (18)	0.59	0.82	0.82	1.02	1.26	0.27	0.38	0.44	0.52	0.55
Higher (16)	1.03	1.49	1.82	2.06	2.37	1.08	1.44	1.57	1.78	2.23
Total (110)	*0.53*	*0.80*	*0.87*	*0.90*	*1.12*	*0.48*	*0.60*	*0.65*	*0.72*	*0.76*

Note: Agricultural Research Intensity ratios, as defined here, measure agricultural research expenditures as a percentage of AgGDP.

[a]Weighted by the respective country's share of aggregate AgGDP.
[b]Countries assigned to income classes based on 1971-75 per capita GDP averages where Low, $600 <; Lower-middle, $600-1500; Middle, $1500-3000; Upper-middle, $3000-6000; and High, > $6000.

local or national, to appropriate fully or be compensated for the research benefits that spillover to areas outside their jurisdiction (i.e., a market failure rationale) leads to less than socially optimal levels of investments in research. Oehmke (1986) suggests that public agencies subject to institutional rigidities or who use imperfect (historical) information when determining research investment levels have a tendency to set *actual* levels of research investments which fall short of (secularly increasing) *optimal* levels of investment. Anderson (chapter 4) raises the further possibility that the substantially high levels of risk and uncertainty surrounding agricultural research endeavors and their potential impact on the agricultural sector could lead governments to shy away from investing in research at the levels which expected (or deterministic) relative rates of return suggest are appropriate.

Two other perspectives challenge the underinvestment hypothesis itself. Taking a public finance perspective Fox (1985) argues that previous rates-of-return studies failed to discount their estimates by the deadweight losses in factor and product markets that occur when government expenditures are financed by distortionary tax collections. When coupled with the notion that the social rate of return to conventional capital is undervalued by neglecting benefits that do not accrue to the private investor, Fox (1985, pp. 810-11) concludes that "agricultural research conducted at public expense in recent years has generated a social rate of return comparable to investments in the corporate sector, and neither under nor overinvestment seems to be the case." Others, such as Hertford and Schmitz (1977) and Pasour and Johnson (1982), focus attention on the validity of the rates-of-return estimates themselves. The inference is that the shortcomings in the analytical framework used to identify the costs and benefits from agricultural research have meant that many prior estimates of the (ex-post, marginal and average) social rate of return to agricultural research have, on balance, been biased upward. One strand of this criticism is that the rates-of-return evidence is heavily biased in favor of the research success stories. While no doubt sample selection bias is a factor in those studies at the research project level, the criticism holds less weight at the research program or commodity level and is not an issue for the 48 aggregate studies (Echeverría 1990b, table 1) that have been carried out at the sectoral level.

The wide disparity in ARI ratios between more- and less-developed countries, when buttressed by a large array of empirical studies suggesting relatively high rates of returns to public investments in agricultural research and accompanying rationales in support of an underinvestment hypothesis, has led to a variety of operational guidelines concerning "desirable" research investment levels. The 1974 UN World Food conference set a 1985 research intensity target of 0.5% (UN 1974, p. 97) while the World Bank (1981a) proffered a widely cited target for 1990 of 2%. Johnson (1982, p. 81) argued that "... the evidence presented on the returns to agricultural research definitely supports the proposition that a given country should spend no less as a percent of the value of its agricultural output than is now being spent by the average of countries with comparable levels of incomes."

The difficulty with these rules of thumb is that the conceptual, empirical, and even practical bases for such generic recommendations are not clearly established. Ruttan's

(1980, p. 53) suggestion that "... a level of expenditure that would push rates of return to below 20% would be in the public interest" is one guideline that comes closer to having some conceptual merit. At a minimum it leaves room for optimal levels of expenditures to be at, below, or above the 2% (or, for that matter, the 0.5%) level, given cross-country and temporal variations in the efficacy of public investments in agricultural research. Certainly the evidence in table 1.6 makes it clear that, with few exceptions, less-developed countries are far from realizing a 2% target and many fell well short of the recommended 0.5% level.

Political Economy Perspectives

To do justice to a debate on the "appropriate" levels of public support for agricultural research goes well beyond our brief here. However, a potentially instructive means to a more complete understanding of the structure of support for agricultural research is to place publicly funded research in the context of the overall level of public support for agriculture. A motivation for this approach lies in the political economy perspective developed earlier in this chapter. This perspective takes public agricultural research expenditure levels to be the outcome, at least in part, of an allocation process subject to constraints imposed by the (possibly countervailing) influences of various interest groups within society. Governments direct funds either to the agricultural sector or the nonagricultural sector and give differential preferences to various public programs within each sector in response to such pressures.

Table 1.7 presents some indicators of public-sector expenditures for 70 countries grouped by per capita income. Both the share of agricultural expenditures and agricultural research expenditures in total government expenditures decline dramatically when moving from low- to high-income countries. Something in the order of 10% of total government expenditures in lower income countries goes directly to agriculture and approximately 0.6% to agricultural research while the corresponding percentages for high-income countries are around 3% and 0.2%, respectively.

Expressing agricultural research expenditures as a percentage of agricultural expenditures provides an indication of the *relative* importance given to research on agriculture within the constraints imposed by overall public spending on agriculture (table 1.7). Countries on average directed about 8% of their 1981-85 agricultural expenditures to agricultural research endeavors. Within the range of tolerances relevant for these data — stemming in large part from the difficulties of generating comparable measures of government expenditures from published data[31] — there are no obvious trends revealed by this

---

31 Comparability across countries would best be served if "consolidated total government expenditures" were used to perform these calculations. Unfortunately, comprehensive data of this sort are available for just a few countries. The practical compromise was to rely on "national government expenditure" data only, which is an acceptable alternative to the extent that national government expenditures constitute a significant (and stable) share of total government expenditures. Juxtaposing agricultural research expenditures against public expenditures on agriculture is subject to misinterpretation if (a) in some countries agricultural research expenditures arise, at least in part, from science and education budgets or

tabulation. Certainly the income-linked pattern of support of agricultural research that many have implied from an inspection of ARI ratios is far less evident in this case.

This gives prima facie support to the view that governments in poor compared with rich countries in general do not give differential (i.e., lower) priority to spending on agricultural research within the overall constraints of spending on agriculture. To more fully comprehend the (political) trade-offs between financial support for agricultural research vis-à-vis other forms of government interventions that impact upon agriculture would require access to detailed case-specific data. But, it appears from this evidence at the aggregate level that fundamental limitations to public support for research in low-income countries may well lie in the financial and political constraints imposed by overall and agricultural-specific levels of public sector spending.[32]

The data in table 1.8 give some insights into the political economy forces at work here. While total government spending on agriculture, indexed over the agricultural population, increases dramatically by a factor of 85 times, from around $21 per capita in the low-income countries to $1800 per capita in the high-income countries, there is only a corresponding 8-fold increase in agricultural spending indexed over the total population. Per capita spending on agricultural research follows a similar pattern. Thus, as one moves from low- to high-income countries the level of per capita "benefits" or transfers accruing to rural-based coalitions may well increase at a disproportionately larger rate than the per capita incidence of "costs" associated with such programs.

While this interpretation is consistent with the political economy perspective discussed earlier in this chapter one runs the danger of over- or mis-interpreting the data. Certainly public expenditures in agriculture have been largely responsible for the long-term decline in the real cost of food world wide that has, in turn, allowed specialization and division of labor to occur in agriculture as well as other sectors of many of the world's economies. Thus, the returns to public agricultural expenditures have been shared by those outside the agricultural sector. However, policies have intervened in this process. In many industrialized market economies, agricultural price and foreign trade policy has protected agriculture so that rents from cost-reducing technologies have, in the short run, largely accrued to the rural sector in terms of higher returns over variable costs and increases in the value of sector specific assets such as land. As the costs of the policies to protect agriculture have risen, either trade liberalization or the subsidization of agricultural exports has allowed food-importing countries to capture some of the gains from public expenditures in agricul-

---

the like and (b) the coverage of agriculture expenditures is subject to variation across countries and time. periods. But, expenditure classifications bear no obvious relationship to per capita income levels so, while care needs to be exercised when using these statistics, the simple averages presented in table 1.7 are likely to be informative.

[32] In fact, to raise agricultural research spending levels for low income countries from their current level of 0.60% of AgGDP to the high income average of 2.13% would require that low-income countries increase, on average, the research component of their agriculture-related expenditures from their current level of 7.1% to 25.2%.

Table 1.7: *Agricultural Research and Public-Sector Expenditure Shares*

Income group[a]	1971-75	1976-80	1981-85
	%	%	%
	Agricultural research intensity ratios[c]		
Low (13)[b]	0.42	0.44	0.60
Lower-middle (18)	0.64	0.65	1.04
Middle (12)	0.56	0.52	0.63
Upper-middle (12)	0.62	0.77	0.95
High (15)	1.63	1.88	2.13
Total sample (70)	0.79	0.88	1.11
	Percentage of agricultural expenditures in total government expenditures		
Low (13)	10.5	11.7	11.2
Lower-middle (18)	7.5	8.1	9.3
Middle (12)	6.5	5.7	5.2
Upper-middle (12)	6.7	4.7	4.3
High (15)	3.0	2.7	2.5
Total sample (70)	7.1	6.9	6.8
	Percentage of agricultural research expenditures in total government expenditures		
Low (13)	0.82	0.72	0.67
Lower-middle (18)	0.67	0.50	0.58
Middle (12)	0.52	0.39	0.36
Upper-middle (12)	0.22	0.20	0.17
High (15)	0.29	0.24	0.24
Total sample (70)	0.52	0.42	0.42
	Percentage of agricultural research expenditures in agricultural expenditures		
Low (13)	7.8	6.5	7.1
Lower-middle (18)	9.7	8.4	8.4
Middle (12)	8.3	7.3	7.7
Upper-middle (12)	6.7	6.3	5.9
High (15)	5.6	6.6	7.0
Total sample (70)	8.2	7.8	7.9

Note: All data represent simple averages across all countries in each income class.

[a] Countries assigned to income classes based on 1971-75 per capita GDP averages where Low, <$600; Lower-middle, $600-1500; Middle, $1500-3000; Upper-middle, $3000-6000; and High, > $6000.

[b] Bracketed figures represent number of countries in each income class.

[c] Measures agricultural research expenditures as a percentage of AgGDP. These figures differ slightly from corresponding figures in table 1.6 due to sample size differences.

Table 1.8: *Public Spending per Capita on Agriculture and Agricultural Research*

Income group[a]	Government expenditure on agriculture			Agricultural research expenditures		
	1971-75	1976-80	1981-85	1971-75	1976-80	1981-85
	(1980 PPP dollars per head of agricultural population)					
Low (13)[b]	14.02	18.90	21.11	0.94	1.11	1.28
Lower-middle (18)	43.95	69.46	102.10	3.66	4.02	5.32
Middle (12)	77.76	94.82	119.19	5.45	6.09	7.55
Upper-middle (12)	218.75	358.67	552.28	12.55	19.79	26.49
High (15)	1338.16	1423.13	1801.02	91.79	113.24	140.63
Total (70)	*362.38*	*404.07*	*531.22*	*23.87*	*29.94*	*37.58*
	(1980 PPP dollars per head of total population)					
Low (13)	10.02	13.35	14.06	0.73	0.83	0.93
Lower-middle (18)	20.93	29.62	38.72	1.47	1.82	2.29
Middle (12)	31.59	35.30	38.11	2.36	2.30	2.60
Upper-middle (12)	66.01	62.06	73.04	2.19	2.49	2.72
High (15)	111.49	112.38	115.02	7.32	8.14	8.46
Total (70)	*47.86*	*50.87*	*56.27*	*2.93*	*3.19*	*3.49*

Note: All data represent simple averages across all countries in each income class.

[a] Countries assigned to income classes based on 1971-75 per capita GDP average where Low, <$600; Lower-middle, $600-1500; Middle, $1500-3000; Upper-middle, $3000-6000; and High, >$6000.

[b] Bracketed figures represent number of countries in each income class.

tural technologies, albeit at the cost of distorted world food prices.

In contrast, as noted earlier, many low-income countries follow price, foreign trade, and exchange rate policies that effectively allow real domestic prices to fall in the presence of an agricultural supply response. The result is that rural households only capture a small portion of the returns to cost-reducing technologies. Effectively, these policies can force the rural sector on to an immiserizing growth path. Presumably, these policies can also discourage households from incurring the cost of experimentation and adaptation associated with the more unfamiliar new technologies.

1.4   CONCLUDING COMMENTS

This chapter focused on the interdependence between economic policy and a country's ability and commitment to increase the productivity of resources in agriculture. Section 1.1 provided evidence to highlight the level and selected consequences for agriculture of distortions in countries that typify those pursuing policies that are often designed to promote import-substitution industrialization while maintaining low and stable food prices for urban households. The impact of these policies on the welfare of the typical rural household and on agriculture's contribution to the growth process is surely deleterious.

In the process of economic growth, rural households can be viewed as undergoing a vertical disintegration — a specialization of production activities with an increasing share of household expenditures on preferred foods, housing, clothing, and other nonfood items. Productivity increases are associated with capital deepening and increased reliance on purchased inputs. As the opportunity cost of time increases, labor is allocated away from labor-intensive activities and more reliance is placed on the market for goods and services formerly produced in the more traditional household. At the sectoral level, labor departs agriculture while capital deepening occurs. Agriculture's contribution to the growth process, following Kuznets (1961), includes (a) the low-cost supply of food and raw materials for processing, (b) a market for producer and consumer goods produced by domestic industry, (c) a source of factor contributions (labor, capital) to the industrial sector, and (d) a source of foreign exchange earnings. The policies mentioned here tend to retard this entire process, with strong implications for the types of technological packages that are most useful to households in an environment of distorted markets and macroeconomic imbalances, compared with those operating in more open economies.

In section 1.2 we addressed the question: why have countries persisted in their pursuit of interventions that yield an inefficient allocation of resources and exacerbate adjustments to external shocks? The answer offered in this section is that it is indeed in the self-interest of individuals to influence policy. Section 1.3 builds on this theme by presenting our initial quantitative efforts to place agricultural research in a political economy perspective.

In the case of public goods in general and agricultural research in patricular, lobbying (defined to include voting, the organization of local interest groups, producer associations, and other means of collective action) adds to the social good. However, these efforts often cause government to be pushed and pulled by special-interest groups. It appears that these groups often find higher returns to their lobbying when their efforts are directed toward policy that redistributes income, typically using second-best policy instruments, rather than lobbying for investment in public goods that fosters growth and development.

A number of options are available to lessen the policy discrimination against the rural sector. An important option is to pursue efforts that lead to rural capital accumulation and technological change. Rural capital accumulation and technological change, broadly defined to include human, biological, chemical, and mechanical technology, not only contribute to economic growth, but they also contribute to the willingness of rural households to lobby in their favor, and hence, to countervail the policies that discriminate against the sector. Thus, projects that attempt to increase agriculture's productivity contribute to this end.

Another option is to pursue policy reform that decreases the opportunity for special interests to press for policy instruments that discriminate against the rural sector. The main areas of interest include foreign trade, exchange rate, and capital market policies. Policy instruments that often lead to economic distortions in foreign trade and that are also targets of special interests are quotas on exports and imports, firm-specific licenses that earn rents from foreign trade, various nontariff barriers to restrict quantities imported or exported, and numerous other quantitative restrictions. Foreign trade and capital market controls include licensing of investment in capacity creation and expansion, controls on foreign investment and multiple exchange-rate regimes that are used to discriminate against foreign trade in rural goods. Associated with these controls are the rationing of foreign exchange to selected groups. The supplementing, and in some cases abandoning, of policy instruments with instruments that are less conducive to specific-interest group lobbying activities decreases the opportunities available to discriminate against the rural sector.

Coincident with reforms that remove or limit the use of these instruments are reforms that seek to lessen industrial concentration. Concentration tends to earn excessive rents to sector-specific assets which, in turn, contribute to the willingness of the sector to lobby on its own behalf. The opening up of the sector to foreign capital and the removal of licenses that control investment and capacity creation should be important steps in this direction.

As mentioned, the obstacle to policy reform is often the resistance of those who risk a decrease in wealth from the decline in the value of sector-specific resources, including the laborers displaced by reform. Multilateral support for stabilization and structural adjustment loans that address the needs of the nutritionally deprived and lower the costs of realigning an economy's capital stock should help to alleviate this resistance.

Longer-term goals might include the design of public institutions to better identify and resolve problems created by market failures, to grant individuals and interest groups more equal representation in public choice, and to address problems of distributive justice in ways that minimize the sacrifice of economic growth. An important component of this design is the development of the institutional capacity to implement and manage policy instruments that yield needed public revenues while minimizing their distortionary effect on the economy and their tendency to induce the lobbying efforts of special interests. If instruments are chosen that favor one group at the expense of others, then direct and countervailing lobbying activity can ensue. If political allocation of resources or wealth is to be undertaken without giving rise to rent seeking, then such an allocation should be done without creating a differential advantage to some groups and, according to Srinivasan (1985, p. 43), undertakings not to depart in the future from such an allocation procedure need be sought. Otherwise, the intervention creates the formation of expectations of returns to lobbying resources by other groups which, in turn, can induce lobbying pressures for additional interventions which benefit them rather than the economy at large.

APPENDIX 1.1: OUTLINE OF A GENERAL EQUILIBRIUM MODEL OF COLLECTIVE ACTION WITH PUBLIC GOODS

Here we sketch the general equilibrium model with rent seeking from which the conditions in table 1.5 are derived. For a detailed specification and proofs of the propositions stated here, see Roe and Graham-Tomasi (1990). The simplest form of the model is the case of an economy with two distinct parts: (a) a small open economy with two households (rural and urban), two goods (food and non-food), and two factors, labor and sector specific inputs; and (b) a government which provides a public good to each sector, and sets the relative price of the two goods in response to lobbying by households.

The Household and the Government's Optimization Problem

Households are indexed by $i = r$ (rural), u (urban). Households choose levels of food (q_{ri}) and non-food (q_{ui}) to consume. They also choose the amount of labor (l_{qi}) allocated to the production of the rural good (q_r) and urban good (q_u); the amount of land (x_r) and plant and equipment (x_u) to rent in or out; and the amount of labor to hire in or to work out side of the sector. They are given endowments of labor ($\bar{L}_i$), and land and plant and equipment ($\bar{x}_i$). Market failure is captured by the presence of a rural and urban public good (G_i) that is supplied by the public sector.

The household's *conditional* indirect utility function is defined as, for $i = r,u$:

$$V_i(p,\Pi_i) \equiv \underset{X_i}{\text{Max}}\ U(q_{ri}, q_{ui}) \tag{A1.1}$$

$$X_i = \{(q_{ri}, q_{ui}, l_{qi}, x_i) \in R_+\ |\ \Pi_i = pq_{ri} + q_{ui}\}$$

Disposable income Π_i depends on profits π_i from the production of the i-th good, returns to the endowments of labor ($\bar{L}_i$) and sector specific factors ($\bar{x}_i$), and the proportion, γ_i, of the tax bill T. Denote relative prices by p, w, and c_i for the rural good, labor, and the sector specific factor, respectively. The price of the urban good is taken as numeraire. Then,

$$\Pi_i = \pi_i(p,w,c_i,G_i) + w\ [\bar{L}_i - l_i^\circ] + c_i\bar{x}_i + \gamma_i T \equiv$$

$$pq_i(l_{qi}, x_i;G_i) + w\ (\bar{L}_i - l_{qi} - l_i^\circ) + c_i(\bar{x}_i - x_i) + \gamma_i T \tag{A1.2}$$

for values in X_i that maximize (A1.1).

The indirect utility function defined in (A1.1) is conditional at this point since we treat the household's lobby level l_i° as a parameter. The direct utility $U(\cdot)$ and production functions $y_i(\cdot)$ are assumed to be continuous, strictly concave, and increasing in the household's choice variable. In this situation, the household's problem is separable (Jorgenson and Lau 1969) so that it can be stated in its dual form, denoted here by the "conditional" indirect functions for utility $V_i(\cdot)$ and $\pi_i(\cdot)$.

Letting E_j denote excess domestic demand, commodity and factor market balances are

$$\Sigma_i \, q_{ji} - q_j = E_j, \qquad j = r, u \tag{A1.3a}$$

$$\Sigma_i \, \overline{L}_i - \Sigma_i \, l_{qi} - \Sigma_i \, l_i^\circ - \Sigma_i \, l_{gi} = 0 \tag{A1.3b}$$

$$\overline{x}_i - x_i = 0 \tag{A1.3c}$$

for the rural and urban goods, and for labor and the sector specific factors, respectively. The $\Sigma_i \, l_{gi}$ term in the labor balance equation accounts for the amount of labor the government allocates to the production of the rural and urban public goods.

Treating the world price, p^w, l_i°, and the government's policy instruments, p, l_{gi}, as exogenous variables, (A1.3b) and (A1.3c) are a system of three equations in the three variables w, c_r, c_u. It is assumed that an equilibrium of the economy exists and is unique. In this case, let $w = w \, (\overrightarrow{e_1})$, and $c_i = c_i \, (\overrightarrow{e_1})$ denote the result, where $\overrightarrow{e_1} = (p, l_r^\circ, l_u^\circ, l_{gr}, l_{gu}, \overline{L}_r, \overline{L}_u, \overline{x}_r, \overline{x}_u)$.

We assume that government forms preferences over the utility of households in the economy, and then chooses policy instruments as though it sought to maximize its preferences subject to the condition that it cannot incur a fiscal, and hence a trade, deficit. The government's policy instruments are the relative price p, and the amount of labor l_{gi} to allocate to the production of the rural and urban public goods (G_r, G_u).

That is, government is assumed to solve

$$\underset{X_g}{\text{Max }} U_g = I_r \, (\rho_r, \rho_u) \, V_r + I_u \, (\rho_r, \rho_u) \, V_u \, , X_g \ = \ [(p, l_{gr}, l_{gu}) \in \text{R+}]. \tag{A1.4}$$

Maximization takes place subject to the production function for public goods,

$$G_i = G_i \, (l_{gi}) \tag{A1.5}$$

and the requirement that fiscal expenditures

$$C = -w \, \Sigma_i \, l_{gi} + (p - p^w) \, E_r \tag{A1.6}$$

equal the lump sum income transfers (T) to households, i.e., $T = C.^1$ We assume G_i is continuous, quasi-concave, and increasing in l_{gi}. The values I_i are weights that define the government's preference ordering. They are specified as influence functions whose arguments are determined by the political pressure (ρ_i). Following Becker (1983), political pressure is in turn a function of lobby levels of the i-th household. Effectively, the political

---

1 Fiscal effects of trade are $(p - p^w) \, E_r = (P_r \, / \, P_u - P_r^w \, / \, P_u^w) \, E_r + (P_u \, / P_u - P_u^w / P_u) \, E_u$, where upper case prices represent their respective nominal prices.

pressure function is akin to a political technology that reflects the ability of the i-th special-interest group to form a coalition to lobby government. This pressure is then amalgamated at the national level to form influence.

Hence, the influence functions represent the end product of pressure generated by special-interest groups. Different countries use different methods to define the power of the state. A fundamental characteristic of virtually all political systems is that they are subject to pressures from special interests. This structure is very much a "reduced form" approach. Details of the institutions for establishing laws, politicians, political parties, mechanisms for enacting laws, and defining policy instruments from a set of possible instruments receive no particular attention. The mathematical properties and conditions imposed on this structure are given in Roe and Graham-Tomasi (1990).

The Government's Decision Rules

Proposition 1: If the Negishi (1960) condition holds, i.e., $I_i = 1/V_{i,\Pi_i}$, and if $E_{r_2,p}$ is non-zero, then a maximum to (A1.4) is characterized by $p = p^w$ and $\pi_{i,G_i} G_{i,l_{gi}} = w.$[2]

This condition basically shows that the model does not restrict a Pareto optimal outcome. For the case of an interior solution to (A1.4), let the government's policy decision rules be denoted by:

$$p = p\,(\vec{e_2})$$

and

$$I_{gi} = l_{gi}\,(\vec{e_2})$$

where

$$\vec{e_2} = (\,p^w, l_r^\circ, l_u^\circ, \bar{L}_r, \bar{L}_u, \tilde{L}_u, \bar{x}_r, \bar{x}_u, z_u\,).^3$$

Proposition 2: If the tax burden is borne by urban households, $\gamma_r = 0$, price distortion is determined by:

$$(p - p^w) = \{(l - I)\,[(\,q_r - q_{rr}\,) + (\tilde{L}_r - l_{qr})\,w_p\,]\}/E_{r,p}$$

when

$$\tilde{L}_r - l_{qr} > 0, \text{ and } \tilde{L}_u - l_{qu} < 0 \text{ and by}$$

---

2 Unless otherwise indicated, notation V_{i,Π_i} denotes $\partial V_i/\partial \Pi_i$ and w_p denotes $\partial w/\partial p$.

3 Note that p and l_{gi} are homogeneous of degree zero in I_i.

$$(p - p^w) = \{(l–I)\ [(q_r - q_{rr}) - (\tilde{L}_u - l_{qu})\ w_p]\} / E_{r,p}$$

when

$$\tilde{L}_r - l_{qr} < 0, \text{ and } \tilde{L}_u - l_{qu} > 0, \text{ where } \tilde{L}_i = \bar{L}_i - l_i^\circ \text{ and } I = I_r V_{r,\Pi_r} / I_u V_{e,\Pi_u}.$$

Proposition 3: If the tax burden is borne by urban households, $\gamma_r = 0$, the difference in the marginal value products of the public good are determined by:

$$\pi_{r,G_r} G_{r,l_{gr}} - \pi_{u,G_u} G_{u,l_{gu}} = (1–I)\ ((\tilde{L}_r - l_{qu})\ (w_{l_{gr}} - w_{l_{gr}}) + \pi_{r,G_r} G_{r,l_{gr}}) -$$
$$(p - p^w)(E_{r,l_{gr}} - E_{u,l_{gu}}),$$

for $\tilde{L}_r - l_{qr} > 0$, and $\tilde{L}_u - l_{qu} < 0$, and

$$\pi_{r,G_r} G_{r,l_{gr}} - \pi_{u,G_u} G_{u,l_{gu}} = (I - 1)\ ((\tilde{L}_u - l_{qu})\ (w_{l_{gr}} - w_{l_{gu}}) - \pi_r,\ G_r G_{r,l_{gr}}) -$$
$$(p - p^w)(E_{r,l_{gr}} - E_{u,l_{gu}}),$$

for $\tilde{L}_r - l_{qr} < 0$, and $\tilde{L}_u - l_{qu} > 0$.

Propositions two and three indicate the directional bias in government price and investment policy as a function of government preferences, I_i, and whether a household is labor surplus or deficit.

The Household's Decision Rules

Assuming that the i-th household takes the actions of the j-th household as given, correctly perceives the objective of government, equation (A1.4), knows the political process through which lobbying is transmitted to influence, the household, in principle, can solve the problem[4]:

$$\underset{l_i}{Max}\ V_i\ (p,\Pi_i),\ l_i \in R_+ \tag{A1.7}$$

subject to the government's decision rules for p and l_{gi}. Substituting the policy decision rules into (A1.7), and assuming differentiability of p $(\vec{e_2})$ and c_i $(\vec{e_2})$, the first-order condition for the rural household is the equation presented in table 1.5.

---

[4] It can be shown from the envelope theorem that to constrain the choice of l_i to the household's budget constraint is redundant.

The Game Component of the Model

The simplest approach at this level is to posit a one-shot game, with Nash behavior, and to search for Nash equilibria in lobbying levels. Even with this simple setup, the existence of a Nash equilibrium is not trivial.

Assuming strict concavity of (A1.7) in l_i, let

$$l_i = l_i (\overrightarrow{e_i}) \tag{A1.8}$$

denote the household's lobbying rule obtained from (A1.5), where $\overrightarrow{e_i} = (p^w, l_j^°, \overline{L}_r, \overline{L}_u, \overline{x}_r, \overline{x}_u, z_r, z_u)$.

Equation (A1.8) is the i-th household's best response to the j-th household's action. Then $l_i^°$ are a Nash solution if, and only if,

$$\partial V_r / \partial l_r \Big|_{l_u = l_u (\overrightarrow{e_u})} = 0 \qquad \partial V_u / \partial l_u \Big|_{l_r = l_r (\overrightarrow{e_r})} = 0$$

Of course, a Nash equilibrium need not exist. The consequences of this result are discussed in Roe and Graham-Tomasi (1990) as are extensions to include voting and bureaucratic behavior.

Chapter 2

Agricultural Research in an International Policy Context

G. Edward Schuh and George W. Norton[1]

The issues we address in this chapter involve the role of agricultural research in the context of international economic policy. The environment which that policy helps to define influences the nature of the technologies demanded by producers as well as the supply of technologies offered by private- and public-sector research systems. It also has a substantial influence on the impact of new technologies on producers, the extent to which consumers benefit directly or indirectly from that technology, and ultimately the manner and extent to which technology contributes to general economic growth. Thus, the international economic policy environment plays an important role in shaping the agricultural science and technology policy of both NARSs as well as international development agencies that seek to strengthen and develop the capacity for agricultural research on the international scene.

In the future, national and international policies toward agricultural research and development (R&D) will have a greater influence on the rate of growth of agricultural output and the contribution agriculture makes to general economic growth. An ever-larger share of the increments of agricultural output on the international scene is accounted for by investments in agricultural research. This share is likely to grow in the future, both because the production and distribution of new technology has been found to be an efficient source of economic growth and because the supply of new land that can be brought into production is available only at a sharply rising supply price. Moreover, nation states must increasingly pay more attention to their science and technology policy in order to remain competitive in international markets and earn the foreign exchange they need to service their international debt and to finance higher rates of economic growth.

The parameters national and international policymakers (and private research decision makers) must consider in shaping their decisions are largely reflected in the system of relative prices that prevail in the international economy. In the past, the structure of those

---

[1] The authors would like to thank Jaime Ortiz for data collection and computer assistance.

relative prices was largely influenced by the trade and exchange rate policies implemented by national governments. These policies influenced both the supply of exports from individual countries and their respective demand for imports. They also influenced the level at which prices in international markets were reflected in domestic economies.

Developments in the international economy these past several decades have significantly broadened the number of policies that must now be taken into account. Domestic monetary and fiscal policies have increasingly important international ramifications, influencing to an ever-larger extent the real exchange rates national governments are able to establish for their domestic economies. Moreover, the growing sensitivity to environmental problems, such as global warming, makes it likely that both national and international environmental policies will continue to play a more important role in shaping the structure of international prices and the market opportunities producers in individual countries face.

Two points are important in establishing the context of the material in this chapter. The first is that the factors influencing the international policy context in which national and international agricultural R&D policymakers must operate have become increasingly complex and far-reaching as a result of developments in the international economy. The second is that such policymakers can no longer afford to take a passive role in relation to these conditions in the international economy. Agricultural science and technology policymakers, both private and public, need to take a more active role in shaping their agricultural research programs if the production and distribution of agricultural technology is to be an efficient source of economic growth and if adequate supplies of food for a rapidly expanding global population are to be assured.

The remainder of this chapter is divided into three main parts. The first part provides the background for the analysis to follow. The second discusses some important policy issues. The third examines the implications of what has preceded for strategic agricultural research policies and for international assistance to agriculture. The underlying assumption of the chapter is that policymakers invest in agricultural research to contribute to the achievement of three major policy goals: increased efficiency in resource use, more equitable distribution of the fruits of economic growth, and a more secure environment for their citizens. The expectation is that a sounder agricultural R&D policy, and one that takes conditions in the international economy into account, will make important contributions to the attainment of those goals.

2.1 BACKGROUND

Four issues are discussed in this section: (a) changes in the structure of the international economy, (b) trends in world agricultural trade, (c) changes in technological capability on the international scene, and (d) the structure of agricultural protection. This background provides the setting for the analysis of policy issues that follows in the next section.

2.1.1 Changes in the Structure of the International Economy

Since the end of World War II, there have been enormous changes in the structure of the international economy. At least four of these changes are important to international commodity markets and thus constitute significant changes in the international context in which agricultural science and technology policymakers must operate. These changes are for the most part rooted in technological developments in sectors that provide the infrastructure for international intercourse. Chief among these are technological developments in the communication and transportation sectors that have significantly lowered the costs of these services. These developments have greatly expanded the scope for international trade and for other economic, political, and social interactions among countries. They have been reinforced by the computer revolution, which has made it possible to assemble and analyze large quantities of information.

The first major change in the international economy in this postwar period has been the growth in international trade relative to the growth in global GNP. Trade has grown at a faster rate than global GNP in every year except five since the end of World War II. The five years of exception have been years of severe economic recession in the international economy. The consequence of this relative growth of trade is that the international economy has become increasingly well integrated and interdependent, with national economies increasingly dependent on trade both for markets and for the raw materials, producer goods and services, and consumer goods and services they need.

The second major change in the international economy has been the emergence of a huge, well-integrated international capital market. Starting from a period at the end of the War when there was no international capital market, the global economy has evolved to a point at which the international capital market now simply dwarfs international trade. In a recent year, international financial flows were on the order of $42 trillion, while total international trade flows were on the order of $2 trillion. International capital markets are now every bit as important in establishing links among national economies as is international trade. More important, they dominate foreign exchange markets and establish important links among the macroeconomic policies of national governments (Schuh 1986).

The third major change in the international economy in the postwar period was the shift in 1973 from the Bretton Woods fixed exchange rate system to what can best be characterized as a bloc-floating exchange rate system. At the end of World War II, the international community established a fixed exchange rate system in which the values of national currencies were fixed in terms of each other and remained fixed except under unusual conditions. An important element of this system was that imbalances in the external accounts of national economies were to be eliminated by changes in domestic economic policies.

This system worked reasonably well until the end of the 1960s when international capital markets had grown so large that they dominated foreign exchange markets. The United States devalued the dollar, the key currency on the international scene, in 1971. When that did not reestablish balance in its external accounts, it devalued it again in 1973 and announced that henceforth the value of the dollar would be determined by foreign exchange markets.

Thus ended, for all practical purposes, the Bretton Woods fixed exchange rate system. The system still has a great deal of "fixity" in it since small countries still peg the value of their currencies to the value of one of the major reserve currencies. However, there is a great deal of implicit flexibility in the system because the values of the major currencies change relative to each other and they take the lesser currencies along with them. Something like 85% of global international trade takes place across flexible exchange rates.

The final change in the international economy that is of importance to agricultural commodity markets is the increase in international monetary instability that started about 1968. Prior to that period, international interest rates were relatively stable. Since that date, they have been relatively unstable, at times experiencing large swings. In light of the other changes in the international economy, this increase in monetary instability is of great importance.

At least four of the implications of these changes in the international economy merit further discussion because of their importance to agricultural science and technology policy. First, contrary to the past, agriculture now bears a significant share of the burden of adjustment created by changes in monetary and fiscal policy. Given the existence of a well-integrated international capital market and the prevalence of flexible exchange rates, changes in domestic monetary and fiscal policy are now reflected in changes in real exchange rates and not in real interest rates. The result is to pass the burden of adjustment of these changes to the trade sectors — export sectors and those sectors that compete with imports. In most countries, agriculture is a trade sector. Many countries either export or import agricultural commodities; most do both. Among other things, this means that agricultural resources need to shift from trade to nontrade activities as monetary and fiscal policies change. This is an important challenge to R&D policymakers.

Second, it isn't just domestic monetary and fiscal policies that matter. The policies of other countries are equally, if not more, important. For example, the unprecedented rise in the value of the US dollar in the first half of the 1980s was as much due to the tight monetary and liberal fiscal policies of the United States as it was to the conservative fiscal and easy monetary policies of Western Europe and Japan. That large rise in the value of the dollar had important implications for other countries trying to compete in international markets.

Third, given the present configuration of the international economy, there are strong linkages between international financial markets and international commodity markets. Developments in international financial markets influence the value of national currencies, and these in turn influence trade flows.

Fourth, national economies are now more dependent on forces in the international economy. This is the obverse of the increased dependence on trade that is a logical consequence of that growth in trade as well as the increased importance of the international capital market. A more open economy is, of course, increasingly beyond the reach of national economic policies. Consequently, policy-making and implementation shift in two disparate directions. They shift to the international level and become part of the codes, rules, and disciplines of international institutions such as the General Agreement on Tariffs and Trade (the GATT). They also shift downward to the state and local level. It is this latter shift that is important to research policymakers because important components of R&D policy need to be made at the state and local level.

2.1.2 Trends in World Agricultural Trade

National agricultural research programs operate in a setting characterized by a unique human, natural, and physical resource base. Differences in relative resource endowments across countries are the fundamental basis of comparative advantage and the source of potential economic gains from international trade. Resource endowments are not static, however, as they change over time in response to investments in individual countries. Moreover, both production and consumption are affected by government interventions in the economy by means of tariffs, nontariff barriers, export subsidies, and interventions in foreign exchange markets — directly or indirectly. With the growing openness of national economies, as noted above, the intervention of governments in other countries is as important as the interventions of domestic governments. These interventions, as well as international monetary developments, can mask underlying comparative advantage and affect a country's competitive advantage for long periods of time.[2] Furthermore, agricultural research can influence both underlying comparative advantages and the nature of government interventions, an issue we will address in a later section.

World trade in agricultural products has grown steadily throughout the post-World War II period. In addition to this growth in trade, the patterns of trade also have changed, in part associated with changes in the aggregate growth rates. For example, agricultural trade grew at historically rapid rates during the 1970s, fueled in part by the rapid growth in US trade as the value of the dollar fell in foreign exchange markets. This was also fueled by the rapid growth in international monetary reserves as well as in borrowing by the less-developed countries during this period. Aggregate growth rates slowed in the first half of the 1980s, however, as the international economy experienced the most severe economic recession since the 1930s. The value of the dollar rose, thus reducing the competitive edge of the United States, and many less-developed countries experienced serious international debt crises as international liquidity dried up.

Important shifts occurred in this period in both country and regional shares of total agricultural exports. For example, grain production grew steadily during this period, but trade as a percent of total grain production peaked in 1980-81 and declined through 1985. Less-developed countries, which experienced a declining share of total world food exports in the 1960s and 1970s, realized more rapid export growth than the more-developed

---

2 Competitive advantage refers to the advantage that remains as underlying comparative advantage is affected by exchange rate and trade policy distortions.

countries in the first half of the 1980s (table 2.1). Among less-developed regions, grain imports grew most in sub-Saharan Africa, China, and West Asia & North Africa, but exports grew most in Asia & Pacific (table 2.2). Many other significant changes occurred in individual countries, not the least of which has been India's shift to being a net grain exporter and the increasing volatility of China's agricultural trade.

Table 2.1: *Average Annual Percentage Growth of Agricultural Exports*

	1965-73	1973-80	1980-85
	%	%	%
Low- and middle-income economies[a]			
Food	2.4	4.2	3.6
Nonfood agriculture	2.1	0.4	1.2
Total reporting economies			
Food	3.6	6.8	1.9
Nonfood agriculture	3.1	0.9	2.1

Source: *World Development Report 1989*, p. 150.

[a] Defined as having a level of per capita GNP below $6000 in 1987.

A variety of factors are responsible for these observed changes in trade patterns. These changes indicate that countries must be continuously concerned with international competition in agricultural markets if they want to retain current sources of foreign exchange and carve out ever-larger markets. Food, agricultural, and research policies must take account of changes in both international comparative and competitive advantages. In a sense, each country finds itself on an international treadmill, with a need for continuous increases in productivity and for policies that do not discriminate against agriculture lest that country loses its share of a particular market. An important implication for research policy is that each country must carefully establish research priorities that are consistent with its resource endowment relative to other countries. It must also take into account any changes in the quality of that endowment that may result from research or investments in other countries.

Production in the less-developed countries has now increased to the point at which roughly the same quantity of cereal and other food is produced in them as in the more-developed countries. However, the extent to which changing production and trade patterns are due to technological changes, as opposed to policy or institutional forces, is difficult to assess without a detailed, in-depth analysis on a country-by-country basis. In addition, there are international as well as domestic factors affecting both technological and economic policy changes and therefore also affecting both comparative and competitive advantages.

Table 2.2: *Cereal Trade and Percentage Change between 1974-76 and 1986-88*

Region	Imports (million metric tons)		Exports (million metric tons)		Total change		
	1974-76	1986-88	1974-76	1986-88	Imports %	Exports %	Net imports %
Sub-Saharan Africa	4.1	8.4	0.4	1.0	103	133	99
China	6.7	18.5	2.8	6.0	178	115	224
Asia & Pacific	19.0	21.4	5.2	9.7	13	87	-16
Latin America & Caribbean	13.6	19.5	11.7	11.6	44	-1	320
West Asia & North Africa	13.6	40.6	0.3	3.0	198	1050	181
Less-Developed Countries	*56.9*	*108.3*	*20.4*	*31.4*	*90*	*54*	*110*
MDC Nonmarket Economies	27.2	39.1	8.2	5.3	44	-35	78
MDC Market Economies	72.2	66.8	128.1	180.2	-7	41	-103
More-Developed Countries	*99.4*	*106.0*	*136.3*	*185.5*	*7*	*36*	*-116*
World	156.3	214.2	156.7	216.8	37	38	—

Source: Constructed on the basis of data reported in FAO (1977, 1990c).

Note: Cereals principally include wheat and meslin, rice, barley, maize, rye, and oats.

2.1.3 Structural or Technological Changes

The structural or technological determinants of agricultural production and trade include such things as public and private investments in agricultural research, investments in education, improvements in physical infrastructure (including that for water control), and population growth rates. Relative differences in public agricultural research expenditures across countries, in particular, have been instrumental in altering agricultural productivity and underlying comparative advantage.

A regional summary of changes in cereal yields from the mid-1970s to the mid-1980s is presented in table 2.3. While yield per hectare is only a partial productivity measure and is affected by many factors (chapter 5), these overall levels and trends reflect, in part, the generation and adoption of research results. Research has proved to be a powerful means of altering comparative advantage in world agriculture. In addition, the CGIAR system of international agricultural research centers has generated valuable tropical technologies for rice, maize, wheat, beans, potatoes, cassava, beef, and other commodities. Many new biotechnologies appear to be potentially transferable to those countries with an adequate indigenous research capacity, but decisions regarding the intellectual property rights for new biotechnologies may influence their transfer (chapter 10, Persley 1990).

Table 2.3: *Cereal Yields and Percentage Change between 1974-76 and 1984-86*

	Yield		
Region	1974-76	1984-86	Total change
	(kg per hectare)		%
Sub-Saharan Africa	806	928	15
China	2479	3891	57
Asia & Pacific	1428	1902	33
Latin America & Caribbean	1640	2093	28
West Asia & North Africa	1390	1553	12
Less-Developed Countries	*1628*	*2206*	*36*
MDC Nonmarket Economies	1735	2032	17
MDC Market Economies	3033	3937	30
More-Developed Countries	*2394*	*3051*	*27*
World	1955	2560	31

Source: Constructed on the basis of data reported in *FAO Production Yearbooks*.

A historical summary of the regional data on research expenditures as a percentage of the value of agricultural production is presented in chapter 1 of this volume. These data indicate the upward trend in research-intensity that occurred in many countries from 1961 to 1985. A substantial number of less-developed countries, however, experienced a decline in their research intensity ratios from the late 1970s to the mid 1980s. Failure to maintain research intensity can have serious consequences because much agricultural research serves to maintain past productivity gains as insects and diseases evolve over time and become more resistant to pesticides. Another serious issue in many less-developed countries is that both the capacity to train agricultural scientists and the capacity for research itself appear to have declined in the 1980s as a consequence of severe economic crises (see chapter 7). If so, these declines illustrate the linkage between national agricultural research capacity and the economic environment.

2.1.4   Policy Interventions and the Structure of Protection

Agricultural production, trade, and domestic prices are affected by a variety of policy interventions by domestic and foreign governments. Among these are tariffs, quotas on imports and exports, consumer and producer subsidies, interventions in foreign exchange markets, and monetary and fiscal policies. Each of these policy instruments can affect competitive advantage and mask or enhance underlying comparative advantage.

The emergence of well-integrated capital markets and bloc-floating exchange rates increases the vulnerability of agricultural sectors to foreign economic events, since these markets link economic policies together in ways they have not been linked in the past. They also provide less-developed countries with new development opportunities since they provide individual countries with access to capital from abroad so long as they pursue sound economic policies and are not currently overburdened with debt. They also force those countries that are heavily in debt to improve their export performance, reduce import subsidies, and reduce the overvaluation of their currencies. The latter has probably been the most common policy distortion among less-developed countries.

The implication of this for research is that research that generates new production technology for export commodities, or for those commodities that have been heavily imported in recent years, may deserve more priority than was previously recognized. Emphasis on imported commodities may be called for if it appears that the country has a comparative advantage in those commodities that has been masked under previous policy distortions. Another implication is that these countries may need more sophisticated priority-setting procedures for research that can assess these changing economic forces and draw the implications for research priorities.

An important feature of the international economy in recent decades is that the values of major currencies have experienced large swings relative to each other. For example, the US dollar experienced a six-year decline from 1973 to 1979, during which period the United States became increasingly competitive in international commodity markets. Then the dollar experienced an almost unprecedented rise through May 1985, and the United States lost a great deal of the competitive edge in international markets that it had gained in the previous decade.

This problem was exacerbated by the commodity policy established in the aftermath of the US embargo on sales to the Soviet Union when that country invaded Afghanistan. This policy preordained increases in commodity prices in succeeding years by legislative mandate. Given the importance of the United States in some international commodity markets, this policy raised prices in those markets and sent misleading signals to producers in other countries. This policy ended in 1986 when US loan levels were lowered significantly.

Both more- and less-developed countries discriminate against their agricultural sectors, although in quite different ways. The results of a USDA survey of producer subsidies and taxes in various countries are presented in table 2.4. An important pattern in the structure of protection is apparent. The more-developed countries, for their part, tend to provide large subsidies to their producers. The less-developed countries, on the other hand, discriminate severely against their producers through a large number of explicit and implicit taxes. The result is that far too much of the world's agricultural output is produced in the high-cost, more-developed countries, and far too little is produced in the low-cost, less-developed countries. The result is the sacrifice of a large amount of global income and welfare due to the inefficient use of the world's agricultural resources. This same pattern of protection has important implications for agricultural research, and this will be discussed below.

The consequence of more-developed countries providing such large subsidies to their producers, largely through interventions in commodity markets, is that domestic prices in those countries are set significantly above border price levels. To protect those prices, the more-developed countries have to discriminate against imports from less-developed and other countries. Examples of nominal protection coefficients, which compare domestic prices with border prices, are presented in table 2.5 for several commodities and more-developed countries.[3] The manner in which protection is provided has changed significantly over the past few years, with tariffs becoming less important and nontariff barriers more significant. More-developed countries need to continue reducing protectionism against imports from less-developed countries if they expect the debts of those countries to be paid, markets for exports from more-developed countries to expand, and all countries to benefit from following their comparative advantage as it evolves through agricultural research and the like.

Another dimension to this structure of protection is that the more-developed countries find themselves using export subsidies to dispose of the surpluses they accumulate from setting their producer prices too high. The use of these subsidies by the United States and the European Community in particular causes the international prices of some commodities to be lower than they would otherwise be. This lowers the returns to investments in research in the less-developed countries, an issue that will also be discussed below.

Finally, each country must decide whether its long-run comparative advantage is primarily in agricultural production or in manufacturing. As less-developed countries continue to expand agricultural production through the generation and adoption of new

---

3   Related nominal protection rates for selected commodities in some less-developed countries are presented in tables 1.1 and 1.2 in chapter 1.

Table 2.4: *Ranking of Producer Subsidy Equivalent (PSE) Levels*

Ranking[a]	United States	Australia	Canada	European Community	Japan	South Korea	India	Argentina	Nigeria	Brazil
High tax									Cocoa[b] Sugar	
Moderate tax					Citrus		Cotton (LS)[b] Wheat	Wheat		
Low tax							Cotton (MS)[b] Peanut meal Rapeseed meal Rice Soybeans Soymeal	Maize[b] Sorghum[b] Soybeans[b]	Cotton Rice	Beef[b] Maize Soybeans[b]
Low subsidy	Barley[b] Beef Pork Poultry[b] Soybeans[b]	Barley[b] Beef[b] Cane sugar[b] Cotton[b] Mfd milk[b] Pork[b] Poultry[b] Rice Wheat[b] Wood[b]	Barley[b] Beef Flaxseed[b] Maize Oats[b] Pork[b] Poultry[b] Rapeseed[b] Soybeans Wheat[b]	Barley[b] Common wheat[b] Maize[b] Pork[b]		Poultry	Peanuts[b] Rapeseed		Maize	Mfd milk Poultry

Table 2.4: *Ranking of Producer Subsidy Equivalent (PSE) Levels (Contd.)*

Ranking[a]	United States	Australia	Canada	European Community	Japan	South Korea	India	Argentina	Nigeria	Brazil
Moderate subsidy	Cotton[b] Dairy[b] Maize[b] Rice[b] Sorghum[b] Wheat[b]	Fld milk	Sugar	Dairy[b] Durum wheat Poultry Rapeseed Rice Sheep Soybeans Wheat[b]	Poultry	Pork	Peanut oil Rape oil Soy oil		Wheat	Cotton[b] Rice
High subsidy	Sugar		Dairy[b]	Beef[b]	Barley Beef Fld milk Mfd milk Pork Rice Soybeans Sugar Wheat	Barley Beef Fld milk Maize Rice Soybeans Wheat				Wheat
Weighted average PSE	22	9	22	33	72	64	8	−22	−9	7

Source: Ballenger, Dunmore, and Lederer (1987).

Note: Fld represents fluid, *Mfd* represents modified, *MS* represents medium staple cotton, and *LS* represents long staple cotton.

[a] Low denotes 0-24%; moderate denotes 25-49%; and high denotes≥ 50%.
[b] Net exporter during 1982-84.

Table 2.5: *Nominal Protection Coefficients for Producer and Consumer Prices of Selected Commodities in More-Developed Countries, 1980-82*

Country or Region	Wheat		Coarse grains		Rice		Beef and lamb	
	Producer	Consumer	Producer	Consumer	Producer	Consumer	Producer	Consumer
Australia	1.04	1.08	1.00	1.00	1.15	1.75	1.00	1.00
Canada	1.15	1.12	1.00	1.00	1.00	1.00	1.00	1.00
European Community[b]	1.25	1.30	1.40	1.40	1.40	1.40	1.90	1.90
Other Europe[c]	1.70	1.70	1.45	1.45	1.00	1.00	2.10	2.10
Japan	3.80	1.25	4.30	1.30	3.30	2.90	4.00	4.00
New Zealand	1.00	1.00	1.00	1.00	1.00	1.00	1.00	1.00
United States	1.15	1.00	1.00	1.00	1.30	1.00	1.00	1.00
Weighted average	1.19	1.20	1.11	1.16	2.49	2.42	1.47	1.51

Country or Region	Pork and poultry		Dairy products		Sugar		Weighted average[a]	
	Producer	Consumer	Producer	Consumer	Producer	Consumer	Producer	Consumer
Australia	1.00	1.00	1.30	1.40	1.00	1.40	1.04	1.09
Canada	1.10	1.10	1.95	1.95	1.30	1.30	1.17	1.16
European Community[b]	1.25	1.26	1.75	1.80	1.50	1.70	1.54	1.56
Other Europe[c]	1.35	1.35	2.40	2.40	1.80	1.80	1.84	1.81
Japan	1.50	1.50	2.90	2.90	3.00	2.60	2.44	2.08
New Zealand	1.00	1.00	1.00	1.00	1.00	1.00	1.00	1.00
United States	1.00	1.00	2.00	2.00	1.40	1.40	1.16	1.17
Weighted average	1.17	1.17	1.88	1.93	1.49	1.68	1.40	1.43

Source: World Development Report 1986, pp. 112-3.

Note: Nominal protection coefficients represent domestic prices divided by border prices.

[a] Averages are weighted by the values of production and consumption at border prices.

[b] Excludes Greece, Portugal, and Spain.

[c] Austria, Finland, Norway, Sweden, and Switzerland.

production technologies, this issue will become increasingly important on the international scene.

2.2 POLICY ISSUES

In this section we discuss a number of policy issues facing less-developed countries in which agricultural research has an important role to play in contributing to solutions. Important as new production technology may be in solving these problems, however, it is no substitute for sound economic policy. What the discussion will indicate is that sound science and technology policy and sound economic policy are often highly complementary.

2.2.1 The International Debt Problem

Many less-developed countries are currently burdened with serious international debt problems. These problems are a legacy, in part, of the flood of petrodollars generated in the aftermath of the quadrupling of petroleum prices in 1973. Commercial banks were encouraged by the international community to recycle those dollars lest the international economy collapse. Their efforts to do so were met with open arms by many less-developed countries, who saw borrowing as an alternative to painful devaluations and other policies that were the prescribed medicine for the change in external terms of trade that the rise in petroleum prices represented.

Several indicators of the magnitude of the debt problems are presented in table 2.6. Total debt as a share of total GNP was only about 20% at the beginning of the 1980s. However, this ratio had almost doubled by 1986. Similarly, by the mid-1980s the ratio of interest service to exports was close to 11%, almost double what it had been at the beginning of the decade. Borrowers have to do more than make interest payments, however. They also have to make payments on the principal. Total debt service payments were thus significantly higher than just interest payments. This was especially important since much of the borrowing of the 1970s was on very short terms. Inability to repay meant that the debt had to be refinanced at interest rates that were much higher in the 1980s than they had been in the 1970s.

International debt is serviced either by running a surplus on the trade account, by borrowing additional money, or by some combination of the two. If additional borrowing is to be avoided, policymakers need to increase their trade surplus by either reducing their imports or increasing their exports. This can be done in a number of different ways. The classic means to increase the trade surplus is currency devaluation, which will operate in the desired way on both exports and imports. Devaluations take time to have the desired effect, however. Thus, policymakers typically find ways to reduce their imports in the short term by other means, such as imposing higher tariffs or import quotas. This often reduces the domestic supply of raw materials or critical producer inputs that are imported, thus slowing down economic growth at the very time the economy needs to grow.

Devaluations are painful medicine since they inherently involve reductions in real

Table 2.6: *Debt Indicators for Less-Developed Countries, 1980-86*

Indicator	1980	1981	1982	1983	1984	1985	1986
	%	%	%	%	%	%	%
Ratio of debt to GNP	20.6	22.4	26.3	31.4	33.0	35.8	38.5
Ratio of debt service to GNP	3.7	4.0	4.6	4.5	4.9	5.3	5.5
Ratio of interest service to exports	6.9	8.3	10.4	10.1	10.3	10.8	10.7
			(billions of dollars)				
Total debt	428.6	490.8	551.1	631.5	673.2	727.7	753.4

Source: *World Development Report 1987*, p. 18.

Note: Data are based on a sample of 90 less-developed countries, excluding China.

income for the country undertaking the devaluation. The problem is that, as the currency is devalued, the country has to give up more in terms of domestic resources to earn each unit of currency to pay for exports or to service international debt. This is why policymakers almost inevitably avoid devaluations if they can. To the extent that negative shifts in the external terms of trade require devaluation, the proper policy is to get on with the devaluation so the adjustment process can start immediately and spread widely in the domestic economy.

The production of new agricultural technology through agricultural research can make important contributions to the solution of this problem, especially if a capacity for this research is in place. The supply of new technology for export sectors will make the country more competitive in international markets and thus generate an increased supply of foreign exchange. Similarly, the supply of new technology for domestic sectors that compete with imports can do the same thing. Success on both sides of the trade balance reduces the amount by which the domestic currency has to be devalued in order to reestablish a balance in the external accounts, and thereby limits the reduction in real income the country would otherwise have to experience.

Ideally, policymakers devalue their currency and increase expenditures on their research and extension services at the same time. The devaluation increases price incentives for domestic producers to adopt available technology at a faster rate. Policymakers often discriminated against agriculture by overvaluing their currency prior to the emergence of the debt crisis, which partly explains why supplies of foreign exchange have been limited in these countries and why they have experienced a flood of competitive imports. Thus, devaluations, when warranted, can foster sounder economic growth, independent of the debt problem whose resolution will also increase the payoff from investments in agricultural research.

The only fly in this ointment is that a devaluation of the currency may raise the price of modern inputs if they are imported. Such inputs typically make up only a small component of factor shares, however, and the adoption of new technology will raise the productivity of these inputs. Thus, the effect of a rise in the price of the inputs may be offset in part, at least, by an increase in productivity.

The solution to the debt problem involves more than the policies of the countries experiencing them. More-developed countries need to reduce their barriers to imports from less-developed countries so that the latter countries will have greater access to markets. Similarly, the elimination of export dumping by the United States and the European Community would increase international prices for these commodities and make it possible for less-developed countries to increase their foreign exchange earnings.

2.2.2 Shifts in the External Terms of Trade

A common complaint of policymakers in less-developed countries is that their external terms of trade shift against them over time. This creates a balance-of-trade problem and makes it difficult for them to finance a high rate of economic growth or to service their international debt.

There are many problems with the declining-terms-of-trade argument, at least as it is typically posed. In the first place, there are serious problems in measuring shifts in the external terms of trade attributed in part to the difficulty in making adjustments for changes in traded goods. Much of the measured increase in the price of manufactured goods is due to the failure to account for improvements in the quality of these goods over time. Second, the external terms of trade are unique to individual countries, and it is thus easy to overgeneralize about the decline in terms of trade. And third, whether the terms of trade are declining is determined in part by the choice of time period in which to make the comparison.

Despite these caveats, declines in the external terms of trade often pose problems. The issue is what to do about it. Just as in the case of the international debt problem, the classic remedy is to devalue one's currency. This will deal with the balance-of-payments problem and increase the supply of foreign exchange to finance a higher rate of economic growth and to service internationally held debts. Just as in the case of the debt problem, however, it will also result in a reduction in national income.

Here again, agricultural research and the introduction of new technology can play an important role in addressing the problem. After all, the real prices of primary commodities often decline because a process of technical change is taking place elsewhere in the international economy. The only way to deal with this problem is to sustain a comparable rate of technical change in the domestic economy. If that is done, the effects of the decline in the terms of trade will be offset by an increase in the domestic economy. If the rate of productivity growth in the domestic economy is higher than that in the international economy, the domestic economy will actually benefit as its supply of foreign exchange grows because of expanded markets.

As in the international debt problem, moreover, there is much to be said for a combination of devaluation and a vitalized research effort. The introduction of new production technology will reduce the extent to which the domestic currency has to be devalued, but devaluation can provide shorter-term relief and put the economy on the road to broad adjustment.

2.2.3 Unstable International Monetary Conditions

The exchange rate is the most important price in an economy (Schuh 1986). As it changes over time, it affects the relative prices in an economy between tradables and nontradables, the allocation of resources between these same two sectors, and the distribution of income in the economy. It also affects the terms on which domestic resources are exchanged for foreign goods and services, the competitiveness of domestic export sectors, and the ability of domestic sectors to compete with imports.

Currently, there are many complaints about the instability in exchange rates as a result of the end of the Bretton Woods fixed exchange rate system. Typically, the complaints refer to short-term, day-to-day instability. From our perspective, this is not the critical issue since the risk associated with this instability can be transferred to other sectors by means of futures markets and other marketing arrangements. The more serious problem involves the long swings in the real value of national currencies referred to above. These long swings can mask the underlying comparative advantage for substantial periods of time, reduce the payoff from investments in agricultural research, and create uncertainty about the supply of foreign exchange.

The issue is what can be done about these swings? Reform of the international monetary system is one thing that can be done about it. But barring that, agricultural research can provide some assistance by having a sufficiently diverse agenda where new technology is available for those sectors that bear the burden of adjustment to the swings in exchange rates. Moreover, research that helps build more flexibility in the production sector can make the adjustment problem less burdensome.

2.2.4 The Persistent Need to Diversify

Agricultural diversification issues arise from economic and technological forces operating at the international level, at the national level, and within regions in individual countries. An important component of the diversification issue is the need to transfer resources out of agriculture as economic development proceeds. We approach this set of issues by considering, first, the nature of the diversification problem within agriculture as development proceeds; then, the need to transfer resources out of agriculture; and then international adjustments. Regional issues are left aside since they tend to be subsets of the above.

Increases in per capita income associated with economic development induce strong pressures for agricultural diversification. In the first place, as income increases, there are relative shifts in the demand for individual commodities that induce changes in consumption patterns, other things being equal. Consumption patterns shift away from a dependence on tubers such as cassava and less-preferred grains such as sorghum and millet, and toward higher-income grains such as rice and wheat. Continued increases in income shift demand further, bringing about increases in the demand for livestock and livestock products,

68 *Schuh and Norton*

poultry, and fruits and vegetables.

An important component of these shifts in consumption patterns on the international scene is the shift out of direct foodgrain consumption and into the indirect consumption of feedgrains in the form of livestock and livestock products. The demand for feedgrains is derived from the demand for these latter products. Although the consumption of feedgrains is at low levels in most less-developed countries, large income elasticities for livestock and livestock products suggest that there will be significant increases in the demand for feed grains in the future. Research programs need to anticipate these shifts.

Current consumption patterns differ significantly among regions in the less-developed world, largely reflecting differences in the stage of economic growth (table 2.7). In sub-Saharan Africa, for example, which has very low per capita incomes (see chapter 6), the shift from food- to feedgrains is proceeding quite slowly, while major substitutions are occurring among the foodgrains. The consumption of rice and wheat has increased rapidly, while consumption of other, more traditional foods has increased very little or declined.

Table 2.7: *Annual Growth Rates in the Consumption of Grains, 1961-83*

Region	Wheat		Rice		Maize		Other coarse grains		Total cereals	
	Feed	Food	Feed	Food	Feed	Food	Feed	Food	Feed	Food
	%	%	%	%	%	%	%	%	%	%
Sub-Saharan Africa	7.7	6.2	3.2	5.1	5.3	3.3	1.6	1.4	3.9	2.9
Asia & Pacific	6.6	5.9	3.4	3.2	1.5	2.9	1.5	0.03	6.0	3.4
Latin America & Caribbean	4.9	3.5	4.3	3.5	4.2	2.8	9.1	3.4	5.4	3.2
West Asia & North Africa	5.5	4.3	1.0	4.5	7.9	2.7	3.4	1.0	4.7	3.6

Source: Pinstrup-Andersen (1986).

Another factor affecting consumption patterns is urbanization. Urban areas consume more wheat (and rice in some cases), vegetables, and livestock products, while rural areas consume more traditional foods. Urbanization is, in effect, a surrogate for a variety of more fundamental economic forces. For example, some of the observed changes in consumption patterns can be attributed to increases in per capita incomes, others to differences in the opportunity cost of women's time, and others to differences in relative prices.

Still another factor is the change in production costs as an economy experiences increases in per capita incomes. Increases in real wages are associated with increases in per capita incomes. Since commodities differ in their labor intensity, shifts in the structure of production will be induced as labor costs rise.

Finally, technical change itself can induce changes in production patterns. It does this by being non-neutral in its effects on resource use and by eventually lowering the price of the commodity — thus bringing about substitution effects.

We now turn to a consideration of the need to diversify resources out of agriculture as economic development proceeds. The income elasticity of demand for agricultural production is inherently less than that for nonagricultural goods and services. This by itself implies the need to shift resources out of agriculture. But if technical change raises the productivity of resources in agriculture, this will still further reduce the demand for resources in agriculture. This issue is important because it flies in the face of so much of conventional wisdom.

Last, there is the issue of diversification on the international scene. As technical change occurs in specific commodities or commodity groups, general equilibrium effects will be induced as shifts in supply outpace the growth in demand and as the prices of affected commodities decline. This causes resources to shift out of the affected sectors and the output of other commodities to expand. The international implications of this can be quite great since comparative advantage could shift on a broad scale.

2.2.5   Environmental and Natural Resource Problems

Although food production is expanding in most areas of the world, deterioration of the natural resource base threatens agricultural and domestic development in a number of countries, with important implications for agricultural research. In some cases, the technical knowledge is currently available on how to sustain the productive natural resource base in the face of expanding population pressures, but the institutional means for encouraging the implementation of that knowledge is not. In other cases, the technical knowledge is also lacking, particularly for the fragile soils in parts of Africa (Lal 1987). The result is that one observes deforestation, overgrazing, desertification, and increasingly severe soil erosion and flooding around the world. In addition, lakes, streams, and rivers are becoming increasingly polluted with industrial wastes and agricultural runoff, weakening their potential as a direct source of food and threatening their potential as a source of water for modern agriculture. Biologists, physical scientists, socioeconomists, and policymakers must cooperate to help solve these environmental problems.

A concerted effort will be required, both on the part of national agricultural research systems and on the part of multilateral and bilateral development assistance agencies. Natural resource or environmental problems have implications both within countries and internationally. The river silting and resultant flooding of northern India as a result of deforestation and soil erosion in Nepal is one example; so are the changes in climate and water quality associated with the cutting of tropical rain forests and with certain types of industrial growth.

Many natural resource problems are associated with the most marginal lands within countries and can be understood only as a subset of the general problem of rural underdevelopment and poverty. Marginal lands become a problem when the socioeconomic factors associated with underdevelopment are combined with a land resource subject to degradation under expanding human use (Bremer et al. 1984). Consequently, part of the solution may lie in creating opportunities for off-farm employment or intensification of the use of nonmarginal lands, while another part lies in intervention on the marginal lands themselves.

National agricultural research systems seeking solutions to problems of environmental degradation must consider on- and off-site technologies and institutions (chapter 3).

Recently, the World Bank has recognized the need to place greater emphasis on issues of environmental degradation associated with development activities. Bilateral assistance agencies are also expanding support for activities in the area of preserving natural resources. Institutional means must be found for (a) creating incentives to reduce regional and international environmental externalities, (b) overcoming rates of time preference that skew consumption toward the present in many less-developed countries, and (c) offsetting income elasticities of demand for environmental improvements that value food consumption over the environment. The latter two factors are closely related to income levels.

Economic policies contribute to many of these environmental and sustainability problems. For example, policies that discriminate against agriculture in poor countries cause land and water to be undervalued, as do the export-dumping policies of the European Community and the United States. When land and water are undervalued, producers have less incentive to protect their value as productive agents, or to invest in enlarging the flow of services from them in the future. Moreover, these same policies, together with the failure to invest in the creation and diffusion of productivity-enhancing technologies, causes output to expand on the extensive margin rather than on the intensive margin. The result is to push agriculture into marginal areas, up hillsides, and onto land with fragile topsoil, where degradation and erosion are almost inevitable.

Thus, once again one observes the complexity between economic policy and science and technology. More nearly optimal policies on both sides can do much to improve sustainability and to strengthen the underlying resource base for the future.

2.2.6   The Role of Foreign Aid to Agriculture in the Context of International Trade

An international policy issue with broad and important ramifications for agriculture and for agricultural research systems in less-developed countries, as well as for policymakers in more-developed countries, is the importance and desirability of foreign economic assistance for agricultural development. Countries are poor in large part because of the relatively small amount of capital per worker or per hectare in their agricultural sectors. Much foreign assistance involves capital transfers and might, therefore, relieve a major development constraint. Technical assistance and food aid are, of course, other major forms of foreign assistance. Agricultural research systems are often recipients of foreign assistance.

Recently, policymakers in the United States (and to a smaller extent in other more-developed countries) have received pressure from agricultural commodity groups to reduce foreign assistance to less-developed countries on the grounds that aid encourages competition with US agricultural exports. Some economists have supported this view (Avery 1985). However, many agricultural economists have argued that agricultural growth stimulates income growth with resulting positive effects on less-developed country imports of US farm products.[4] This debate is of importance to agricultural research institutions in less-developed countries if the following apply: (a) if in fact aid to agricultural research systems has borne positive results, (b) if aid to agriculture in general (including policy reform, credit programs, infrastructure, etc.) has influenced adoption of new technologies, and (c) if (assuming aid has had positive results) a misunderstanding of the impact of aid on agricultural trade has resulted in the curtailment of aid, thereby hindering agricultural and overall development.

The studies cited above presuppose that foreign aid increases agricultural productivity. The authors then examine the impact of agricultural productivity on trade. Surprisingly little empirical analysis of the linkage between foreign aid and agricultural productivity actually exists. However, Peterson (1989), who fitted a Cobb-Douglas production function to data from 113 countries, does provide evidence that capital transfers from rich to poor nations increase output per worker (including nonagricultural output) in less-developed countries.

In an attempt to examine the impact of foreign assistance on agricultural productivity more closely, we estimated the parameters of an aggregate agricultural production function for a sample of 98 less-developed countries using cross-sectional time-series data. Official development assistance (ODA) was included as a variable in the analysis. The basic model was as follows, with all variables measured in logs:

$$Q = \alpha_0 + \alpha_1 X_1 + \alpha_2 X_2 + \alpha_3 X_3 + \alpha_4 X_4 + \alpha_5 X_5 + \alpha_6 X_6 + \alpha_7 X_7 \tag{2.1}$$

where Q measures the real value of agricultural gross domestic product; X_1 is livestock measured in number of cattle equivalents; X_2 is labor measured as economically active population in agriculture; X_3 is a land quality index; X_4 measures tractor power; X_5 is a schooling variable measured as the number of pupils enrolled in primary and secondary levels; X_6 is a level-of-technology variable proxied by the number of pupils enrolled in the third level of schooling; X_7 is the real value of foreign aid (ODA); and $\alpha_0, \ldots, \alpha_7$ is a set of coefficients to be estimated.

The output and input variables represent annual data from 1975-1985, while the foreign aid variable — included as a quadratic distributed lag of a three-year moving average of ODA receipts — was lagged six years back to 1970. The output variable measured in nominal local currency units was first deflated to 1980 currency units using country-specific implicit agricultural GDP deflators and then converted to an "international" dollar using 1980 purchasing power parity indices obtained from Summers and Heston (1988). To reduce problems with heteroskedasticity due to large differences in country size, all outputs and inputs were measured on a per hectare basis.

The results of the estimation are presented in table 2.8. Seven models were estimated.

---

[4] See for instance, Kellogg, Kodl, and Garcia (1986), Lee and Shane (1985), de Janvry and Sadoulet (1986b, 1988) Houck (1987), Vocke (1987), and Christiansen (1987).

Table 2.8:   Agricultural Production Function with Foreign Aid Variables for 98 Less-Developed Countries

Explanatory variable	Model 1	Model 2	Model 3	Model 4	Model 5	Model 6	Model 7
Constant	-1.421 (-9.60)[a]	-3.325 (-20.05)	-3.704 (-20.95)	-3.270 (-12.34)	-2.462 (-13.27)	-0.819 (-4.24)	-1.250 (-6.02)
Labor[b]	0.463 (20.06)	0.366 (21.85)	0.442 (23.95)	0.518 (29.05)	0.518 (29.05)	0.518 (29.05)	0.518 (29.05)
Land quality	0.375 (9.76)	0.901 (20.06)	0.855 (19.25)	0.673 (15.75)	0.673 (15.75)	0.673 (15.75)	0.673 (15.75)
Livestock[c]	0.275 (17.69)	0.336 (23.39)	0.297 (20.24)	0.213 (14.60)	0.213 (14.62)	0.213 (14.62)	0.213 (14.62)
Tractor horse power[d]	0.177 (22.52)	0.104 (12.68)	0.089 (10.79)	0.087 (11.38)	0.087 (11.38)	0.087 (11.38)	0.087 (11.38)
Primary and secondary education[e]	0.070 (6.28)	0.081 (8.06)	0.081 (7.85)	0.069 (7.21)	0.069 (7.21)	0.069 (7.21)	0.069 (7.21)
Higher education[e]	0.073 (11.20)	0.052 (8.64)	0.040 (6.58)	0.020 (3.48)	0.020 (3.48)	0.020 (3.48)	0.019 (3.48)
Foreign aid	0.005[f] (0.68)	0.027[f] (3.50)	0.051[f] (5.50)	0.127[f] (8.25)	0.030[f] (3.05)	-0.010[f] (-0.87)	-0.011[f] (-1.03)
Sub-Saharan Africa intercept dummy	—	—	—	0.808 (3.18)	—	-1.640 (-8.40)	-1.212 (-6.34)
Sub-Saharan Africa slope dummy on aid	—	—	—	-0.970[f] (-5.60)	—	0.390[f] (7.43)	0.041[f] (7.27)
West Asia & North Africa intercept dummy	—	1.859 (9.96)	2.200 (11.01)	2.450 (9.10)	1.642 (8.40)	—	0.430 (2.00)
West Asia &North Africa slope dummy on aid	—	-0.077[f] (-5.25)	-0.091[f] (-6.33)	-0.137[f] (-9.10)	-0.039[f] (-2.77)	—	0.013[f] (0.09)

Table 2.8: *Agricultural Production Function with Foreign Aid Variables for 98 Less-Developed Countries (Contd.)*

Explanatory variable	Model 1	Model 2	Model 3	Model 4	Model 5	Model 6	Model 7
Asia & Pacific intercept dummy	—	—	—	—	-8.080 (-3.17)	-2.450 (-9.10)	-2.020 (-7.25)
Asia & Pacific slope dummy on aid	—	—	—	—	0.097[f] (5.60)	0.137[f] (7.43)	0.138[f] (7.27)
Latin America & Caribbean intercept dummy	—	—	1.154 (5.77)	2.020 (7.25)	1.212 (6.34)	-4.300 (-2.00)	—
Latin America & Caribbean slope dummy on aid	—	—	-0.062[f] (-4.40)	-0.138[f] (-7.27)	0.041[f] (-3.03)	-0.00002[f] (0.09)	—
$\bar{R}^2$ (n = 98)	0.905	0.924	0.927	0.938	0.938	0.938	0.938

Source: Output, World Bank (1989); Land Quality, Peterson (1987); Labor, Livestock, Machinery and, Land, *FAO Production Yearbooks*; Education, *UNESCO Statistical Yearbooks*; Foreign Aid, *OECD Development Cooperation.*

[a] Figures in parantheses are *t*-ratios.

[b] Defined as the number of economically active population in agriculture.

[c] A measure of "animal unit aggregates" with weights of: 0.8 for cattle and asses; 1.0 for horses, mules, and buffalos; 1.1 for camels; 0.2 for pigs; 0.1 for sheep and goats; and 0.01 for chickens, ducks, and turkeys.

[d] Represents number of tractors weighted by a time-varying (country-invariant) weight of average tractor horsepower where weights represent a linear interpolation and extrapolation of Hayami and Ruttan's (1971, 1985) 1970 and 1980 estimates of 35 hp and 40 hp, respectively.

[e] Measures the number of students enrolled in primary plus second and third levels of education.

[f] Calculation based on coefficient taken from the distributed lag variable.

[g] Measures net receipts by individual less-developed countries of total net official development assistance from DAC countries and territories. Grants, loans, and credit for military purposes are excluded by definition.

The coefficients of all variables were significant to at least the 5% level, except for foreign aid. The coefficient for foreign aid was significant in some regressions but not in others. In model 1, which included all 98 countries in the aid variable, the coefficient on foreign aid was positive but nonsignificant at the 5% level. However, in model 2, where an intercept dummy variable and a slope dummy variable on foreign aid were included for West Asia & North African countries, the foreign aid variable was highly significant for the remaining countries.

In model 3, the countries in both West Asia & North Africa and Latin America & Caribbean were excluded. As a result, the coefficient on foreign aid for the remaining countries became larger and more significant than in model 2. In model 4, all countries except those in Asia & Pacific were excluded and the foreign aid coefficient became larger and still more significant. In model 5, all countries except those in sub-Saharan Africa were excluded and the aid coefficient, although smaller and less significant than for Asia & Pacific, was still significant at the 5% level. However, when all countries except for those in West Asia & North Africa were excluded, foreign aid was nonsignificant. This was also the case when all countries except those in Latin America & Caribbean were excluded.

The conclusion that can be drawn from this is that foreign aid had a positive and significant impact on agricultural output in sub-Saharan Africa and particularly in Asia & Pacific, from 1975-1985. Impacts on agriculture in West Asia & North Africa and Latin America & Caribbean were, on average, nonsignificant. The agricultural marginal value product (MVP) of foreign aid in Asia & Pacific was around $10.40 per dollar of aid. The aid MVP in sub-Saharan Africa was $0.40, and for the world as a whole, except for West Asia and North Africa, it was $0.85. While these MVPs may at first appear to be a small return on the dollar, remember that the measure of official development assistance used as the foreign aid variable in the analysis was directed at nonagricultural as well as agricultural development. The agricultural impact is, therefore, an underestimate of the total impact.

The results of the analysis are time-period specific and clearly vary by region. The effects of foreign aid in Latin America & Caribbean may have been masked by the effects of the debt crisis in several countries of that region. A high proportion of the aid in West Asia & North Africa may well have been directed at nonagricultural programs. It appears that aid has had a positive impact on agriculture in the most populous region of the world (Asia & Pacific) and in the poorest region (sub-Saharan Africa). This evidence that aid has had at least some of its intended economic effects should encourage recipients and donors alike.

If aid has, in fact, improved agricultural productivity, the next question is whether increased agricultural productivity has spurred overall economic growth, consumption, and additional imports. Several recent studies have examined different aspects of this question, as noted earlier. One of the studies that examined each of these pieces was by de Janvry and Sadoulet (1986b, 1988). The estimated model included the following equations based on growth rates between 1970 to 1980 for 60 less-developed countries:

Growth rate of manufacturing:

$$\dot{I} = \alpha_0 + \alpha_1 \dot{A} + \alpha_2 \dot{X} + \alpha_3 \dot{P} \tag{2.2a}$$

Income equation:

$$\dot{Y} = \beta_0 + \beta_1 \dot{A} + \beta_2 \dot{I} \tag{2.2b}$$

Consumption equation:

$$\dot{C}_i = \gamma_{0i} + \gamma_{1i} \dot{Y} = \gamma_{2i} \dot{U} + \gamma_{3i} \dot{POP} \tag{2.2c}$$

Import equation for product i:

$$\dot{M}_i = (C_i / M_i) \dot{C}_i - (Q_i / M_i) \varepsilon_i \dot{A} \tag{2.2d}$$

such that

Agricultural growth structure equation:

$$\varepsilon_i = \dot{Q}_i / \dot{A}$$

and where $\dot{Y}$ represents the growth rate of GDP; $\dot{C}_i$ is the growth rate of consumption of agricultural product i; $\dot{U}$ is the growth rate of urbanization; $\dot{POP}$ is the growth rate of population; $\dot{Q}_i$ is the growth rate of agricultural product i; $\dot{M}_i$ is the growth rate of net imports of agricultural product i; $\dot{P}$ is the inflation rate; and $\dot{A}$ is the growth rate of agriculture.

From the above model, they derive the elasticity of import demand for product i with respect to agricultural growth:

$$d\dot{M}_i / d\dot{A} = \varepsilon_i = (g_i - \varepsilon_i)/D_i$$

where $g_i = \gamma_{1i}(\beta_1 + \alpha_1 \beta_2) = (d\dot{C}_i / d\dot{A})$ represents the elasticity of consumption with respect to agricultural output, and $D_i = (M_i / C_i)$ represents a "dependency" ratio for product i.

Their estimated parameters from this model for all 60 countries as well as the results for the least-developed countries (GDP per capita <$600) are shown in table 2.9. They found the elasticities of manufacturing output with respect to agricultural output and of consumption with respect to income to be positive and highly significant. Their derived elasticity of consumption with respect to agricultural output was highest for wheat in the least-developed countries and for maize in the newly industrialized countries, as one might expect.

The resulting elasticity of import demand with respect to agricultural growth is positive or negative depending on g_i, on the level of dependency for product i ($D_i = M_i/C_i$), and on the growth rate of product i relative to that of agriculture in general (ε_i). Moreover, de Janvry and Sadoulet found that agricultural growth led to growth in the demand for

Table 2.9:   *Parameter Estimates from Growth Model, 1970–80*

Per capita GDP (1965 US$)	Number of countries	α_1	β_1	β_2	δ_{li}			g_i			ε_i		
					cereal	wheat	maize	cereal	wheat	maize	cereal	wheat	maize
<600	37	0.94*	0.56*	0.31	0.26*	1.01*	0.27	0.22	0.86	0.23	1.20	1.43	1.39
>600	23	0.56*	0.50*	0.46	0.35*	0.36	1.34*	0.26	0.27	1.01	1.62	-0.35	1.24
all	60	0.82*	0.53*	0.37	0.33*	0.80	0.93	0.28	0.67	0.78	1.33	0.79	0.78

Source: de Janvry and Sadoulet (1988, p. 10).

* Significant at the 5% level.

imports of cereals in 27% of the countries in their sample, of wheat in 90%, and of maize in 48%. The countries that had positive growth rates of agricultural output per capita, high growth in per capita GNP, non-negative growth rates of product i relative to the growth rate of agriculture (ε_i), and positive elasticities of import demand for product i with respect to agricultural growth, included South Korea (cereals and maize), Brazil (wheat and maize), Malaysia (cereals and maize), Egypt (wheat), Tunisia (cereals and maize), Kenya (wheat), Guatemala (wheat), Colombia (maize), and Paraguay (wheat).

These examples suggest a strong relationship in many cases between increased agricultural production and increased agricultural imports. The primary reasons for this relationship are, first, that agriculture is an important sector in most of the less-developed countries and consequently overall economic growth depends on agricultural growth. Second, people in less-developed countries have high income elasticities of demand for food, often 0.5 or higher, which means that a high proportion of every extra dollar is spent on food. Population growth rates are still relatively high in most of the less-developed countries and when a high population growth rate is combined with a high income elasticity of demand for food, even modest per capita income growth can cause the demand for food to exceed increases in domestic production. In addition, diets shift to livestock products with resulting increases in the derived demand for feedgrains, as discussed earlier.

The above scenario depends to a major extent on (a) the size of the income elasticity of demand for food and (b) the degree to which agricultural growth is translated into overall income growth. Income elasticities depend on the stage of development, which implies that, in the long run, as development occurs and income elasticities of demand for food decline, then the less-developed and more-developed countries will increasingly compete for world markets. The degree to which agricultural growth is translated into overall income growth depends on the growth paths that countries choose (e.g., heavy reliance on industrial exports, cash crop exports, oil and primary product exports, food production increases through technical change, or some other path). Furthermore, those countries that follow a path of increasing food production through technical change may, with the help of public policies, translate that growth in food production into broad-based, employment-generating growth or perhaps into narrower capital-intensive industrial growth.

2.3   IMPLICATIONS FOR AGRICULTURAL RESEARCH

Considering agricultural research in the context of international policy leads to a substantial broadening of the research agenda from what it is conventionally taken to be. This section draws the implications from the two previous sections for this broadened agenda.

1.   The development of more flexible agricultural sectors should receive priority in research programs. Not only do resources need to flow in and out of agriculture as international monetary conditions change, but they also need to shift on the margin back and forth between the tradable and nontradable sectors. An important corollary is that more efficient adjustment policies and programs need to be

designed for agriculture.

2. Assessments of competitive advantages must take into account configurations of monetary and fiscal policy on the international scene.

3. Strategies are needed for dealing with long swings (five to seven years) in real exchange rates which mask underlying comparative advantage.

4. Because of the decentralization of economic policymaking and implementation, decision making about agricultural research also needs to be decentralized. It also needs to reflect local resource endowments and to give more attention to local resource problems. An important corollary is that expanded research efforts in the social sciences are needed to design local institutional arrangements to deal with agricultural poverty.

5. National research systems need to give more attention to identifying the constant changes in the comparative and competitive advantage of agriculture in their country as the basis for sharpening their research priorities. However, their priorities should not reflect a passive acceptance of changing comparative advantage, but rather, should seek to identify potential niches in the emerging pattern for their producers and to commit resources to help them realize their potential advantages.

6. An important related challenge is to identify those cases in which a country's external terms of trade are shifting against it because of technological developments in other parts of the world. If the potential for catching up in the domestic economy can be realized efficiently, resources should be committed to this end. This will help the country deal with its longer-term problems with balance of payments.

7. An international effort is needed to verify the extent of declines in national postgraduate training in the agricultural sciences and agricultural research capacity in the less-developed countries. Programs to arrest this decline and restore growth should be developed.

8. Research is needed that identifies the extent to which international trade patterns are a reflection of existing trade and exchange rate distortions. An important part of establishing national agricultural research priorities is to determine whether national research efforts have a higher social payoff if domestic export sectors are made more competitive or if domestic sectors that compete with imports are made more competitive. More generally, research priorities need to be established in terms of the reality of existing or probable future trade and exchange rate distortions.

9. Social science research that helps reduce existing barriers to trade among both the more- and less-developed countries can lead to significant efficiency gains on the international scene and a more equitable distribution of global income. It can also raise the social rate of return from investments in agricultural research and thus induce a larger flow of resources to this important source of economic

growth.

10. Social science research is needed for a better understanding of international labor markets and for designing policies and institutional arrangements to facilitate the transfer of labor out of agriculture as economic development proceeds. This will help to realize the full benefits of agricultural research and thus raise the social rate of returns to investments in such research.

11. The research agenda to deal with environmental and sustainability problems is growing rapidly. Because of the long lags involved, more resources need to be committed to assessing and dealing with the potential problems of global warming. Of particular importance are studies that assess the regional impact of global warming around the world. To date, such studies have focused primarily on the United States. On deforestation and sustainability issues, the highest priority should go to evaluating existing policies that motivate such counterproductive activities and to designing policies and institutional arrangements that lead to more socially rational behavior.

12. Careful assessments are needed of the extent to which agricultural modernization leads to more general economic growth by means of the production and dissemination of new productive technologies, as well as the extent to which the benefits of economic development that this generates rebound to the benefit of low-income groups. Such research should help justify expenditures on agricultural research by both national governments and the international community.

13. Research is also needed to better understand how such broad-based economic growth translates into import demand and the structure of international trade.

14. Social science research needs to receive much higher priority on both the national and international scene. This research is needed to guide domestic economic policy in directions that minimize distortions to underlying comparative advantage, to understand policies in other countries that affect comparative and competitive advantages, and to assist in establishing research priorities domestically. It is also needed to better understand the linkages between economic policies and science and technology policies.

2.4   CONCLUDING COMMENT

The days are gone when highly segmented national economies could develop agricultural research agendas in isolation of the rest of the world and without taking into consideration the effects of domestic and international economic policies. The research agenda that results from the consideration of international policies and the changing structure of the international economy is far more complex than the agenda from the more segmented world, as is the problem of establishing research priorities. Moreover, an important part of the broadened agenda is the need for a stronger social science research agenda and a more sensitive interaction between the social sciences on one side and the biological, natural, and physical sciences on the other.

Chapter 3

Sustainability: Concepts and Implications for Agricultural Research Policy

Theodore Graham-Tomasi[1]

The agricultural research community is placing increasing emphasis on a somewhat vague notion called "sustainability." The concept of sustainability encompasses a wide variety of concerns regarding the potential for economic development to run into resource and environmental constraints, which act to retard future progress. The incorporation of these potential constraints into the analysis of current systems in general, and into agricultural research programs in particular, is a major undertaking. But it is one that many view as critical to improving the quality of life of the world's burgeoning population.

The purpose of this chapter is to provide a framework for analyzing the implications of sustainability concerns for agricultural research; hence, it is quite limited in scope. It addresses some issues in the definition and measurement of sustainability and suggests some implications of sustainability for research policy.

The basic conclusion drawn here is that, at this juncture, sustainability is a broad set of concepts which should serve to guide research in all of its facets. It is not a set of technologies that can be recommended for adoption, nor is it close to being so. Even achieving an operational definition of sustainability is problematical. Perhaps the notion of sustainability will never move beyond being an implicit framework for organizing a set of reactions to environmental and resource concerns. However, given the potential long-term importance of these concerns, the sustainability concept is likely to play a role in research policy and management for some time to come.

---

[1] This chapter was originally prepared as a discussion paper for ISNAR; the author acknowledges ISNAR's support, while absolving it of any responsibility for the product. The author thanks Vernon Ruttan, G. Edward Schuh, Hans Gregerson, C. Ford Runge, and the staff of ISNAR, especially Krishan Jain, Willem Stoop, Howard Elliott, and Philip Pardey, for helpful conversations on this chapter.

3.1   SUSTAINABILITY: BASIC ISSUES

3.1.1   Concepts of Sustainable Development

The concept of sustainable development can be traced to the debates of the early 1970s concerning the limits to growth, a discussion spurred by the widely read work of Meadows and co-workers (1972). The ability to maintain the pace of economic development, while accounting for changes in the resource base upon which development depends, was given a central focus in the report of the World Commission on Environment and Development (1987), known as the "Brundtland Report." On page 43 of this report sustainable development is defined as "... development that meets the needs of the present without compromising the ability of future generations to meet their own needs." In a related vein, the TAC/CGIAR (1989, p. 3) defines sustainability as "... the successful management of resources for agriculture to satisfy changing human needs while maintaining or enhancing the quality of the environment and conserving natural resources."

These days, just about everyone is on the sustainability bandwagon, and sustainability has come to mean all things to all the riders on this bandwagon![2] But few have tackled the difficult chore of translating the idea of sustainability, which has taken on the features of myth, into a set of practical evaluative criteria.

In providing an operational definition of sustainability, it must be recognized that, for any given system, one may wish to sustain more than one aspect of the system, and conflicts may arise. Also, an approach to sustainability in one geographic area may conflict with or enhance sustainability in others. There are other social goals besides sustainability that are relevant, and indeed, we shall argue that a sustainability criterion is not operational without reference to other social objectives. Thus, the concept of "sustainability" without further specification of what is to be sustained, at what levels, over what geographic area, and without elaboration of the relationship between sustainability objectives and other objectives, is devoid of content and not useful for scientific discourse and serious policy analysis.

Definitions

In the literature on renewable resources, as well as in common usage, sustainability refers to the ability to maintain a given flow over time from the base upon which that flow depends. A simple analogy of a stock of funds in a bank is useful. A level of consumption is sustainable if it costs no more than the interest earned from those funds; in this way the base of wealth in the bank remains intact, allowing for at least as much consumption in the future as today.

Swindale (1988) remarks that this is a static notion of sustainability and suggests that a more dynamic formulation is needed. The more dynamic definitions that have appeared

---

2 See the myriad definitions in Pezzy (1989) and the discussion by Batie (1989).

arise either from further specification of what is to be sustained or from combining sustainability objectives with others. For example, if one wishes to sustain levels of consumption per person and the population is growing, then growth in the output of consumption goods is necessary. Similarly, if the prices of inputs and outputs are changing, because, for example, of increasing energy costs and/or changing final demands, then sustaining the profitability of farming requires that the production systems be continuously altered. Finally, one might be interested in improving the quality of life over time and not just in maintaining the status quo.

In all of these situations and in others, the basic idea of sustainability becomes more dynamic. In this more elaborated form, sustainability becomes "sustainable development." As Ruttan (1988) argued, sustainability is "not enough."

In defining sustainability, *what* is to be sustained must be specified. It is proposed here that the object to be sustained is overall, aggregate well-being in individual countries. Short of a global definition, this approach allows the broadest consideration of effects on well-being with the broadest geographic and sectoral scope. Of great importance is the manner in which aggregation over sectors, areas, and individuals is to be achieved, and this depends on country-specific internal policy considerations. Hence, it is not possible to offer even an operational definition of sustainability without including additional social objectives; analyses of sustainability cannot be based on sustainability considerations alone.

As emphasized by Lynam and Herdt (1989), in practice systems that are smaller than the aggregate welfare of entire countries must be specified. Sustainability of these smaller systems then becomes an instrumental sub-goal for the sustainability of the larger one. A useful specification of a hierarchy of systems in agriculture is Conway's (1984) agroecosystems approach. But focusing on a lower system level does not mean that higher ones may be ignored. In order to avoid having sustainability at a higher level being undermined by activities at lower levels, the sustainability criteria must be consistent across system levels, with measurement at one level reflecting concern for linkages to the next.

By way of example, Lynam and Herdt (1989) suggest that, at the farming system level, the natural indicator of output is total factor productivity (value of output divided by value of inputs). If the evaluation of sustainability at the farming system level is to be consistent with sustainability at a higher level, then the values attached to the inputs and outputs must reflect linkages to other systems. Many such linkages exist, such as downstream siltation of reservoirs from erosion and water pollution due to pesticides and fertilizers. These linkages are not always incorporated into readily measured indicators of sustainability at lower levels, which raises severe difficulties for the measurement of sustainability of subsystems.

Substitution

It is vital to note that, by focusing on overall well-being as the object of sustainability, it is not necessary, and may even be counterproductive, to insist on the sustainability of every component subsystem. Resorting to the earlier analogy, there may be several ways of holding funds to provide interest income, e.g., in domestic banks, in foreign banks, in government issued bonds, and in private company stock. One does not need to sustain each source of funds separately, but one does need to sustain the productivity of the overall complex.

It is possible to employ some resources in excess of sustainable levels while maintaining the overall productivity of the resource base, and this takes advantage of the substitution possibilities among resources. Many analysts have been excessively concerned about the sustainability of particular components of an overall system, while ignoring substitution possibilities among components. Such substitution for and among natural resources is similar to the ability of machinery to substitute for labor in crop production.

At a farming system level, increased knowledge about the varieties of soils within a producing unit can be used to better tailor fertilizer applications to soil needs, thereby reducing overall fertilizer inputs without sacrificing crop outputs, as well as (perhaps) increasing total factor productivity. At a higher level of aggregation, the forests of North America have been diminished greatly, and they have been replaced with agricultural land. While the forest resource has not been sustained, the overall capability of the system to provide social well-being has been enhanced.

The key concept of substitution has been used by resource economists to point out that sustainability can be achieved where it would seem to be impossible. Suppose, for example, that there are three inputs to the production of a single good: a nonrenewable resource, labor, and manufactured capital. Suppose further that the input is drawn from a finite resource and that it is essential to production, i.e., if this input is zero, then output is zero as well. The single good produced can be either consumed or invested in new capital. Then, as long as capital and/or labor can provide a sufficient substitute for the resource, it is possible to maintain a positive level of per capita consumption for a fixed population forever, despite the resource constraint.[3] While this is obviously a highly stylized result of limited practical importance, it does serve to point out that substitution possibilities should not be ignored.

3.1.2 Alternative Concepts of Sustainability

A number of alternative concepts have been put forward under the rubric of sustainability. In this section some of these concepts and their relationship to the definition proposed here are discussed; for further analysis, see Pezzy (1989).

Before proceeding, let us dispense with one issue. The approach taken in this chapter is unabashedly anthropocentric. Some reject this approach altogether and seek alternative evaluative schemes, based, for example, solely on considerations of energy flows in ecosystems. Here, the concern is with enhancing the well-being of the human species, with

---

3 For details, see Dasgupta and Heal (1979, pp. 193-207).

appropriate concern given to the relationship between our species and the natural world.

Stability

Conway (1985) has proposed defining sustainability in terms of the stability of the system. There are two aspects of stability that appear to be relevant.

First, if stability means an absence of fluctuations, it is not necessarily the case that instability is opposed to social well-being. A tight distribution about a low and constant average might not be preferred to a wide distribution around a high and growing average. Since fluctuations will occur, one must confront the possible trade-offs between averages and variability. The willingness to make such trade offs may differ substantially according to the level at which the trade-offs are assessed — whether it is at the level of the individual farmer or some higher level of social aggregation. For the purposes of this chapter, attention is directed to averages; further analysis of the implications of fluctuations for agricultural research is provided in chapter 4 of this volume.

Second, this notion of fluctuation does not really capture the concerns raised by Conway (1985), which seem more directed to stability as the way of withstanding shocks to the system. This view is in close accord with the definition of sustainability offered above. If it is believed that a series of shocks to the system will occur and that the system will move progressively into degraded states from which it cannot recover, then the issue is not one of random fluctuation about some mean, but of a deteriorating mean itself. The chances of shocks occurring, the rate of degradation of the system, and the time horizon for planning all interact when one is assessing the impact of the stability of the system on sustainability.

Clearly, a resilient system is preferred to one that is adversely and irreversibly affected by shocks. However, the current resilience of a system that is changing may be difficult to assess using data from the recent past. This is because, in the absence of the kinds of harmful shocks that might occur in the future, no degradation of the system may be seen. Hence, more theoretical aspects of system operation need to be incorporated into assessments of sustainability.

In the theoretical literature, a great deal of attention has been devoted to the possibilities of catastrophic jumps in nonlinear dynamic systems, rather than to smooth and continuous development. The recognition of possible threshold effects is of key concern here. One wishes to keep the system well away from any possible thresholds that may exist, and this must be done in the face of random fluctuations. The recognition of potential catastrophes in dynamic systems increases the concern one must have for risk, since we should be extremely sensitive to the very large, potential risks such catastrophes may entail.

This idea of maintenance of a "safe minimum distance" from such "bifurcation" points needs to be given operational meaning in particular contexts by biologists and other scientists, so that such a distance can be incorporated into policy analyses.

Intergenerational Equity

Some authors have equated sustainability with achievement of equity between generations. If current high levels of well-being are not sustainable, achievement of sustainability will promote intergenerational equity. Conversely, if systems are degraded only over long time horizons and evaluation procedures do not recognize future demands, then emphasizing intergenerational equity in evaluations will automatically lead to greater emphasis on sustainability.

However, the two ideas are not necessarily identical. To see this, consider a dryland agricultural system with possible future access to a confined fossil aquifer (i.e., one incapable of being recharged) for irrigation. Suppose that the current rate of agricultural production is sustainable, but augmented levels associated with irrigation cannot be, since the groundwater stock is finite. Alternative patterns of use of the groundwater over time are the object of equity and economic efficiency discussions; the concept of sustainability alone gives little guidance here, since decreasing consumption over time is not possible. Does a sustainability criterion require that the necessarily temporary benefits of irrigation be foregone and the aquifer forever unexploited? Strict adherence to sustainability in this case seems to place excessive emphasis on the status quo. This example highlights once again the idea that the sustainability criterion needs to be combined with other criteria if reasonable decisions are to be made.

The Discount Rate

Closely related to concerns about intergenerational equity is the debate over discounting. A discount rate is an adjustment factor used in economic analysis to adjust the benefits and costs of actions occurring in the future so that they are comparable to those occurring now; the discount rate is closely related to interest rates that prevail in an economy.[4] If a discount rate is used, a future outcome is of less importance in the analysis than is a current one. Does use of a positive discount rate mean that sustainability is necessarily undermined? The answer to this question is quite complex, but basically, there is not necessarily any conflict between sustainability and discounting.

Both high and low interest rates are consistent with high rates of resource use in certain contexts. For example, consider the possibility of conversion of natural forests with long-lived benefits to industrial land use with only near-term payoffs. On the one hand, if interest rates are very high, many investments in industrial capacity may appear unprofitable, and natural forests may be preserved. On the other hand, if interest rates are very low, the future benefits of the natural forest may outweigh the more temporary profits of industrial development.[5]

---

4 For an introduction to discounting, see Markandya and Pearce (1988).

5 For details regarding this argument, see Porter (1982).

Part of the objection to the use of private discount rates in evaluating patterns of resource use stems from models that are not sufficiently specified. It is well known (Clark 1976) that, if the rate of interest is higher than the natural reproduction rates of renewable resources at low levels of stock, then it may be "socially optimal" to harvest the resource to extinction, where *optimal* means maximization of discounted utility from consumption. The reason is clear: harvest of the resource to extinction and investment of the proceeds in capital is a more effective way to create future consumption than is sustainable exploitation of the resource. If this is at odds with one's sensibility, it is because either the value of the resource itself (and not just consumption) was left out of the model, or the role of the resource in an ecosystem that generally sustains consumption was neglected. It is not discounting per se that causes the problems, but rather poor economic theorizing.

There do exist many compelling arguments where, in evaluating public investments, it is inappropriate to use discount rates that reflect interest rates determined in private markets. In this view, private interest rates tend to be determined by decisions that are insufficiently forward-looking, and hence, a lower *social* rate of discount should be used for public evaluations.[6]

While it is well recognized that social rates of discount should be lower than private interest rates for a variety of uncontroversial reasons (such as taxation), the use of "artificially distorted" public discount rates causes problems. This is because discount rates serve to ration the allocation of capital. If the discount rates used in one sector are lower than those used in another, investment in the first sector will expand at the expense of the sector with the higher discount rates. Thus, using an "artificially low" public discount rate will result in a tilt of investment to the public sector, which perhaps will lead to a lowering of overall social output and well-being, as highly productive private projects are passed over in favor of less productive public ones. This may serve to undermine sustainable development.

This said, it must be recognized that, in general, the use of high discount rates may tend to undermine the interest of the future. There is a need, then, to develop methods other than adjustments to discount rates to account for sustainability in policy analysis. If intergenerational justice is a social concern to be reflected in decisions, as well it might be, it should be incorporated directly into evaluation tools on its own terms. If this is done, the reasons for adjusting discount factors are undermined. Of course, if no practical ways of building sustainability and equity concerns into analyses exist, then adjusting interest rates may be a reasonable second-best strategy. But, as will be discussed briefly below, such methods are under development; it is strongly recommended that further research be devoted to refining practical approaches for including these ideas in decision-making and evaluation procedures.

---

[6] For one interesting analysis, see Sen (1967).

Self-Sufficiency

Some writers in the area of sustainability seem to imply that self-sufficiency is a defining feature of sustainable systems. As a general proposition, self-sufficiency undermines the prospects for sustainable development (although not necessarily *sustainability* per se). Economists have made great efforts to show that there are gains from trade and specialization. In the absence of trade, the level of a system's overall output will generally be quite low, with little potential for development.

Several authors (e.g., Carter 1988) have equated sustainability in agriculture with systems that use few external inputs, such as pesticides and manufactured fertilizers. This emphasis on reduction of external inputs (self-sufficiency at the farming system level) largely reflects two concerns. First, many external inputs are derived in part from energy, and energy prices will exhibit a continual upward trend in the absence of sufficient technological change in the energy sector. In this case, reliance on external inputs will lead to ever-increasing production costs, which may undermine the sustainability of factor productivity. Second, external inputs are often associated with pollution.

More will be said about these concerns in section 3.1.3; suffice it to say here that there is no reason to conclude a priori that sustainable agriculture is low-external-input agriculture. Indeed, for many resource-poor, marginal lands, sustaining even low levels of output requires significant applications of external inputs. We might also note that "low-input agriculture" is not really low-input; most such systems specify a substitution of large doses of knowledge and human capital for manufactured inputs.

Sustainability and Other Objectives

As has been stressed, a sustainability criterion gives little guidance for social decision making on its own — it needs to be combined with other objectives when decisions are made. In this section, further evidence of this is presented.

In renewable resource systems, there are a large number of alternative levels of output that can be sustained, with different levels of sustainable extraction corresponding to different levels of availability of that resource. The choice among these different levels requires additional considerations that typically involve linkages to higher-level systems. One might, for example, counsel maximum sustained yield, or the sustained yield that maximizes economic efficiency. Taking the more narrow, subsystem view, maximum sustained yield has great appeal, but this ignores the fact that a move to a lower level of output in the subsystem frees up scarce capital and labor inputs for use elsewhere; recognition of the latter is embodied in the economic efficiency criterion.

The existence of different geographical areas within a country raises the issue of assessing possible trade-offs between them in terms of sustainability. For example, if there is a high-potential area that responds well to intensive management using external inputs that cause pollution, and a marginal area that may undergo resource degradation under increasing production levels, it may be possible to decrease the rate of degradation in the marginal area by increasing pollution in the high-potential area. The aggregation of these effects into an assessment of the overall situation must take place on grounds other than pure sustainability.

Attention has been given in the literature to physical and biological sustainability, to socioeconomic sustainability, and to institutional sustainability. Some of these may be sub-goals to sustaining overall well-being, but in some circumstances, they may be separate and conflicting goals. For example, it may seem generally desirable to sustain traditional institutions. But traditional institutions for the management of common property resources may not be sustainable in the face of external shocks and a growing population, and hence, may not fit into a pattern of sustainable development focused on other goals. Alternatively, the ability to sustain output levels from common property resources may require sustaining traditional institutions: imposition of a private property regime may be so at odds with the local culture that the system becomes untenable.[7]

This example serves the additional purpose of emphasizing that one should not focus on single attributes of systems, such as biophysical measures of system health. Whatever recommendations are made, they should be compatible with local cultures as well as a wide array of social and economic influences on behavior if they are to be adopted in a sustainable fashion. One should also recognize the potential feedback mechanisms from groups of individuals to the institutions attempting to enforce sustainability, e.g., via the political system. A program with incentives built in to alter it will not last long, and many sustainability recommendations are faulty in this regard, since there may be enormous gains to circumventing them.

The inability to focus on biophysical measures is also underscored by pollution, which introduces a quality element to consumption streams. In the context of factor productivity in agriculture, adjustments to the values of physical inputs and outputs to reflect their quality is vital. Thus, an input (such as fertilizer), which has a market price of $\$p$ and which causes $\$x$-worth of environmental damage per unit, should be evaluated at its "quality-adjusted" price of $\$(p+x)$.

In summary, we have proposed assessing sustainability in terms of some measure of aggregate well-being. How the adding-up over people and places occurs requires attention to other objectives. Discussion of aggregation issues over time, space, and individuals has received woefully inadequate attention in much of the sustainability literature.

3.1.3 Irreversibility and Uncertainty

It was stated earlier that one does not need to sustain every one of the components of a system for the overall system to be judged sustainable. If one component is degraded, then other compensating actions can be taken to substitute for it. The ability to substitute one

---

[7] For further discussion of common property see Magrath (1989).

component for another is crucial to attaining a high overall level of output in the system. Similarly, it was stated that self-sufficiency and sustainable development in general are largely incompatible at a national level, since specialization and trade undergird many of the potential sources of maintenance of well-being.

These statements presume, of course, that one has sufficient understanding of the system to make confident assessments of the substitutability of resources and the continued availability of traded goods. Several authors have severely criticized traditional economic models for being excessively sanguine about substitution possibilities between resources and other inputs to production; this criticism has been leveled by both biologists and economists alike.[8] The earlier conclusion was that constant per capita consumption can be sustained in the face of a finite resource base if capital can substitute for resources. And it is a simple matter to show that this conclusion must be altered if any strictly positive level of resource input is required, even if it is just to run the capital, due, say, to the laws of thermodynamics (Dasgupta and Heal 1979, p. 208).

In addition to the concern that policy recommendations will be based on inappropriate, simplistic models of natural systems, we must also recognize that our understanding of natural and human systems in general is fraught with uncertainties (chapter 4). While these uncertainties are of little concern if previous decisions can be reversed costlessly, the depletion of natural capital is in many instances effectively irreversible. It would be most unfortunate if we degraded certain crucial resource or environmental systems based on the expectation that substitutes exist either internal or external to the system — only to find that we are incorrect, cannot alter what we have done, and are facing the collapse of the system.

There are many examples of such possibilities. Consider, for example, the case of conversion of tropical forests to agricultural land. It might be believed that there exist sustainable agricultural systems for tropical forest soils, based on inputs of nutrients, and that sustainable development can be enhanced by this conversion. An unanticipated shock to energy prices could render the use of manufactured inputs unprofitable for farmers, or the proposed system might not work because of unforeseen biophysical reasons; the substitute system is not adopted, and the forest is irretrievably lost.

Perhaps the most enduring loss is that of genetic diversity. The extinction of species and the loss of diversity of gene pools is of growing concern, and arguments of uncertainty are quite cogent in this context. This concern has been particularly strong in connection to the depletion of tropical forest resources. Naturally, there is long-standing interest in the maintenance of genetic diversity in plants associated with agriculture.

Recognition of uncertainty and irreversibility has led some analysts to question basic manipulations of natural systems. They hold that maintenance of the status quo, or even a "rolling-back" of our intervention in the natural world is prudent. This argument is not without foundation. However, it also must be recognized that such an approach risks

---

[8] See, for example, the papers by Erlich (1989) and Christensen (1989) in the inaugural issue of *Ecological Economics*.

foregoing a large number of perfectly sustainable activities that greatly enhance human well-being.

In many respects, how much weight is given to uncertainty and irreversibility, and the degree of aversion to the implied risks, is what the sustainability debate is all about, once the more obvious fallacies are stripped away. Much "traditional" economic development theory might be characterized as reacting to such risks via "technological optimism," while the more extreme sustainability critiques might be characterized as reacting via "technological pessimism."

What is needed more than rhetoric is a careful assessment of the likely losses from finding oneself in the undesirable situation in which one regrets past decisions. What is also needed is an assessment of the chances of these situations occurring. The expected losses from possible incorrect decisions can then act as an additional cost of actions that might result in regret. The presence of this additional cost then creates a disincentive to irreversible actions and a positive incentive for flexibility to delay irreversible actions while additional information is sought. This approach implies that we consider the arguments of uncertainty and irreversibility, but these do not necessarily rule out all deviations from the status quo, or from states that are sure to be sustainable. The approach is outlined in more detail by Graham-Tomasi (1985) and by Hanemann (1989), while Perrings (1989) offers a proposal for a system of "environmental bonds" which is similar in intent.

One clear implication of this discussion is the need to identify, for different agroecological zones and sociocultural situations, possible irreversible activities and the associated indicators of impending irreversibility.

3.2   SUSTAINABILITY OF AGRICULTURAL SYSTEMS

Having discussed some general issues of sustainability, we now turn to specific concerns raised by the sustainable development of agricultural systems in the developing world. The list of potential problems is familiar: erosion, soil compaction, reduced soil nutrients, salinization, waterlogging, lowering of groundwater tables, pollution of ground and surface water, pest-related problems, and loss of genetic diversity. Since the basic concerns have been well elaborated elsewhere (TAC/CGIAR 1989; Stoop 1990), we will focus here on those with direct implications for agricultural research.

3.2.1   Lack of Sustainability and the Marginal Lands Hypothesis

Concern for sustainability implies a need to identify areas of potential degradation and impending irreversible harm to natural and environmental systems. There are three basic sources of concern: (a) areas experiencing rapid change, at a pace sufficient to outstrip the capacity of farmers and others to adapt; (b) areas under intense resource management resulting in environmental pollution, overreliance on exhaustible inputs, such as energy, and overexploitation of renewable resources, such as irrigation water; and (c) marginal areas of low potential, such as mountainous areas, areas at the edges of plant ranges, or those with poor soils or climates.

Also of concern is sustainable agricultural development, where the primary focus is the ability of the agricultural system to expand output to meet the demands of a growing population. Here, the need is to identify future constraints to growth in the total output of the system. There is ample evidence that the rapid growth in aggregate agricultural output that arose over the past few decades from the green revolution is now slowing (e.g., Byerlee 1990; Pingali, Moya, and Velasco 1990).

The dual concerns of maintaining the resource base and increasing crop yields have lead some analysts to the conclusion that the majority of research attention should be directed to the marginal lands and to technologies other than those of the green revolution (Stoop 1990). If this view is an indication of future trends, it has far-reaching implications for research policy, organization, and management. It is worthwhile exploring further this line of reasoning, which we will call the marginal lands hypothesis.

The gains in productivity arising from the green revolution have occurred in more favorable areas and have relied on high levels of external, nonorganic inputs. Since the supply of such favorable areas has largely been exhausted, low-cost replication of past successes in new areas will not be possible. Thus, one must assess the opportunities for expanding the output from favorable lands. But limitations occur here as well. We may be reaching a ceiling beyond which changes in the genetic composition of plants will fail to provide yield gains on currently cultivated lands. Even highly managed experimental plots may be experiencing a reduction in yield per hectare for important crops (Byerlee 1990; Pingali, Moya, and Velasco 1990). Additionally, heavy applications of external inputs are not only expensive but also cause environmental pollution, both of which serve to curtail increased application rates.

If past gains in favorable areas will not be sustainable in the future, then marginal lands must meet the demands of growing populations. Plant breeding strategies spurred the green revolution and remain the mainstay of the agricultural research system, but the ability to use these strategies on marginal lands is questionable. The green revolution required favorable areas, and therefore this approach cannot be applied to marginal lands. In this case, the agricultural research system must be altered to reflect the more fundamental soil/water/pathogen aspects of production on marginal lands. Let us examine these arguments in more detail.

Limited Gains per Hectare from Favorable Areas

First, the marginal lands hypothesis is predicated on the idea that future gains from favorable areas using green revolution technologies are limited. We take this as given for the crops at issue (for fixed levels of inputs of water and nutrients). However, it may be that input rates can continue to rise on favorable lands and on other lands that might be brought into the green revolution system. There is evidence that significant yield gaps exist for favorable lands, i.e., that yields obtained by farmers are falling short of potential yields, but the size of such gaps appears to be shrinking (Byerlee 1990; Pingali, Moya, and Velasco 1990).

There are two concerns raised by expanded reliance on manufactured inputs. First, these inputs are heavily dependent on energy, and under plausible scenarios, energy prices will rise in the future. However, to abandon these inputs in the name of sustainability because of possible increases in future energy prices seems premature. There are near- and far-term issues here. Energy supplies, as with other goods, are described via a supply curve, which shows that increasing energy inputs can be obtained at higher prices. Certain adjustments can be made to higher energy prices over the near term, if one is interested in sustaining production levels. (If one is interested in sustaining *aggregate* factor productivity, it is not clear whether use of marginal lands would be any more sustainable than use of favorable areas.) Over the longer term, clearly the availability of energy will be a limiting factor if the pace of technological change in the energy sector and in the production of agricultural inputs is not sufficient. But again, to abandon current activities because of the prospect of insufficient technological change in input markets in the future appears premature.

Second, the use of external inputs may cause pollution. However, the costs of pollution must be weighed against the costs of pollution control. The appropriate level of pollution is generally not zero, and it increases with increases in the benefits associated with the activities that cause pollution. Moreover, the supply of chemicals to the environment is quite heterogeneous. Lands vary tremendously in terms of the risk of pollution for any given level of inputs. There is a considerable ability to recognize the gains from applications of inputs on some lands, while applications are restricted on others, even under the same cropping system. This kind of targeted approach represents a substitution of knowledge for environmental capital. Research is needed on the extent to which this substitution can be achieved across a variety of agroecological zones and cropping systems.

Limited Area of Favorable Lands

In most areas of the world, the geographic limit of favorable lands has been reached. Thus, under the supposition that future gains per hectare on these areas are limited, the marginal lands hypothesis is implied. However, it must be recognized that land capabilities can be altered via investments of capital. Thus, previously favorable locations that have been degraded can be restored. Unfavorable, or marginal, lands can be improved as well: terraces can be built to reduce slopes, access to inputs and markets can be enhanced with roads and other investments in infrastructure, and water resources can be developed.

To the extent that these investments are possible, marginal lands can be made favorable, albeit at a price. The resource base upon which technology can be applied is itself the subject of change; thus, it is not *necessarily* the case that only technologies other than those of the green revolution can be used to make some currently marginal lands usable —

it does appear that limited gains could be made on marginal lands with improved cultivars and without other kinds of investments. But the sustainable use of marginal land requires substantial investments of some sort in either event.

3.2.2 The Marginal Lands Hypothesis Revisited

The above remarks are not intended to be definitive arguments against the marginal lands hypothesis. The overall point of this section is that it is not clear at the current time that the marginal lands hypothesis is so obviously true in such a wide array of situations that the program it implies should be embarked upon without question and without need for empirical research.

More important, it must be recognized that all courses of action have opportunity costs. An appropriate approach balances the gains to be obtained across all of the available courses of action. The incremental costs of obtaining increments of output from the system should be equated across additions from the three basic sources: the existing favorable sector, investments to extend the favorable sector, and the creation of mechanisms for obtaining sustainable production in marginal areas using new agronomic approaches.

These arguments are based on criteria of sustainability and economic efficiency. There are compelling issues of fairness that arise in discussions of the marginal lands hypothesis (e.g., TAC/CGIAR 1989). In this view, enhancing the ability of marginal lands to produce income from agricultural uses is a mechanism for the redistribution of social products to these regions. There is little doubt that extreme poverty and inequality are often highly correlated with impending unsustainability of the resource base, and that the causal mechanism for this association runs in both directions. However, it should be recognized that there may be alternative ways to achieve more equitable outcomes, and research expenditures may be an inefficient tool for redistributing social outputs.

While the marginal lands hypothesis itself may be questioned, many of its implications for research policy and organization apply as well to the ability to make continued gains from favorable areas, further closing the yield gap, without excessive pollution. To do so requires careful consideration of highly diverse biophysical and socioeconomic environments. The requirement that research programs recognize this heterogeneity implies considerable alteration of the current systems of research and technology transfer (Lynam and Herdt 1989; Stoop 1990). For example, fertilizer needs vary across soil types and aspects even within an individual farm field. Given sufficient understanding of this variation and incentives to act and by tailoring application patterns to soil needs, fertilizer use can be reduced substantially with no diminution in yields. But this potential will go unrealized unless research and "technology transfer" resources are directed to management skills rather than crop improvement.

3.3   SUSTAINABILITY AND AGRICULTURAL RESEARCH

In this section, some implications of sustainability for the conduct of agricultural research are discussed. Many of these are addressed in greater detail by Stoop (1990).[9] Here, the discussion is organized around issues of research policy, organization, and management.

3.3.1   Research Policy Issues

There are two sets of issues regarding the relationship between research policy and sustainability. The first set concerns the impact of agricultural policies in general and other broader policies of governments on sustainability. The second set concerns the implications of sustainability for policies regarding research on agriculture per se and the setting of research priorities and plans.

Linkages between Sustainability and Agricultural Policies

It was argued above that a sustainability criterion does not have any clear implications for research unless it is combined with other policy objectives. Thus, in addition to influencing policies that shape overall funding levels (chapter 1), the conduct of agricultural research in terms of priorities and planning must be heavily influenced by other social goals.

The interaction between the national agricultural research communities and other components of government is of vital importance, especially at administrative levels, where funding is on the basis of institutional support rather than research projects. In the absence of such communication, the research program will reflect the opinions of agricultural researchers rather than the opinions of the broader society regarding what is to be sustained and how sustainability is to be assessed. Ultimately, this will lead to reduced funding levels for research. These considerations imply that the boundaries of the agricultural research system must be explicitly broadened in the consideration of research organization.

In addition to the overall influence that development policies have had on the definition of sustainability, these policies also have important implications for the success of research programs. These linkages must be understood and incorporated into agricultural research policy. Individual agents within a social system respond to behavioral incentives, which are at once cultural, social, psychological, and narrowly economic in nature. If sustainable agricultural systems are to be adopted, they must be compatible with the incentives individuals face. Hence, the general policies of governments that affect agriculture have a profound influence on sustainability. Understanding the relationships among policies regarding taxes, subsidies, tariffs, exchange rates, interest rates, pollution control, population growth, land tenure, and sustainability is a complex task. These policies and a host of others promote the use of unsustainable agricultural practices and inhibit the

---

[9] See also Lynam and Herdt (1989) and TAC/CGIAR (1989).

adoption of sustainable ones.

Since many of the proffered remedies for unsustainable agriculture are considerably more complex in design than, say, planting one crop variety versus another, the concerns regarding incentives and linkages to other national policies are of substantial importance for planning and evaluating agricultural research. Returns to efforts at developing sustainable agricultural production systems may be negligible if these constraints are not recognized and built into the design and evaluation of alternative systems.

The current capability to address these concerns is limited within the international and, especially, the national agricultural research communities. The development of a social science capability within them is of paramount concern if research is to be effective in addressing either the sustainability of agricultural systems in marginal areas or the sustainable development of agriculture in more favorable areas. A basic concern for research policy, then, is that it must reflect linkages to other policies.

Agricultural Research Policy

The previous section briefly addressed the relationships between sustainability and broader social policies. Regarding agricultural research policy per se, the need is to appropriately allocate resources among alternative research activities.

A sustainability perspective in agricultural research policy implies a need to identify potential future constraints to increases in system productivity and the rate of change of these constraints. Systems that are changing rapidly are at risk of being degraded as existing approaches to resource management and the capacity to adapt to change are outstripped. It is the ability of farmers and others to adapt to such changes, as well as the recognition of the impending irreversible consequences of insufficient adaptation, that must be incorporated into research evaluation procedures and research policy deliberations.

As the object of sustainability is defined more broadly, difficulties for research policy expand. Agricultural research policy must be anticipatory, across a wide array of agroecological zones and the sources of constraints upon them. Due to linkages to other systems and sectors, the information on emerging constraints may not come solely from the agricultural sector. Declining fish yields far downstream of agricultural production areas, the impact of water pollution on health, decreasing demands in export markets due to food contamination — all of these have implications for altering agricultural research priorities.

The severity of such spillover effects is highly diverse across specific settings and will not be signalled by any research findings that x cropping system practiced on y soil results in erosion at rate z. Lack of sustainability is not solely an attribute of a farming system applied in a given physical environment. The inability to rely on biological and physical measurements alone is also evidenced by the key role local institutions play in resource management. This includes, for example, institutions for restricting usage rights in common property as well as organizations for local irrigation management. The need for social science capability is again raised by this discussion.

Even if a more narrow focus on output or factor productivity from agricultural systems is taken, there are several implications for sustainability in the allocation of research effort. The discussion of the marginal lands hypothesis indicates that in addition to decisions on which commodities and geographic areas will receive attention, the research system must assess trade-offs between expenditures of research effort on the following:

(a) commodities and crop improvement;
(b) the control of pollution from intensively managed lands;
(c) the ability to invest in infrastructure or in other ways to extend the area of land amenable to intensification using green revolution or other technologies;
(d) appropriate levels of and means for sustainable intensification of production on marginal lands.

Given the wide variety of biophysical and socioeconomic conditions that have an effect on the system, each of the above must be allocated across agroecological zones and/or other delineations of geographic areas.

All of the implications of the sustainability perspective elaborated above are true enough, and they point towards a highly systems-oriented approach. However, it is dangerous to insist on too comprehensive a view. All of the above considerations, while of obvious relevance, could lead to a paralysis of activity. A full systems view, which is logical, is also daunting, especially in light of current research capabilities within the international agricultural research centers and, more particularly, within the NARSs. It is unreasonable to require that all analyses take account of all possible interactions before they are deemed useful. Moreover, it is not necessary to understand all aspects of every linked system before coming up with a reasonable basis for action. These considerations are offered as an indication of the difficult task posed by sustainability, not as a set of minimal specifications for a research system that addresses sustainability. Balance and practicality must be achieved at the same time that one strives for more comprehensive understanding.

Research Priority-Setting Tools

One major goal of the research system will be the ability to identify emerging constraints to sustainable development, ex ante. The capacity to monitor and evaluate the various aspects of sustainability requires a longer-term commitment than is possible under project-oriented assistance. The implications of sustainability for setting research policy will require the development of more comprehensive tools for priority setting.

The essential difficulty is that when one considers sustainability, the feedback from research to observable, measurable results within a narrowly defined system that can guide future allocation of effort is disrupted. In the absence of sustainability research, the base case outcome is much less easily defined than it is for commodity-based research. Thus, the payoffs to research are less easily measured. This places severe demands on the development of priority-setting tools that will reflect sustainability concerns.

The approaches for setting research priorities can be distinguished by their degree of measurement. The scoring methods[10] require a decision maker to subjectively assess the importance of various outcomes and the likelihood of achieving them under alternative research activities, and then to assign scores on this basis. These informal, but structured, scores are then aggregated to achieve an overall score for the different research activities. While existing categories in scoring methods are not well adapted to sustainability, perhaps they could be altered to be more useful in this regard. However, it seems very unlikely that any simple procedure for introducing a sustainability score into such tools would be adequate or even useful.

Economic surplus techniques are more demanding. They attempt to formally measure the impact of outcomes of agricultural research on market prices and thereby on social well-being throughout the economy and over time. They can also incorporate linkages across sectors. The economic surplus approaches to setting research priorities are not currently designed to incorporate sustainability concerns, but they could readily be adapted to consider sustainability, at least conceptually; the availability of data to implement the expanded model is another matter at this juncture.

The basic strategy would be to alter measures of the values of inputs and outputs of subsystems to reflect linkages to higher-level systems. This would require adjusting the market prices used in the valuation of inputs to agriculture and the crops produced, since the observed market prices often do not include all of the costs incurred and benefits accrued by society. The use of prices for evaluation that are not equal to market prices is fully consistent with the sustainability objective and with the attainment of efficient economies in general.

If sustainability research could reduce pollution from agrochemicals while maintaining crop yields, it could be incorporated into research evaluations as a reduction in the cost of production of the crop. The amount of cost reduction is difficult to specify, of course, and the benefits of pollution reduction in the specific context being studied would have to be measured. Similarly, charges could be developed against agricultural production systems for resource depletion, and these would reflect the social cost of reducing stocks of natural resources. Any research outcome that altered the pace of this depletion could then alter the cost of production, and this could be evaluated using the surplus approach. Again, the key issue is measurement.

3.3.2   Research Organization and Management

Clearly, the consideration of sustainability places great demands on the research system beyond its current capability. The issue is how the existing capacity, and any future enhancement of it, should be organized to make maximum effective use of limited resources. It is impossible to make detailed recommendations about research organization as

---

[10]See Norton and Pardey (1987) or Stoop (1990) for a brief treatment.

a matter of general principle. Each individual research system has its own existing structure and capability, and recommendations must be made on an individual basis. There is no single best prescription for organizing sustainability research except that it should be infused across the existing program rather than added on via new "sustainability units" (Lynam and Herdt 1989; Stoop 1990). This is particularly evident when one considers the fact that changes in the institutional structure involve their own costs, with radical departures from the existing structure being untenable unless the benefits of change are high.

A sustainability perspective implies that much more diversity should be considered in agricultural research. Diversity of commodities, agroecological zones, sociocultural situations, rates of migration, etc., looms large in determining where resource and environmental constraints will be in effect. This would seem to call for a decentralized approach. However, given current capabilities in the NARSs, it clearly is not possible to engage in research in a highly decentralized fashion on all of the potential concerns. This is an extension of the "small-country problem" in existing research capability, and it implies that a system based on a reasonably centralized or networked approach is in order. A centralized sampling perspective may be in order to avoid replicating research on similar areas and issues across a large number of research units.

Equally clearly, however, the diversity of goals and constraints militates against too much centralization. The highly diverse nature of both biophysical and socioeconomic determinants of sustainability and the appropriateness of alternative treatments calls for a much greater emphasis on bottom-up information flows. Consider also the need for the products of research to be adopted by farmers and the increasing complexity of this concern given the farm-level management tasks required to close existing yield gaps without causing excessive pollution. This implies a need for more farm-level research than is currently the case, as well as station research that is more aware of the adoption of research. This sensitivity to adoption implies in turn that more social science information is needed at the research design stage, especially from sociologists and anthropologists.

The information needed to account more fully for heterogeneity could be generated at least partially by greater use of feedback mechanisms to research evaluation and priority setting from the technology transfer system. The current program of specifying a crop package by the international agricultural research centers is an untenable model for achieving sustained growth in output. In this approach, the package is designed to be reasonably productive across systems, and it is then transferred to the NARSs for minimal adaptive research. Its untenability is especially evident if substantial attention is devoted to the use of low-input/high-output systems on marginal lands and to gains from favorable lands from increased use of external inputs without pollution. It will not be possible to design highly specific farming systems that will be readily adopted by farmers and which meet the needs of countries with diverse goals. The NARSs, then, must engage in more research development, as opposed to research adaptation. And the enhancement of technology transfer systems should be of major concern.

Sustainability research is necessarily long term. This indicates the need for external support for NARSs themselves and not just for projects, as well as the conduct of some on-station research devoted to sustainability issues at the international agricultural research centers. For each sustainability issue for a commodity or farming system in an agroecological zone, a NARS must identify opportunities for long-term evaluation research. This means that continuity across specific donor-funded or country-funded research efforts at these locations must be achieved. In this sense, work on sustainability must be insulated from the short-term vagaries of alterations in research programs. This need must be recognized not only by broader research policy, but also by more specific work plans and budgeting.

3.4   TOWARDS MEASUREMENT OF SUSTAINABILITY

The translation of sustainability from concept to action requires a careful definition of sustainability and an appropriate means of measuring it. In this way, sustainability can be brought into both formal and informal decision making.

Unfortunately, sustainability is difficult to measure. At current levels of knowledge, it is not possible to identify a few variables at the farm level, such as soil pH or soil moisture, which adequately gauge sustainability. These kinds of measures provide useful information, but they are too narrowly confined to the farm level and to productivity and output concerns to measure sustainability as conceived here. A substantial research effort is needed to determine a set of standardized procedures for measuring sustainability. However, it is clear that some theoretical measures can be defined, and this will help to guide more practical empirical procedures. Two types of measures are needed. One type would include categories such as the depletion of natural capital (i.e., fossil fuels, forests, groundwater, soil, or fisheries), and the second type would include environmental pollution and the damages caused by it. We will very briefly discuss approaches to measuring these; a full analysis is beyond the scope of this chapter.

3.4.1   Measurement of Resource Degradation

The measurement of resource degradation should be undertaken in such a way that, as a resource is increasingly exploited above a sustainable level, the "sustainability cost" assessed against the activity should go up. Additionally, this cost should reflect the uniqueness of the resource, i.e., the substitution possibilities for it. The measure should be equally applicable to exhaustible and renewable resources and should be consistent with the measurement of other economic variables.

Such measures have been proposed in the literature on adjustment of national income accounts to reflect sustainability concerns (Peskin 1989) and in the literature that proposes "compensating projects." The basic idea is to set up a charge to reflect changes in natural capital in a way that is similar to the accounting procedures used with manufactured capital. As human-made capital stocks depreciate, a charge is levied to reflect the reduction in future production possibilities, equal to the cost of investment required to replace the capital stock as it wears down. A similar charge should be levied against depletion of natural capital.

In principle, such charges could be assessed via so-called compensating projects. That is, one could determine the magnitude of a fund that would have to be set aside in order to substitute for the losses in well-being due to services lost from the natural environment. Thus, a system of "resource depletion bonds" might be instituted, either actually or hypothetically, in determining the costs and benefits of research programs.

This is fairly straightforward under conditions of certainty. Naturally, the key considerations here are those of substitution possibilities and the treatment of uncertainty. Suppose, for example, that a fund is set up from the proceeds of current resource depletion and that this fund is used to invest in infrastructure. This is a very different compensating project from one where the fund is used for investments in artistic achievement or religious monuments. To date, these issues have not been adequately resolved in the context of measuring the sustainability of natural resources.

The development of techniques to measure such capital depletion charges is an important topic for further research.

3.4.2   Measurement of Environmental Pollution Costs

A wide variety of techniques have been developed for measuring the costs of environmental degradation due to pollution.[11] These techniques have been developed and applied primarily in the United States, and less so in Western Europe. They are based on the notion of measuring the willingness of individuals to make exchanges of income for environmental quality (willingness to pay). They are fully consistent with methods of measuring gains from agricultural research and the techniques employed in economic surplus methods for setting research priorities.

There are two basic sets of methods that exist. The first set is direct techniques, where individuals are queried directly in sample surveys in order to elicit willingness to pay. The second set is indirect techniques, which employ observations of behavior regarding market goods that are closely related to environmental goods. The demand for environmental goods can then be inferred from the manner in which the demand for the market good changes in response to changes in environmental quality.

These techniques hold some promise for application in less-developed countries, but they would need considerable refinement before practical empirical research could be conducted that would be useful for policy analysis.

Although a complete set of rigorous methods that could be applied in less-developed countries does not exist, progress can still be made toward incorporating pollution costs into research evaluation tools. In essence, a proxy for pollution costs associated with agricultural production can be obtained by increasing the costs of inputs associated with pollution in

---

[11] There is a large body of literature in which these techniques are discussed; see, in particular, Freeman (1979) and in the context of forest resources in less-developed countries, Graham-Tomasi (1990a,b).

existing economic surplus evaluations.

3.5   CONCLUSION

In this chapter some basic issues related to sustainability and its implications for agricultural research were set forth. The treatment of this issue was largely conceptual, with few very specific recommendations made. In part, this reflects the current status of the concern for sustainability: a set of broad principles advanced by a diverse set of interested parties, each with their own approach. Here, a modest attempt was made to focus the discussion and to provide a framework, rooted in the discipline of resource economics, for examining more detailed proposals.

Further progress toward building a sustainability perspective into the conduct of agricultural research will take place in the context of specific geographic areas and resource concerns. It is difficult to make generic recommendations; however, a few do emerge. First, efforts should be made to measure the economic costs of external effects of agricultural production in less-developed countries, so that prices of inputs and costs of production can adequately reflect their full cost to society. Second, techniques should be developed for accounting for resource depletion in economic terms, i.e., for adding a "natural capital depletion cost" to technologies that are heavy users of resources. Third, primary attention should be given to uncertainty and impending irreversibility in determining priorities for action. Fourth, adoption and management skills should be given more emphasis, and the effects of broad social policies analyzed. And fifth, monitoring capability needs to be enhanced so that the status of the resource base can be assessed.

In all of this, it should be recognized that sustaining the current resource base is one issue among many; discussions of sustainability should reflect a backdrop of rapidly increasing demand for food. Neglect of substitution possibilities and innovative adaptation to scarcities leads to an overly conservative approach to resource management at a time when productivity is of central concern. At the same time, a hands-off attitude, reflecting the notion that things will take care of themselves with the current institutions and decision-making approaches, is a formula for a dire future.

Chapter 4

Agricultural Research in a Variable and Unpredictable World

Jock R. Anderson

Most of the formal literature on the agricultural research process per se, whether of a managerial or evaluative orientation, implicitly treats research and its setting as being deterministic. In fact, of course, the process is intrinsically uncertain. Most agricultural sectors are highly variable and much of the observed variability is extremely unpredictable, so that it is, technically speaking, risky. The conjunction of an uncertain research process with an uncertain physical and economic environment is the reality of agriculture that makes it all an extremely risky business.

There is thus a considerable mismatch between nearly all the formal literature on research resource allocation and that on decisions about investing in research in the risky environment in which this takes place. It is the purpose of the present chapter to describe this environment and how decisions are made in it, and how consideration of the uncertainties involved may lead to decisions that differ from those that might otherwise be made. Risk and uncertainty are so pervasive in the system that the overall situation might well be described as "turbulent." Recognition of this turbulence may help to explain the sometimes seemingly cautious behavior of potential investors in agricultural research.

4.1 UNCERTAINTY SURROUNDING AGRICULTURAL RESEARCH

4.1.1 Sources of Variability in the Agricultural Sector

There are many sources of variability in the agricultural sectors of the world. Only in uninteresting tropical paradises with benign climates and governments that extensively interfere in cushioning the sources of natural and other variability does this generalization not hold good. Specific sources of environmental variability that add to the challenge of agricultural research administrators are elaborated in this section.

The Natural Environment

The variability of a natural environment over time is widely appreciated by observers from many different perspectives. As a generalization, variability tends to be exceptionally high in less-developed countries, as opposed to the sometimes harsh but generally rather more predictable climates and environmental circumstances of many more-developed countries. Climate is usually the major driving force behind natural environmental variation. There are many aspects of climate that cause variation and, depending on the particular geographical circumstances, some of these climatic factors may be more or less important (Anderson 1979).

In many less-developed countries, the major climatic driving force is the precipitation regime, with the important overriding influence of the temperature regime. Temperature conditions tend to be more predictable, although some extreme events such as severe frosts do not fall within this generalization. Precipitation, especially rainfall, tends to be less predictable and is usually of overwhelming importance in determining the ultimate performance of crop and livestock productivity and production. Needless to say, the influence of human managers in such systems can be great in moderating the effects of natural variation. Farm managers have a considerable influence in dealing with, say, droughts through cautious stocking decisions or selection of appropriate planting density in crops.

Beyond specific enterprise management, however, there is also considerable scope for decision making in order to help to endure the consequences of an unstable environment. Different farm enterprises can be combined in different proportions over different seasons so that there is a portfolio of diversified enterprises that, in combination, may be considerably less variable than a more specialized operation (Heady 1952, ch. 17). Some agricultural activities are inherently more stable in their performance than others. A classic case of this is cassava growing. The standing cassava crop serves as a store of food that can be harvested with flexible timing and that is fairly safely preserved while in the soil.

Other devices for managing natural variability can operate at higher levels of aggregation than that of the individual farm. One example of such an intervention is crop insurance, usually based on the physical performance of crops. The regrettable thing to report about this particular form of insurance is that it has been singularly unsuccessful, except in those few countries that have been able to afford to underwrite the insurance heavily from the public purse (Hazell, Pomareda, and Valdés 1986). Examples that can be described as reasonably successful are to be found in countries such as Sweden and the USA but are essentially unknown in the less-developed countries.

Needless to say, some of the effects of natural disasters can be quite long-lived in their impact on farm households and on those who depend on such households. One has only to consider the devastating impact of hurricanes, typhoons, etc., on tree crops ranging from short-cycle ones such as bananas to long-cycle ones such as coffee and rubber. Some of the effects of such disasters can be moderated through disaster-relief measures, perhaps from international sources, but inevitably, small-scale producers are severely disadvantaged through such mishaps.

The Economic Environment

Many aspects of the economic environment faced by farmers are subject to considerable variability. Much of this is reflected in variable prices (OECD 1980; Blandford 1983). Prices may vary from year to year and also greatly within a year. The root cause of such variability depends, in turn, on many other factors, sometimes including government intervention (see next subsection) but, more fundamentally, it usually devolves to changes in incomes of consumers of farm products, and largely unpredictable changes in their tastes and preferences.

Another fundamental source of variation in price relates to the natural variability that is reflected in varying quantities of commodities supplied to markets. The nature of the demand functions faced by agricultural suppliers leads to the varying price effects of such natural variability.

Less-developed countries, to the extent that broad generalizations may be made, typically have many small-scale producers of their major export commodities, and national groups of producers together make up a relatively small portion of global production and trade. This means they are effectively price takers on world markets for their major exports and the price regime that they face is virtually uninfluenced by their individual decisions. Nations can seek to differentiate their products on quality grounds and thus move to face less perfectly elastic demand schedules. To the extent that demand is somewhat elastic, there may be some natural cushioning through the demand function for variations in the supply, which in turn leads to a negative correlation between yields and prices so that, in the event of, say, a disastrous coffee harvest resulting from unfavorable weather or wide-scale pest and disease attack, farmers enjoy relatively higher prices than would otherwise have been the case. Thus the variation in their incomes is less than that in either prices or yields (Anderson 1985a).

Apart from the natural buffering effect of downward-sloping demand curves, other possibilities for attempting to intervene in the economic environment are various stabilization schemes. These can take many forms, ranging from buffer stock operations, such as that operated by the Australian Wool Corporation, to voluntary export controls and local stock management, such as that attempted under the International Coffee Agreement when it is in effect, or other more financially managed schemes such as buffer funds (Newbery and Stiglitz 1981; Scandizzo, Hazell, and Anderson 1984).

A considerable degree of sophistication is required to manage successfully any sort of agricultural stabilization attempt. For schemes requiring international cooperation, considerable goodwill as well as good management is required not to ʌead to even more problems than were set out to be solved by the intervention. By the nature of things, many less-developed countries are not well supplied with the requisite management skills to handle such attempts at making the economic environment of their farmers less risky. There can even be considerable macroeconomic consequences of the management of stabilization funds. For one such example, consider the considerable volume of funds (relative to gross national product) tied up in the Papua New Guinea national coffee stabilization reserves during much of the 1980s (Brogan and Remenyi 1987).

It is widely acknowledged that research is a rather time-consuming process. Sometimes research discoveries may arise after only a short period from the initial investment in research facilities but typically there is a lag of several years. Most econometric studies addressed to this issue have yielded mean research lags of four to ten years. In less successful cases of research endeavor, the lag may be much longer.

Since a major component of the valuation of research benefits is the price received for the products that are subject to the technical innovations arising from research, planners need to make forecasts of the prices ahead of their realization. Price forecasting long into the future is notoriously difficult, at least as judged by "errors" assessed by comparing prices actually experienced with those that were forecast many periods ahead.[1] Public research bodies in their research-planning activities need to use the best possible information about long-term trends in the prices of the commodities subject to research investigation. Sources of this information include international agencies such as the World Bank, and national and regional commodity price-forecasting agencies.[2] No matter what the source of a forecast, errors are inevitable.

All this translates into price uncertainty being a significant factor in any research-planning activity. An answer to the question of whether or not this uncertainty matters is something that probably cannot be generalized. The issue is addressed below in section 4.2.2 where the conditions under which uncertainty can safely be ignored and mere expected prices used are discussed.

The Political Environment

A special aspect of the economic environment that is deserving of separate attention is the influence over economic matters that arises through political intervention. Policy concerning variation in the agricultural environment can take many forms. The role of public agencies in managing stabilization schemes has been noted. If the rules of a stabilization scheme are well thought-out, firmly established, and consistently adhered to, much stability may well be achieved. If the contrary is the case in any of these respects, uncertainties surrounding the rules of the game can easily add to the uncertainty effectively faced by people dealing with the commodities in question. This type of policy uncertainty is a greatly neglected aspect of the unstable environment faced by farmers around the world (MacLaren 1980, 1983). Governments come and go and bring with them new slants on policy which may make the ultimate task of farm managers, and those concerned with planning resource

---

[1] See, for example, Freebairn (1978), Cornelius, Ikerd, and Nelson (1981), and Lee and Bui-lan (1982).

[2] For example, World Bank (1986).

allocation within the agricultural sector, awkward and possibly self-defeating (Hobbs et al. 1988).

One common field of policy intervention is in dealing with severe droughts (and sometimes other natural hazards). During such stressful times, governments are inevitably under much pressure to be seen to be doing something to help people in various states of distress and plight. Typically, well-intentioned but somewhat rash decisions are taken in the heat of the moment, but such decisions may well serve ultimately to disadvantage those careful managers who have organized their farming activities so that they themselves have moderated the effects of the disaster. Incompletely thought-out and cavalier decisions by policymakers can lead to most regrettable and inequitable redistributions of public re-sources. This seems to be a phenomenon endemic to all agricultures, including those of many supposedly sophisticated more-developed countries (Freebairn 1983). In short, analysts designing and appraising programs and projects for nations facing severe national risks, such as war or revolution, must be mindful of the grave consequences of such "downside" risks. The imperative needs will usually mean that development initiatives must still be taken even when there may be high probabilities of failure and when it proves impossible to design flexible plans that would mitigate such risks.

A further aspect of political consideration is the effect of policy on the distribution of research benefits. For internationally traded products, the net national or world benefits may variously be reduced, left unchanged, or increased depending on the nature of a policy and the significance of a particular country in the world market for the commodity being considered. Some of the complexities of these issues have been discussed by Alston, Edwards, and Freebairn (1988). Their results are drawn from a deterministic setting, and it is surely the case that even more opaque, but potentially significant, results may be drawn from an appropriate stochastic casting of their trade-model view of the impact of national versus rest-of-world research, as well as more appropriate shifts in supply curves. Present-ing such a conceptualization of models in a multicommodity as well as multicountry setting, with spillover effects between countries such as are noted by Davis, Oram, and Ryan (1987) and Evenson (1989) would add further to the complexities of such an analysis. A criticism of both these sets of models is that all the work has been done thus far with linear supply curves. Since most such assumptions are more or less gross simplifications, the reality of the distribution of benefits to different parties under a realistic setting of distorted international markets remains a considerable uncertainty in itself (Anderson 1989a).

4.1.2   Sources of Uncertainty in Agricultural Research

Against this background of pervasive variability within an agricultural sector, the less-than-certain functioning of agricultural research itself can be sketched. An essential feature of research is the fact that it is a chancy process of discovering new knowledge. If an investigative activity is certain in its outcome, it is hardly describable as research. Investors in agricultural research systems, private or public, commit resources without knowing exactly what will be discovered through the work. Historically, the rewards to investment have tended to be high on average. Oehmke (1986) argues that slow response to changes in the optimal level of research largely explains the often persistent high returns and the associated underinvestment. The reality of research investments is that they are, in fact, highly diverse in their economic effects (Ruttan 1982; Pinstrup-Andersen 1982). The significant successes tend to be given rather more publicity, analysis, and attention than do the many failures — large and small. This is how research is, and in essence it cannot be changed. Good managers may be able to increase the probability of relatively successful projects being undertaken and of the results ultimately being implemented, but even superlative managers must have their mistakes or, in geological parlance, dry holes (Arnon 1975).

Agricultural research is seldom a single-stage process and, in most cases, involves several sequential steps with uncertainty encountered at every one. In the first place, there is the uncertainty of whether a planned research activity can actually be effectively implemented. Resources, both human and physical, need to be brought together and, in many of the less-favored parts of the world, even this stage has its profound difficulties and consequential uncertainties. Finding and then encouraging skilled research workers to engage in research work for little personal financial reward, in remote and difficult circumstances, can be quite awkward, not to mention expensive.

When research personnel are at least in place and are appropriately equipped, there is then the uncertainty that they can make worthwhile discoveries that add usefully to the body of knowledge. This process is one that is littered with risks and, even when things seem to be discovered, there may be difficulty in having the findings accepted and made available through, say, the mainstream scientific literature. Authors of research papers everywhere know the difficulties of convincing their peers that what has been discovered is really new and worthy of publication. Analogous difficulties of acceptance are experienced by the creators of other research products, such as new plant cultivars or new machine designs, for which scientific papers are here treated as surrogates.

Once new knowledge has been claimed to have been gained through some such form of publication, there follows the issue of how it is picked up and eventually used by innovative farmers. Sometimes this process is aided and facilitated through an extension service that may variously be closely linked or more distantly related to the research or sales service. Whatever may be the nature of such links, the net result is again surely uncertain. Some findings are readily translated into cost-reducing efficiency gains by early-innovating farmers and subsequently adopted widely within the farming community. These may be quite mundane innovations but highly profitable. Others that are highly specific to particular locations may not be so readily implemented or adopted and there are some scientifically exciting findings that may not be nearly so profitable and, in spite of possibly considerable investment in related extension or selling services, may never really be taken up in a widespread manner, and so have little observable effect in market structures. These various adoption effects are usually depicted by economists as rightward shifts of supply curves as a net response to the new knowledge gained.

There is a large literature on how supply curves shift in response to investment in research. The literature is cluttered with controversy over the nature of such shifts, the market context of such shifts, and thus of the nature of the distribution of the benefits from research (Lindner and Jarrett 1978, 1980; Davis 1981). The gainers and losers from agricultural research are determined by such theoretical matters. In a planning sense, only limited attention has been given to the distribution of benefits in allocating resources to research.[3] It seems that virtually no attention has been given to the effects of the inherent uncertainty in this process on the worthwhileness of the risky investment by public or private agencies.[4] This is not to say that uncertainty itself has been ignored by research analysts (Binswanger and Ryan 1977). Indeed, stemming from the work of industrial operations researchers such as Sprow (1967), considerable application has been made of digital simulation methods in quantifying the risks associated with returns from agricultural research.[5]

4.2   INVESTING IN RESEARCH IN A RISKY AGRICULTURE

The riskiness associated with agriculture in general, and with agricultural research in particular, has now been overviewed and the question now addressed is: does it matter? The issue involved is just how quantitatively important is the uncertainty that enters at so many points, especially in its cumulative effects. This can be tackled and modeled either as a single-commodity (illustrated below) or as a more complex multi-commodity case involving essentially portfolio management methods. General research strategies, such as Nelson's (1961) parallel research strategy, are clearly important in influencing research achievements but are beyond the scope of this chapter.

The ensuant puzzle, once any quantitative importance of risk has been established, is the extent to which the risk dimension is important in public decision making. This raises questions of the degree to which a project is statistically independent of other sources of public income and of the relative size of the risk in relation to other risks in the economy. These are explored in section 4.2.2 with illustrative reference to the particular situation of cocoa in Papua New Guinea.

Some of the special issues in research planning relate to the targeting of research endeavors to specific groups — perhaps those that are relatively impoverished or that can be seen to be in great need. These matters, and further ones concerning risk pooling through

---

[3] Duncan and Tisdell (1971), Hayami and Herdt (1977), Edwards and Freebairn (1982, 1984), and Freebairn, Davis, and Edwards (1982).

[4] An exception is the work of Brennan (1988).

[5] Fishel (1971), Parton, Anderson, and Makeham (1984), Dyer, Scobie, and Davis (1984), and Dyer and Scobie (1984).

a mixed portfolio of research projects, are considered in the final two sections of this part.

4.2.1 kesearch Benefits under Risk

The literature on evaluating benefits from agricultural research is vast and is far from being internally consistent. There has been considerable controversy about the analytical frame-work used, even when analysts confine themselves to the rather restrictive framework of economic surplus. Economic surplus measures are used in the present context as they seem to be the only workable and reasonable way to attempt to quantify the extent of benefits. Even with this simplification, there is yet little agreement about the "best" approach to representing supply-and-demand relationships and, most importantly, the nature of the shifts in supply or demand curves that are induced through the adoption of research results by farmers.[6] It is not the intention in this review to elaborate these theoretical and empirical controversies. Rather, the approach taken is to consider relationships that seem reasonable and to recognize the difficulties inherent in them through modeling explicit random components to represent the unexplainable uncertainties.

A logical starting point in conceptualizing the research process is to posit elements of what, in the literature, is usually described as the research production function. The way in which this is tackled here is to introduce first a function representing the contributions made to knowledge through specified research investments (Weaver 1986; Pardey 1989). This is done in equation 4.1 in such a way as to highlight the lags that inevitably prevail between investing in research resources and in gaining the knowledge, and also to recognize the uncertainty in this process, namely,

$$\Delta K_t = f_1 (R_{t-i}) + f_2 (\ )u_1 \tag{4.1}$$

where ΔK_t represents the increment to knowledge associated with current and prior investments in research R_{t-i}, $i = 0,1,2, \ldots , T$, and f_1 and f_2 (with arguments analogous to but not necessarily identical to those of f_1) are functions, the latter leading to a heteroscedastic transformation of a disturbance u_1 (Just and Pope 1978).

Just how the increments to knowledge are measured in this formulation is a good research question in itself. It is not at all clear that any single measure does justice to the subtlety of what knowledge really consists of. Of the various measures that could be used, the one that has received most attention in the literature is some counting of scientific publications of various qualities (Evenson and Kislev 1975b, p. 29). Sometimes these are corrected in various ways to make the measure more descriptive of the phenomenon, but this is usually done with due regard to the difficulties in actually measuring such performance indicators.

An operational concept of a research production function may well combine with a

---

[6] Rose (1980), Wise and Fell (1980), Wise (1984), and Anderson (1989a).

relationship such as (4.1) a further one that includes the transfcrmation of new formal knowledge into productivity and production gains by producers. In the present formulation, this is represented as equation 4.2, which translates (stochastically) such incremental additions to knowledge into shifts of the supply curve, which is to say, an aggregate response by agricultural managers to the potential changes in technology revealed through research findings (Stefanou 1987). This particular relationship seems to be one of the greatly underresearched ones in the economics of agriculture. Any attempt to specify such a relationship is necessarily highly speculative. Apart from being related to the body of knowledge at large, and the seemingly unknown response of producers to it and to changes in it, some of the factors that must necessarily influence any such association are those relating to the effectiveness of communication between the research system and the production system, such as the formal extension (selling) service of a government (firm), the degree of edaphoclimatic homogeneity within the domain of the research, and the price regime (for inputs and outputs) faced by producers that influences the profitability of particular innovations. Just how all these factors might be combined in an ultimate shifting of the supply schedule is an important but seemingly unstudied phenomenon. It seems to be a fertile field for research. Some of the work on adoption and related models of Bayesian learning is pertinent here.[7] In the large, however, the relationship must be perceived as essentially a black box, encompassed here in equation 4.2:

$$k_t = g_1 (\Delta K_{t-j}, X_t, H, V) + g_2 (\ )u_2 \qquad (4.2)$$

where k_t is the proportional new vertical shift in the supply schedule in the region of current price P_t; g_1, g_2, and u_2 are analogous to their counterparts in (4.1); and the arguments of the functions include increments to knowledge lagged j periods, investment in extension services X_t, an index H of agroecological diversity relevant to the commodity in question,[8] and a vector of other relevant prices V. The new annual shifts accumulate progressively over time as

$$ck_t = \sum_{i=0}^{t} k_i$$

dating from the initial impacts of research.

Moving from a given shift of the supply curve to an evaluation of the benefits, there are several key components, some of which are well established and others much less well established or quantitatively defined. The core of any such relationship is the first bracketed term in equation 4.3 below. The variables in this term consist of (a) the proportional shift

---

[7] See Lindner and Fischer (1980), Lindner, Pardey, and Jarrett (1982), Feder and O'Mara (1982), Feder and Slade (1984), and Feder, Just, and Zilberman (1986).

[8] Judd, Boyce, and Evenson (1986), Evenson (1987), and Pardey and Craig (1989).

in the supply schedule, variously represented vertically or horizontally but here by the preferable vertical (price) shift using the parameter ck, (b) the price prevailing before any research-induced shift, and (c) the relevant (i.e., pre-shift) quantity produced. Cutting across the controversial literature on evaluation of benefits using surplus measures, these three factors in their multiplicative form stand out as the uncomplicated and dominant elements of the benefit side of research evaluation. The next term in equation 4.3 is one that is a function of the relevant market parameters, particularly the elasticities of demand and supply and, most importantly, the type of shift of the supply function, and the shape and location of both the supply and demand functions. The latter depend on concepts of the market and attempts to estimate econometrically the relevant parameters of such markets. All the features and parameters are either judgmental or largely unknown and thus are also appropriately represented by a considerable degree of uncertainty that is captured here through a multiplicative disturbance in equation 4.3:

$$B_t = 0.5(ck_t \, P_t \, Q_t) \; M \, (E_d \, , E_s \, , S) \, u_3 \qquad (4.3)$$

where gross research benefit B_t depends on previously defined variables, the expected quantity Q_t produced in the absence of the accumulated supply shift ck_t but not necessarily in the absence of its anticipation, and a relationship M that links the price elasticity of demand E_d, the price elasticity of supply E_s, and S representing the type of shift and the shape of the supply schedule.

One source of variation that may often be important in an equation such as (4.3) is what could be called exogenous variation in market prices. This is particularly the case for the more or less perfectly elastic demand structures faced by small exporting countries, where the results of occurrences in the rest of the world generate variable product prices. The uncertainty in such price regimes is beyond the influence of the exporting country but the consequences, particularly when interpreted through the national exchange rates with major trading partners, can be profoundly important to the economic viability of production. For most traded commodities there is considerable price variation (OECD 1980). Such variation can be represented through stochastic processes that may or may not be stationary or parametrically unchanging over time.

Under the simplifying assumption of a small exporting country, equation 4.3 can be simplified to a representation such as equation 4.4, wherein the demand parameters vanish and, in this particular representation, the uncertainty regarding the nature of the supply shift is captured in a random and multiplicative error term:

$$B_t = 0.5 \, ck_t \, P_t \, Q_t \, E_s \, u_4 \qquad (4.4)$$

where u_4 is a multiplicative disturbance term of unit mean and possibly more or less constant variance.

To make any such equation operational, it is necessary to invoke a supply relationship that expresses quantities expected to be produced as functions of relevant planning variables, including appropriate, probably "expected" or anticipated, prices. In most econometric representations, this is done through postulated lag structures, which can vary greatly in type and complexity.

For the present purpose, the realized quantity supplied can thus be thought of as the product of three components, namely (a) an otherwise anticipated quantity Q_t supplied that depends through a supply relation $q_1(\cdot)$ on previously observed prices P_{t-m} and quantities Q_{t-n}, (b) a correction factor for the short-run adjustment arising from the new incremental and proportional cost reduction k_t which is $k_t E_{ss}$ (Pinstrup-Andersen, Ruiz de Londoño, and Hoover 1976), where E_{ss} is the short-run price elasticity of supply, and (c) a multiplicative disturbance term u_{5t} to represent the riskiness of supply (Scandizzo, Hazell, and Anderson 1984, pp. 7-8):

$$Q_t = q_1 (P_{t-m}, Q_{t-n})(1 + k_t E_{ss}) u_{5t} \tag{4.5}$$

The cumulative effects of new technology are captured through the lagged quantity produced.

The optimizing problem involved in research allocation can now be stated more formally. Employing the simplest plausible economic indicator for social choice, namely the expected value of the net present value of benefits from the research, this can be defined as in expression 4.6. This provides the valuation function that can be manipulated to optimize resource allocation in the research process. Formally, expression 4.6 should be maximized with respect to the main decision variables, namely the investments over time in research (R_t). Given the uncertainties that abound and that have been variously modeled in the several equations, such a maximization is more readily said than done. In practice, it will also involve careful judgment by research administrators to select the particular research tasks that make the basic knowledge generation function 4.1 operational and effective.

The valuation criterion can thus be written, for brevity as

$$E [PV (B_t - R_t)] \tag{4.6}$$

where E[·] denotes the expected value operator, and PV the present value operator that, in turn, depends on the social opportunity cost of capital r_t, that may well not be constant over time or, indeed, even known.

To make this framework more operational, it may help to look at some simple relationships in order to see how the abstract models outlined in these equations might apply. These are described in the following section, including the related stochastic specifications, in order to provide an illustration of the likely orders of magnitude of both the risks and the expected values of the relationships.

4.2.2 Public Investment in Research under Risk

Whether risk really matters is a nontrivial issue that depends on several potentially awkward questions. The first of these is how to measure risk. Many different measures have been proposed and, indeed, several are widely used in practical risk analyses. The least unsatisfactory and most widely used is variance of the performance indicator (e.g., present value), particularly as it relates to the mean performance indicator via the mean-variance (E, V) criterion. It has been widely castigated in the literature for being crude and simplistic (Anderson, Dillon, and Hardaker 1977) but its popularity remains undiminished (Tsiang 1972). Another measure of risk that has received some attention is the standardized measure of variability, the coefficient of variation, although, unfortunately, it is not a criterion that is particularly defensible in theoretical terms. Its virtue is that it is widely comprehended among professionals of many different disciplines, is unit-free, and is readily measured (Anderson and Hazell 1989, p. 10).

More theoretically acceptable measures of risk generally depend on more comprehensive descriptions of the probability distribution functions of performance variables. The most obvious candidate is the application of principles of stochastic dominance to the distribution function of uncertain present value (Anderson 1974). In recent years, development of concepts of stochastic efficiency has led to more powerful ordering techniques. Particularly noteworthy are those in the family of rules for "stochastic dominance with respect to a function," where reference can be made to specified ranges of risk aversion on the part of concerned decision makers.[9]

These introductory remarks somewhat beg the question of whether decision makers really need to be concerned about risk. If they are private decision makers who have some natural level of aversion to risk, perhaps quite small if they have ample financial resources, the issue raises no new economic questions. If, on the other hand, the decision maker is working on behalf of public entities, it is a moot point as to whether an appropriate level of societal risk aversion should be involved in the appraisal.

The topic of risky public investment appraisal has been the subject of considerable debate. The classic contribution is by Arrow and Lind (1970) who put forward persuasive arguments that, in typical public-investment situations, all that is important is that public decision makers use carefully assessed expected values of economic performance as their guide to merit.

The key to arguments about dealing with risk in public investments is the size of the uncertain project relative to the overall context in which it is judged. If a public agency is working on behalf of a national government, the project should probably be considered relative to the magnitude of national income. It is rare that individual projects are large relative to national income, and, in the typical small-project situation, the risks associated with the project are essentially diluted in the large pool and, for all practical purposes, can

---

[9] Meyer (1977), King and Robison (1981), and Cochran, Robison, and Lodwick (1985).

be ignored. The exceptional cases, where explicit account of risk may need to be taken, arise when the project is large relative to the corresponding aggregate income or, what is somewhat more unusual, where the project is highly correlated with the uncertain aggregate income.

Such situations are rare in public project appraisal and, accordingly, the general guideline is for decision makers to use merely the expected value of relevant (appropriately discounted) net present values of returns on a public investment. Where the exceptional situations are encountered and dealt with, accounting for risk is potentially a complex matter and perhaps this explains why it has mostly been ignored in public investment analysis generally and in research evaluation in particular. Some simplifying approaches to dealing with risk in such exceptional situations are, however, available (Anderson 1983, 1989b, c). These consist of approximating formulae for determining appropriate risk deductions to otherwise carefully measured expected values of the present worth of projects.

Such procedures can be applied to investment in risky agricultural research. Accordingly, a concrete example is presented below in order to explore the likely dimensions of risk deduction and to point to where research decision makers can conveniently and safely ignore risk.

Another issue in accounting for risk, even in the simplest case where only the expected value of a performance criterion is required for guidance, is the way that different contributing sources of uncertainty interact together. In the most straightforward case, independent sources of risk enter the analysis additively and lead to rather simple representations of how the combined risks can be assessed. When the summary measure of performance is a simple summation of many different sources of variation, even if these happen to be statistically interdependent, the expected value is still readily computed as a simple function of the relevant expected values of all the component parts. When the risks do not enter additively and linearly into the overall assessment, however, things are not nearly so straightforward, and merely combining simple functions of expected values of random variables does not lead to unbiased estimates of the expected value of the overall criterion. In such cases, it may be necessary to take explicit account of the extent of uncertainty and the stochastic interdependencies in order to produce unbiased estimates of the expected value of the performance criterion. There has been very little attention to these somewhat subtle matters in the literature of public project appraisal. Most of the limited work available has been rather abstract and of a cautionary nature (Anderson 1976).

Decision making about portfolios of research projects is unquestionably more difficult than about individual projects. That research managers tend to focus on particular commodities on a one-at-a-time basis raises issues of the decision criteria that they may wish to use. One of the traditional measures through which research is described is a research intensity or congruence measure (Fox 1987), which might be approximated, for example, by research expenditures divided by total industry revenue for the commodity in question (Boyce and Evenson 1975). Various authorities have suggested pragmatic (which

is to say, convenient but almost totally arbitrary) indicators of appropriate research intensities such as the 0.5% guideline 1985 target of the World Food Conference of 1974 (UN
1974). There is also, for example, the infamous 2% (1990 target) rule of the World Bank
(1981a, p. 8).

Small-project Case

The conceptual structure introduced above is too abstract to be illustrative of likely practical
implications, and thus, an empirical simulation of a small-project case is introduced for this
purpose. The illustration is cocoa research in Papua New Guinea. To initialize a model, it
is necessary to make concrete assumptions about all the relationships introduced. The first
is a highly speculative empirical counterpart to equation 4.1. The deterministic core of the
hypothetical representative knowledge production function is presented in figure 4.1.
Scientists clearly differ in their inherent productivity, and part of the art of the successful
research manager is to attract productive scientific workers to a project. Figure 4.1 is
representative in the sense that it abstracts from individual variation. The orders of
magnitude of scientific productivity used here are based loosely on the data reported by
Evenson and Kislev (1975b, pp. 29-31). Aspects of such variation are captured in the
stochastic specification of equation 4.7:

$$\Delta K_t = \sum_{i=0}^{5} (d_{t-i} + e_{1t}) \, s_{t-i} \tag{4.7}$$

where ΔK_t is the number of relevant new research publications in year t, d is specified in
figure 4.1 as a function of prior years of scientific endeavor, s is the number of scientist
years in any given period, which is determined by the research budget R, and e_1 is a random
component with a five-point discrete probability distribution: with values -0.4, -0.2, 0, 0.2,
0.4 occuring with equal probability.

The second challenge is to specify an empirical counterpart to equation 4.2. This is
done here by positing a relationship between publications and new productivity gains
summarized in figure 4.2 and expressed as

$$k_t = \sum_{j=1}^{5} v_{t-j} + u_{8t} \tag{4.8}$$

where k_t is as defined for (4.2), v_{t-j} is the function of ΔK_{t-i} specified in figure 4.2, and u_{8t}
is a normally distributed additive error term with mean zero and standard deviation 0.001.
The relationship depicted in figure 4.2 features diminishing marginal returns to numbers of
publications in any previous period and an appreciating, peaking, and depreciating effect
over time of the productive impact of new knowledge captured in scientific publications
(Wise 1986; Stefanou 1987).

Figure 4.1: *A knowledge (publications) production function for equation 4.7*

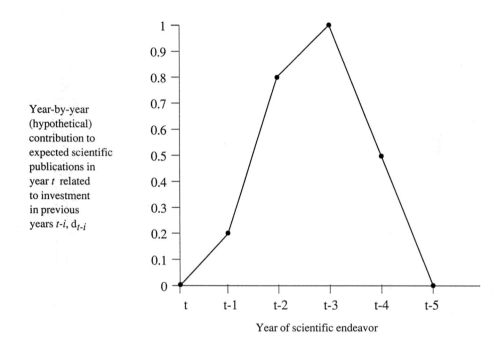

In several senses, these hypothetical relationships are unit free. To continue with the development of a concrete exemplification, it is now necessary to specify a particular commodity and country context. Cocoa in Papua New Guinea (PNG) is the case chosen. PNG supplies about 3% of the world trade in cocoa, and this accounts for about 3% of its gross domestic product, although about 10% of its exports. The national currency, the kina (K) is approximately at parity with the US dollar.

This market is quite volatile, with a mean export price (in 1987 values) of about 2800 K/ton and a year-to-year standard deviation of about 1800 K/ton. The quantity exported (in recent years) averages 30,000 ton with a standard deviation of about 1000 ton. Some econometric investigation of the market has been made by Akiyama and Duncan (1982). They elected to employ log-linear functions for the supply specification of the PNG cocoa market and proceeded to determine the lagged relationships involved. For simplicity, a linear, Nerlovian adaptive-expectation supply specification is used here in conjunction with lag structures and (mean) elasticities consistent with those found by Akiyama and Duncan (1982, pp.18, 53).

The short-run anticipated price is given by

$$P_t^* = (0.1/0.09) \, P_{t-1} + (0.17/0.09) \, P_{t-2} - (0.18/0.09) \, P_{t-3} \tag{4.9}$$

Figure 4.2: *A relationship between publications and supply-curve shifts for equation 4.8*

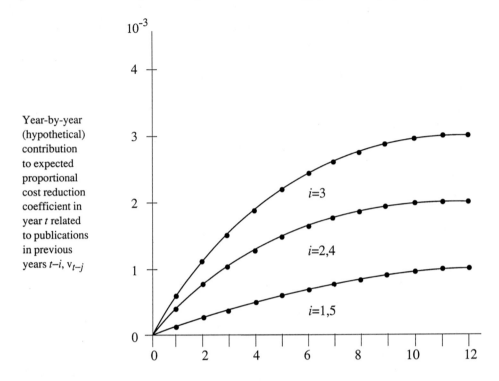

Publications in year t–i relating to the commodity, ΔK_{t-i}

and the dynamic counterpart to equation 4.5 is

$$Q_t^* = 5.0765 + 0.0084\, P_t^* + 0.83\, Q_{t-1} \qquad (4.10)$$

where the Q_t^* are in 10^3 tons per year, P_t^* is in 10^3K/ton, and the asterisks denote anticipated values.

The above empirical relationships provide nearly enough information to implement the model. The additional assumptions required are for the exogenously determined export prices P_{t-m}; the unit cost of the research scientists C, so that $R_t = C s_t$; the short-run elasticity of supply in equation 4.4 $E_s = 0.09$; and a stochastic specification for the disturbance in equation 4.4. The distribution used here is triangular, with mean 1.0 and range 0.9 to 1.1. Initial conditions including lagged values are required for some variables, and these are set at mean levels.

A trial value of the allocation of research resources is also needed and this is set initially at three scientist-years per year for all 40 years of the initially analyzed situation.

This base case is simulated a number of times (NS) to investigate the nature and distribution of returns from investing in research under these several assumptions. The riskiness of the investment can be summarized in many different ways. Two of the most widely used measures are the net present value, NPV, and the internal rate of return, IRR. A discount rate is required for the first measure and the one used here for the 30-year accounting period is 7.5% per annum, the assumed real, expected social discount rate for PNG. This rate is only required for the second measure if it is computed using the modified method to avoid multiple solutions in rather unusual situations (those that are not encountered in the present cases). It is assumed that the present value of the pre-investment-period activities (costs) is K0.5 million.

The base case (others are compared in the next subsection) is seemingly quite profitable. The data presented here (and below) are for NS=99, a sample of size sufficient to yield fairly reliable estimates of both the average performance and the riskiness of performance. The mean NPV is K1.660 million with a standard deviation K0.448 million. The corresponding data for IRR are 15.7% and 2.2%. In both cases the distributions are slightly positively skewed, as indicated by the statistics based on the standardized-third moments (α_3 in figure 4.3). The risk is described most comprehensively by the complete sample distribution functions, as reported in figure 4.3.

These cumulative distribution functions (CDFs) enable any critical value to be readily read, e.g., P(NPV $\leq$ K1.0 million) = 0.07, or P(IRR > 15%) = 0.6.

The richness of such probabilistic data can be contrasted with the poverty of data from a deterministic analysis. Recomputing the financial performance under the very strong assumption that all random variables take on their mean values at every realization, the summary measures to be compared with those for the mean stochastic performance are NPVdet = K2.387 million and IRRdet = 18.5%. Given the highly nonlinear structure of the model and the multiplicative nature of several of the uncertain elements, it is not surprising that there is considerable difference in the stochastic means and the deterministic estimates. Indeed, the net present value assessed by deterministic methods, even using appropriately assessed mean values of the component random variables, overestimates average net present value by 43.8%.[10] It thus seems that, notwithstanding any social attitudes towards risk, the appropriate accounting for uncertainty might well be important in accurate financial assessment of risky research investments, at least in absolute terms. It may be, however, that relativities, especially in ex ante work, are both more important and less compromised by uncertainty.

Large-project Case

In a large-project case, the only ready solution for practical implementation is to use something like the approach developed by Wilson (1982) and made pragmatically operational by Anderson (1983, 1989b). In such an approach, there is a heavy demand for

---

[10]Calculated as 100 ((2.387 - 1.660)/1.660) = 43.8%.

information that may not always be readily forthcoming. First it is necessary to define the
size of the project relative to the appropriate aggregate. For most public research situations,
the relevant divisor will be something approximating gross national income. In essentially
agrarian economies, particularly those that do not have a very diversified commodity
orientation, it is not inconceivable that some agricultural commodities will occupy a large
fraction of the national earnings. If the same countries are also of a relatively homogeneous

Figure 4.3: *Financial performance probability distributions for base-case cocoa research*

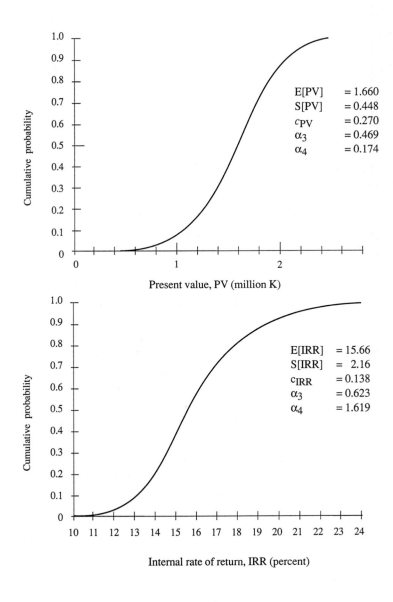

edaphoclimatic composition, individual research activities may well have potentially significant effects on national income. In this case it is necessary to pin down another rather information-demanding relationship, namely, the stochastic dependence between returns from investment in research in the major commodity and returns in the rest of the national income. The classic cases of this type of economy are the sugar-based economies of the Caribbean and elsewhere. The issues come to a head during times of severe downturn in the international market for the commodity.

It is no mean feat to estimate all the parameters that are required for applying the large-project formula of Anderson (1989b), namely,

$$D = A \cdot c_{NPV} \left(c_{NPV} \cdot F/2 + g \cdot c_{GNP} \right) \tag{4.11}$$

where D is the proportional deduction to be applied to mean project net present value NPV, A is the coefficient of society's relative risk aversion, c denotes a coefficient of variation, F is the fraction of mean project return relative to mean GNP, and g is the correlation between project (i.e., research) returns and GNP. It seems that, if analysts are prepared to make some informed judgments, reasonable ball-park estimates can be made. The procedure is then quite mechanical. With some assumptions about the size of the economy and its relationship to a particular major commodity, the following illustration demonstrates the application of this formula. The importance of accounting for the riskiness of returns from the research investment can thus be roughly assessed.

To return to the PNG cocoa example for the sake of concreteness, the question of whether the risk itself matters to the economy depends on the factors noted. In general, research is not a strong contender for risk discounting because it is a relatively small component of the agricultural industry concerned. Most agricultural industries are individually only small parts of the totality of agriculture, and the agricultural sector is but a small part of most economies, approximating only about 5% of gross national product (GNP), for example, in most more-developed countries, but it is substantially more, ranging between 10% to 50%, for most less-developed countries.

In the PNG cocoa case, cocoa represents about 3% of GNP, and research gains in the industry are but a small fraction of this — approximately the cumulative proportional cost reduction over a designated period of observation, say, about 10% over 15 years or so. The main driving force in a socially desirable, cautious approach to research investment is, naturally, the research itself. Social desirability, from a broad perspective, depends, in turn, on the significance of the industry concerned. In this regard, the present case is not very significant as, applying equation 4.11 and using the recent historical variability of the PNG economy and an appropriately high level of relative risk aversion (A=2),[11] the base case

---

[11] Newbery and Stiglitz (1981) argue that the appropriate "normal" range for A is from one to two, while Anderson (1989c) suggests that, for a country of Papua New Guinea's present income level and distribution, a value of two is reasonably applicable.

leads to a minuscule proportional risk deduction (table 4.1). Analogously, some potential levels of risk deduction, were cocoa to be a more important commodity in PNG (e.g., if gold and copper were to diminish greatly in importance, for whatever reason), are also reported in table 4.1.

Table 4.1: *Proportional Risk Deduction and Size of Industry,*
PNG Cocoa Research

Industry revenue as a percentage of GNP	Proportional risk deduction
%	%
3[a]	0.022
10	0.073
20	0.146
50	0.364

Note: Assuming that (1) representative research benefits are, in mid-project, approximately 10% of industry revenue, (2) relative risk aversion is 2, and (3) correlation between research benefits and GNP is zero.

[a] Base case.

Even when cocoa (very hypothetically) accounts for one-half of the economy, a level it approaches in some West African economies, the risk deduction only reaches approximately one-third of one percent. This surely is rather inconsequential and, in the context of the many uncertainties surrounding the research investment decision, could safely be ignored.

Now that the base case has been set in context, the question of more efficient research resource allocations can be addressed, albeit in a crude and partial way. This is tackled here through a small simulation experiment with the resource vector over time being the key experimental variable. The base case consists of a three-person research team constantly in place ($R3$). Variants of this examined here are one, two, and four research scientists ($R1$, $R2$, and $R4$). Further variants with similar mean resource commitments are also explored. These include (a) a step-up arrangement with two scientists for the first fifteen years of the investment period and four for the remainder ($R2,4$), and (b) a reversed scheme ($R4,2$). Two other "plans" are investigated, namely, (c) a constant absolute growth rate in research investment — increasing linearly from two researchers in the base period of t_0-10 to four in t_0+30 years, and (d) a random arrangement with the research effort determined annually as a triangular distribution with range two to four and mean (and mode) three researchers. The results, including mean present value, coefficient of variation of present value, and risk deduction (as a fraction) are shown in table 4.2.

It is not surprising that the risk deductions are all of the same order of magnitude as that for the base case of table 4.1. They can thus be dismissed as trivial, providing that the

Table 4.2: *Alternative Investment Experiment, PNG Cocoa Research*

Research resource situation	Mean return E(NPV)	Coefficient of variation of NPV	Risk deduction, D	Risk-adjusted return[a]
	(million K)			*(million K)*
Constant 1 researcher	0.981	0.319	0.017	0.9808[b]
Constant 2 researchers	1.620	0.265	0.021	1.6197
Constant 3 researchers	1.660	0.270	0.022	1.6596
Constant 4 researchers	1.428	0.302	0.024	1.4277
Step-up 2 to 4 researchers	1.602	0.268	0.021	1.6017
Step-down 4 to 2 researchers	1.529	0.291	0.024	1.5286
Steady increase, 2 to 4 researchers	1.611	0.278	0.023	1.6106
Random (stochastically independent) annual allocations[c]	1.659	0.272	0.022	1.6586

[a] Certainty equivalent return = E[NPV](1–D), from first and third data columns.

[b] The reporting of these data to five digits is not to imply that they are statistically significant but rather to illustrate the effect of the tiny risk adjustments.

[c] Triangular distribution, range 2-4, mode 3.

industry remains a small part of the economy (table 4.1), or they may not be so readily dismissed, as is now explored in a more regionally confined context. Given the hypothetical nature of the embedded relationships, not too much can or should be made of the comparisons possible in table 4.2, but a couple of observations can be made. First, the risk adjustments are sufficiently small and uniform that the ranking of the resource situations compared is unaffected by the adjustments. Second, a constant staffing of three researchers is suggested as economically superior, rather than more or fewer. The third situation, and one that is perhaps not very realistic, although it is relevant given the fluidity of expatriate research staffing in PNG, of rather unstable staffing around a mean of three researchers is found in this particular model specification to be quite insignificantly different from the steady-state staffing of three.

4.2.3 Targeting Research to Groups with Special Needs

The focusing of research to a particular group is really a special case of the issues broached above. The size of the project may be large relative to the relevant aggregate because the latter becomes a more targeted aggregate — for example, the social welfare of a particular disadvantaged tribal group in a remote region of a small part of a less-developed country. Addressing research to attempt to lift the welfare of such disadvantaged groups means that the project becomes a very specialized case of a large project, as discussed in an earlier section.

Often various other dimensions beyond mere economically assessable performance may be important. One such example would be a research program with nutritional objectives for a group that is perceived to be facing either chronic or periodic food insecurity. Nutritional aspects of research performance must then surely take a predominant role in the measures used to assess research benefits. It is conceivable that almost any financial accounting at all becomes irrelevant in such cases.

Some accounting can still be done, as illustrated, for example, by the work of authors such as Pinstrup-Andersen, Ruiz de Londoño, and Hoover (1976) and more recently other work managed by Pinstrup-Andersen at IFPRI. The analytical difficulties become progressively more intractable as a move is made towards more purely subsistence forms of agriculture (Hayami and Herdt 1977). Consider, for example, the case of attempting to improve the productivity of sweet potato culture in the highland subsistence areas of PNG (Antony and Anderson 1988). Here markets are extremely thin to nonexistent for this now rather traditional commodity. Any surplus is basically fed to pigs which, in turn, have rather specialized valuation characteristics. It makes little sense to grow sweet potatoes for pig consumption per se. The crop is essentially one for direct household consumption. When it is in short supply it is sorely missed. When there is a surplus, it is an embarrassment that has to be disposed of. Attempting to improve productivity from this form of cropping is fraught with difficulties of both a technical and economic nature. If significant productivity-enhancing cultivars or practices are discovered, the likely main effect will be displacement of labor. This labor happens to be highly gender differentiated since it is firmly within the province of the females of the households to manage the sweet potatoes and to provide nearly all the labor for their cultivation. This then provides an example of highly targeted research opportunities that, although the economic attributes may be rather questionable, may have profound social dimensions if successes are achieved.

Rather than introduce a new example to illustrate some of the points about a project that is large relative to the regional economy, the previous example dealing with cocoa is given a further twist. Reference is made to the formula for proportional risk deductions of Anderson (1989b). As a project becomes larger, its size can be reflected in the relative size variable, F, in equation (4.11). This, in principle causes no real problem for the analyst. A more subtle and difficult aspect to try to pin down is the likely degree of correlation between the project return and regional income. Concentrating on an important product for a local region means that, to the extent that it is variable, it will move in close association with the regional aggregate and this will typically be reflected in high correlation coefficients. Such a situation is depicted with the rather speculative funnel in figure 4.4, which illustrates likely feasible combinations of different degrees of positive correlation as the size of an industry's research benefits in an economy increases.

Iso-risk deductions that would be applicable for different combinations of the size of the industry and correlation are also presented in figure 4.4. It can be seen that, as the proportionate size of the industry grows, there is a steady (linear) increase in risk deduction, reflecting the unit elastic situation noted in equation 4.11. As correlation increases, ceteris paribus, there is also a linear rate of increase of proportional risk deduction. Needless to say, as the previous discussion indicates, the case of national PNG cocoa research falls low

Figure 4.4: *Illustrative iso-proportional risk deductions; A* = 2, *c*~NPV~ = 0.27, *c*~GNP~ = 0.04

Figure 4.4: *Illustrative iso-proportional risk deductions;* $A = 2$, $c_{NPV} = 0.27$, $c_{GNP} = 0.04$

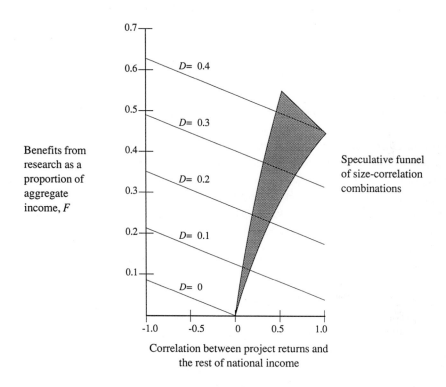

Benefits from research as a proportion of aggregate income, F

Speculative funnel of size-correlation combinations

Correlation between project returns and the rest of national income

in the cone of the funnel, with minuscule deductions. A pragmatist might well say that these deductions are so small that other inevitable errors in assessing future research returns will surely overwhelm any such subtle accounting for risk.

The same may not, of course, hold for some smaller aggregates. Suppose that, for some reason, concern in a research project was focused on a particular region. Consider, for example, the PNG province, East New Britain, which is an important producer of cocoa (some 22% of national production with some twelve thousand small holders and some 120 large holders, each group producing about 3000 tons of dried cocoa beans annually). Cocoa thus provides about 11% of the aggregate income of East New Britain, and a research project of the type considered in section 4.2.2 ($c_{NPV} = 0.27$) could represent about 2% (F = 0.02) of provincial income. Such returns would be highly correlated with cocoa export returns and thus strongly correlated with provincial income — say, $g = 0.3$ — assumed to have the same coefficient of variation as national income ($c_{GNP} = 0.04$). Substituting these data into equation (4.11) yields

$$D = 2[(0.27^2)(0.02)/2+(0.3)(0.27)(0.04)] = 0.008 \equiv 0.8\%$$

which, by any account, is still a small proportional deduction. Evidently, the regional or geographic focus would need to be very localized for the deduction for research projects to climb to really significant proportions.

On the basis of the variants of this constructed example, it seems that the main advantage of accounting for risk in investment analysis of research projects is for the accurate determination of mean project performance rather than to account for social risk aversion per se.

4.2.4 Portfolio Management in Research

The portfolio aspects of resource allocation to research can be considered at several levels. At one of the broadest levels, for instance, there is inevitably a problem for research administrators to allocate their budgets across disciplines. In many cases research agencies are organized on largely disciplinary lines and thus this allocation is a central problem for resource managers. There is no well-established literature on the complementarity of different disciplines, but it is clear that, in most cases, there are benefits to be gained not only from individual disciplinary activities but also through their positive joint effects. The issue becomes a significant one for rather small research systems such as might be found in, say, a small island economy (Hardaker and Fleming 1989). There may be just so few professional resources available that not all disciplines can be adequately represented in the organization. A common case is for a small research organization not to have any significant representation from the social sciences in its professional staffing. In research organizations that involve multidisciplinary work, such as many farming systems research programs, for instance, there is at least implicit accounting for the probable complementary effects. In general, however, it seems that much more careful research is required to identify the nature of complementarity between a range of disciplines, and the importance of alternative administrative arrangements for tapping such interactive benefits most effectively.

At a lower level of organizational structure, there is rather more in the way of research findings that deal with portfolio aspects. There is a large literature on portfolio management in enterprises that involve risk. This features diverse approaches such as risky whole-farm planning and investment analysis involving efficient portfolios of risky enterprises. The latter may be viewed from an individual or institutional level. The analytic framework usually used is one of mathematical programming in which defined objective functions are maximized subject to sets of constraints. The decision-making problem of a research administrator can readily enough be conceptualized in this framework, although the interdependence between research activities often complicates analysis beyond the simple accounting procedures that are embodied in many of the models used.

This is not the place to broach the technicalities of portfolio management under risk, but it should be noted that, particularly in the industrial research literature, much attention has been given to these matters (Anderson 1972a), although it remains rather questionable as to the extent to which the procedures are used by practical research managers. Large-scale firms with sophisticated operations-research departments seemingly have a proclivity for developing elaborate procedures for conceptualizing the problems without necessarily giving full regard to the practicalities and information requirements surrounding the application of the models to practical problems (Anderson and Parton 1983).

4.3   CONCLUSIONS

There are two broad sets of conclusions that can be drawn from the foregoing considerations. The discriminating factor seems to be the temporal point of reference to the research activity. Research is surely a risky activity, whether viewed retrospectively or prospectively. In retrospective evaluations, it seems that a sympathetic accounting for the riskiness of the activity can be a useful adjunct to understanding the success achieved and to the worthwhileness of the activity. This is the topic tackled in section 4.2.1. What has not been established in this paper or, indeed, anywhere else as far as can be determined, is a reconciliation of the inherent uncertainty in research with the regularities that seem to guide the invisible hand in research investment. The induced-innovation hypothesis of Binswanger and Ruttan (1987), for instance, implies that technical change and institutions are responsive to changes in resource endowments and prices of factors of production and products. Presumably what this means is that these theoretical notions apply to measures of central tendency such as average research benefits. It may be that more complex hypotheses can be advanced for addressing jointly the influence of changes in factor prices as well as the variability of these changes and measures of the uncertainty of research results themselves. Such future theoretical work would seem to be much more complex than the existing literature on the induced-innovation hypothesis and may not be especially worthwhile if most of the important trends in averages are satisfactorily explained by movements in variables that are adequately described through their means.

The more important conclusions are, however, for prospective planning. The role of risk in such decision making is a sadly neglected field, which could be quite important in some situations. This has been the main thrust of the present chapter, and the main messages for operational research planners are summarized below.

Estimating future returns to agricultural research investment is something between a challenging task in applied economic analysis and a fledgling art form. Uncertainty is intrinsic to the phenomenon of research and must thus be dealt with in some way. The simplest way is to collapse all implicit probability distributions to point estimates of relevant parameters and thus undertake what could be described as a degenerated or deterministic analysis. This will yield reliable estimates if (a) the degeneration is to good mean estimates and (b) the random variables so degenerated enter the assessment linearly and additively. Otherwise, as has been illustrated herein by reference to a simple model of research on cocoa in PNG, the resulting estimate of the mean economic return, which is required for good research decision making, may be seriously awry.

Individuals and society at large can safely be described as (technically speaking) risk averse. At the individual level, this is often manifested in somewhat cautious behavior in action and investment. At the community level, as in public investment appraisal such as ex ante research assessment, the risks of individual projects are typically so diluted in the overall economy and so widely shared by the members of society that the influence of risk aversion pales into insignificance. Such an influence can be approximated by a proportional risk deduction that should be applied to mean project worth (such as its net present value). For research projects on particular commodities, these deductions are, as is also illustrated by the PNG cocoa case, so small that they can be safely ignored, except where a very localized perspective is being taken in the assessment. In this special case, some illustrated simple procedures can be used to make an appropriate adjustment.

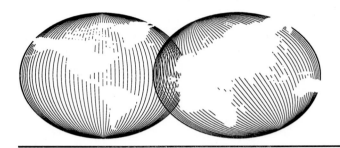

PART II

MEASURING AGRICULTURAL RESEARCH
AND ECONOMIC DEVELOPMENT

Chapter 5

Internationally Comparable Growth, Development, and Research Measures

Barbara J. Craig, Philip G. Pardey, and Johannes Roseboom

Constructing measures of real economic activity for the purpose of making international comparisons is a useful but tricky exercise. Unless data are collected specifically for the problem at hand, the resulting series may only poorly measure the variable of interest. Available data may not provide uniform coverage of the countries or periods of interest or may be too broadly or narrowly defined. In the analysis of agricultural research and development patterns, problems of data availability and quality are compounded by the need to recast value and volume aggregates into units that can be meaningfully compared over time and across countries.

When the data on a series of interest are aggregate values measured in local currency units, the aggregates must typically be deflated to take account of changes in the local price level and converted to a common currency in such a way as to provide an accurate picture of their real value or volume. Both the choice of appropriate converters and deflators and the order in which these two operations are performed matter and thus will, in many instances, have substantial effects on the interpretation of the resulting real-value series.

Data series that are reported directly as quantities or volumes often appear to relieve analysts of the problems of turning nominal values into real ones; nevertheless, subtle but substantive issues of comparability are likely to persist. With volume aggregates, some unweighted and perhaps undesirable aggregation is implicit in the data collection process.

In this chapter the strengths and weaknesses of data used throughout this volume are discussed, and the systematic approach taken to enhance comparability is described. Since there are some insurmountable obstacles in these data sets, evidence on the effects of using less than ideal aggregation procedures is presented. This evidence will both aid the interpretation of imperfect data and provide some sense of the boundaries within which the truth lies.

In section 5.1 we describe ideal aggregation procedures. Some practical options for translating preaggregated data into real value or volume measures are discussed in section 5.2 along with selected evidence on the consequences of using different conversion, deflation, and scaling procedures. In section 5.3 we discuss the specific concepts and measurement issues that underpin the agricultural statistics used throughout this volume and conclude in section 5.4 with a brief review of the definitional, measurement, and practical issues involved in constructing agricultural research indicators.

5.1   AGGREGATION

In a multidimensional framework, the construction of comparable aggregates measuring real economic activity always involves two distinct steps. Index number theory tells us to begin with disaggregated data on prices and quantities in each country to calculate directly a real quantity index.[1] To translate the resulting index into an aggregate that can be compared over time and across countries, some scaling factor — based again on disaggregated prices and quantities — must be applied to the base country and/or time period. One of the major problems of making international comparisons lies in the shortage of systematically disaggregated data.

Index number theory, informed by neoclassical models of economic behavior, argues for aggregating real quantities using price weights that are most specific to the economic activity and agents whose behavior is being summarized (Drechsler 1973). For constructing indices, representative or characteristic price vectors need not replicate local *absolute* price levels, but they should reflect local *relative* prices. Otherwise, one may fail to distinguish between changes in the size of the real commodity basket and changes in the composition of the basket. Even when analyzing sectors or entire countries with badly distorted prices — whether due to trade restrictions, price controls, subsidies, or the like — it is important to use the prices actually faced by economic agents when forming the real aggregate. When constructing comparable international aggregates, it is still desirable to use value weights that are representative but, in this context, accounting for differences in absolute price levels is necessary as well.

5.1.1   Temporal Indices

Contemporary international data sets span years of high price volatility, so the pitfalls of using value aggregates denominated in current local currency units are obvious. To compare commodity baskets produced in different periods, index number theory provides arguments for using timely local prices as weights in the construction of indices that have changing rather than fixed price weights. Changing weights allow one to capture shifts over time in local relative prices, which influence changes in the composition of local commodity baskets. Consequently, discrete approximations of the Divisia index (Divisia 1928) are to be preferred to the more commonly used fixed-weight Laspeyres index; they are less likely

---

[1]   For a useful discussion of index number issues in the context of international comparisons, see Caves, Christensen, and Diewert (1982) or Craig and Pardey (1990a).

to confound changes in the size of the commodity basket with changes in its composition.

There are several possible discrete approximations of the Divisia index. The most commonly used are the Laspeyres and the Törnqvist-Theil approximations:

$$\text{Laspeyres:} \quad I_t^{DL} = I_{t-1}^{DL} \left[\frac{1 + P_{t-1}'(Q_t - Q_{t-1})}{P_{t-1}' Q_{t-1}} \right] = I_{t-1}^{DL} \frac{P_{t-1}' Q_t}{P_{t-1}' Q_{t-1}} \tag{5.1a}$$

$$\text{Törnqvist-Theil:} \quad I_t^{DT} = I_{t-1}^{DT} \prod_{i=1}^{m} \left[\frac{Q_{it}}{Q_{it-1}} \right]^{\overline{w}_i} \tag{5.1b}$$

where

$$\overline{w}_i = \frac{1}{2} \left(\frac{P_{it} Q_{it}}{P_t' Q_t} + \frac{P_{it-1} Q_{it-1}}{P_{t-1}' Q_{t-1}} \right)$$

Here P is an m-dimensional vector of commodity prices and Q is an m-dimensional vector of the corresponding quantities. The transpose of a vector is indicated by a prime, so that $P'Q$ is the sum of the products of the respective elements of P and Q. The t subscripts indicate the time period. The choice between alternative approximations of the Divisia index depends on the nature of the data on hand and the functional form deemed most appropriate for aggregating the quantities of interest (Diewert 1978; Craig and Pardey 1990a).

5.1.2 Spatial Indices

Economic theory gives us less guidance in constructing indices of real aggregates in the cross-sectional dimension since there is no single vector of price weights that is representative for all countries to the extent that international markets for goods and factors are not entirely integrated. As argued elsewhere (Craig and Pardey 1990a), a chained index is of less use in cross section because there is not the same behavioral notion that prices and output evolve over space, i.e., across countries, as they do over time.

If one is forced to resort to fixed-weight indices, attention is focused on the construction of value weights that can be used to calculate real aggregates expressed in common units. The two options most frequently used in international comparisons are the conversion of commodity prices to common currency units or the conversion of all commodities to a common physical unit such as wheat equivalents (Hayami and Ruttan 1985).[2]

---

2 The problem of currency conversion can be avoided if one uses the Törnqvist-Theil approximation of the Divisia index to construct multilateral indices. In these indices, local prices only enter the calculation in the construction of local value *shares*. Since it is only local and base-country value shares that are averaged, one need never employ an exchange rate. If only a single cross section is being considered, this index method has a lot to recommend it. However, with panel data, i.e., cross sections for several years, the

When choosing an exchange rate series to convert local currency to a common or numeraire currency, the goal is to find a converter that correctly translates the purchasing power of the local currency in the particular sector of the economy being analyzed. This is typically not the same problem as searching for an equilibrium exchange rate. Market-determined exchange rates reflect the relative purchasing power of a currency in trade and are thus influenced by a fairly narrow set of real and financial transactions that may or may not be directly related to the aggregates of interest. The managed or fixed exchange rates common in less-developed countries may be even less useful for translating real purchasing power. There is ample empirical evidence that neither market nor managed exchange rates vary in the short run in a way that reflects differences in average price levels across countries (Levich 1985), yet this is exactly what the ideal converter would do.

World Bank staff developed one converter, the Atlas exchange rate, that uses both official exchange rates and short-run changes in relative price levels (World Bank 1983).[3] For some countries, trade restrictions, government exchange rate policies, and the like cause official market exchange rates to deviate flagrantly from the rate that applies to the foreign transactions effectively taking place. In such cases, the Atlas exchange rate is adjusted using secondary data concerning the nature and estimated impact of these distortions.

The International Comparisons Project (Kravis, Heston, and Summers 1982) has generated an alternative series of synthetic exchange rates called purchasing power parities (PPPs) using the Geary-Khamis procedure.[4] These PPPs are an attempt to get a broader measure of relative currency values by comparing the relative costs in local currencies of a detailed basket of traded and nontraded goods and services. One feature of the Geary-Khamis procedure is that it actually performs two steps at once. The set of n country PPPs are calculated at the same time as the m-dimensional "international" price vector by solving a system of $m+n-1$ equations:

$$\Pi_i = \sum_{j=1}^{n} \frac{P_{ij} Q_{ij}}{PPP_j \sum_{k=1}^{n} Q_{ik}} \qquad i = 1, \ldots, m \qquad (5.2a)$$

$$PPP_j = (P_j' Q_j) / (\Pi' Q_j) \qquad j = 1, \ldots, n-1 \qquad (5.2b)$$

---

implied time series for each country in the cross section will not be calculated using only local prices and so may yield a biased picture cf real local growth rates.

3 World Bank (1983) gives details of two earlier versions of the Atlas method, while *World Development Report 1985* (p. 244) describes the current Atlas method, which uses a simple average of the official market exchange rate for the current year and two predicted exchange rates for the current year that are based on observed exchange rates and relative inflation rates of the two previous years. Specifically, $e_t^* = 1/3 \, [e_{t-2} (P_t/P_{t-2})/(\$P_t/\$P_{t-2}) + e_{t-1} (P_t/P_{t-1})/(\$P_t/\$P_{t-1}) + e_t]$, where e_{t-j}, P_{t-j}, and $\$P_{t-j}$ are, respectively, the official market exchange rate, a local general price index, and the US general price index in year $t-j$.

4 For a comprehensive discussion of PPP indices, see Kravis et al. (1975), Kravis, Heston, and Summers (1978, 1982), Summers and Heston (1984, 1988), Kravis (1986), and EUROSTAT (1982).

The PPP for the numeraire country, say n, is set to unity by definition:

$$PPP_n = (P_n' Q_n) / (\Pi' Q_n) = 1 \tag{5.2c}$$

In these formulas, Q_{ij} is the quantity of commodity i produced in country j. The international price of commodity i, Π_i, in equation set 5.2a is the weighted average price of the n country-specific prices, P_{ij}, where country prices are converted to a common currency using implicit exchange rates and then weighted by the physical share of country j in the total quantity of commodity i. The implicit exchange rate or purchasing power parity for country j, PPP_j, is defined in equation 5.2b as the value of its commodity bundle evaluated at international prices relative to that same bundle's value when evaluated at domestic prices.

There is empirical evidence that official exchange rates vary from PPPs in a significant and systematic manner (Heston and Summers 1988). A ratio of annual, average, official exchange rates to PPPs is generally greater than unity for low-income countries and often slightly less than unity for high-income countries. This pattern is due in large measure to differences across countries in the relative prices and quantities of tradable versus nontradable goods and services. Nontradables are generally more labor intensive than tradables, and productivity differences between low- and high-income countries tend to be lower in nontradables. When combined with the fact that labor is relatively cheap in low-income countries, these structural factors lead to lower relative prices of nontradables in low- versus high-income countries.

One advantage of PPPs is that they are not unduly influenced by policy shifts in exchange rates or by sudden swings in financial transactions. They may also be constructed to reflect differences in average prices for a very specific segment of an economy and for a particular set of countries. If the aggregates of interest are dominated by nontraded goods, the PPPs are likely to be more accurate converters than official exchange rates. A practical disadvantage of PPPs is the need to collect detailed data on local prices and comparable quantities in all countries and years in the sample.

Official exchange rates, Atlas exchange rates, and PPPs are converters that have been used in a variety of ways to construct cross-sectional indices. If we use P_j^* to represent the price vector of country j which has been converted to a common currency, say dollars, and let Q_j represent the corresponding quantity vector of country j, then one possible cross-sectional index with base country b is given by

$$I_j^{CS} = (P_j^{*\prime} Q_j) / (P_b^{*\prime} Q_b) \tag{5.3}$$

In this index each country's quantities are aggregated using corresponding prices expressed in a common numeraire which maintains the *local* relative price structure. A more commonly used cross-sectional index formula applies an identical set of value weights to aggregate quantities in all countries, using

$$I_j^{\overline{CS}} = (\overline{P}' Q_j) / (\overline{P}' Q_b) \tag{5.4}$$

The single price vector $\overline{P}$ may be the price vector of the base country or of an arbitrarily chosen third country, a simple average of sample price vectors, or a weighted average of sample price vectors as in the Geary-Khamis procedure.

If the relative price structure differs across countries, each of these different ways of defining the common price vector used in (5.4) will typically result in different indices. If the converters have, in fact, translated all local prices into comparable currency units, there is no obvious need to tamper with the local *relative* price structures when constructing cross-sectional indices. Lack of data on individual country prices may force one to use (5.4), but (5.3) comes closer to the ideal of using relative prices that represent those faced by local agents when summarizing the economic outcomes that are the consequences of their actions.

The use of a common numeraire commodity, rather than a common currency, requires converting each real quantity into units of the numeraire using relative prices to make the translation. For example, a wheat equivalent index can be formed using

$$I_j^{WE} = (R_j' Q_j) / (R_b' Q_b)$$ (5.5)

where R_j is the vector of relative prices in country j, or

$$I_j^{WE} = (\overline{R}' Q_j) / (\overline{R}' Q_b)$$ (5.6)

where $\overline{R}$ represents a *common* vector of wheat relativities applied to aggregate quantities of both the base country b and country j.[5]

Equation 5.6 can be criticized for the same reasons given for (5.4). The use of a common vector of price relativities, however they are chosen or constructed, amounts to imposing an artificial relative price structure on all or most of the countries in the sample. If the units of both the base and comparison country aggregates can, in fact, be converted to comparable units of wheat using local prices, there is no need to impose a synthetic or nonrepresentative set of value weights on either aggregate.

The problem with both of these wheat equivalent indices is that the choice of the numeraire commodity is critical. As shown in figure 5.1, the value of the commodity bundle represented by output point Q_j can be measured in either tons of wheat (on the vertical axis) or dozen eggs (on the horizontal axis). If one uses the local relative price vector P_j the value of country j's output is either A tons of wheat or B dozen eggs. If an alternative relative price vector such as $\overline{P}$ is used, the total value of the country's output as measured in wheat rises from A to C or alternatively falls from B to D when output is measured in eggs. Cardinal and even ordinal rankings of countries may be altered by the choice of the numeraire commodity.

---

5   Each element i of relative price vector R is the ratio P_{ij}/P_{wj} where P_{ij} is the price of commodity i in country j and P_{wj} is the local price of wheat. Each price is expressed in units of the currency of country j.

Figure 5.1:   *Wheat- versus egg-equivalent output measures*

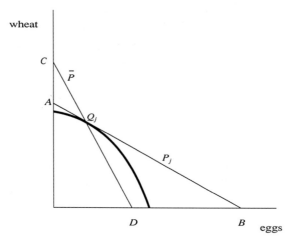

In practice, wheat equivalent aggregates have been manipulated even further. For instance, in Hayami and Ruttan's (1971, 1985) work, the total value of output as measured in wheat is calculated for each country using several *nonlocal* relative price vectors. A geometric average of the resulting aggregates is then taken to be the final measure of the aggregate. This procedure mimics the bilateral Fisher ideal index in a multilateral context.[6] Since different relative price vectors imply different aggregate volumes, the geometric average will tend to provide some ad hoc smoothing of these differences. Referring again to figure 5.1, the geometric average of aggregates would fall somewhere between points *C* and *A* if wheat is the numeraire commodity, or between *D* and *B* if eggs are the numeraire.

5.1.3   Multidimensional Indices and Comparable Aggregates

When we want to compare aggregates *both* across countries *and* over time, sticking to aggregation procedures that use representative relative price weights is still a guiding principle. An effective way to accomplish this is to produce chained time-series indices for each country using local prices, and then scale the resulting series for each country with volume aggregates that have comparable units for all countries in the base year.

As it happens, the construction of these cross-sectional scaling factors involves precisely the same issues discussed above for spatial indices. One need only calculate a comparably measured aggregate for each country in the base year and use it to multiply each observation in a country's time series. The numerators of the spatial indices in equations 5.3, 5.4, 5.5, and 5.6, calculated with base-year data, all provide reasonable scaling factors

---

6   Hayami and Inagi's (1969) wheat-equivalent procedure does not yield a true Fisher ideal index unless the price relativities employed include a local price relativity for each country.

if the objective is obtaining comparable *aggregates*. These same spatial indices can be used to recalibrate country-specific time-series indices if the object is a multidimensional *index*. Once again the desirability of using representative prices points to the use of scaling factors (equations 5.3 and 5.5) that preserve local relative price structures.

We find this procedure more appealing than other procedures for multilateral international indices that have been advocated. Caves, Christensen, and Diewert (1982) have suggested using the Törnqvist-Theil approximation of the Divisia index (equation 5.1b) that compares all country observations to one specific (perhaps synthetic) country and time period. It is, however, more natural to link prices and quantities in neighboring periods of time than to blend value weights from disconnected periods and countries. Khamis (1988) suggests constructing a single set of international prices that are averages of prices over all years and countries in the sample. This may impose a large computational burden while remaining in essence an index method based on fixed, nonrepresentative value weights.

5.2 AGGREGATION IN INTERNATIONAL DATA SETS

We have discussed direct calculations of comparable quantity aggregates, but most international data sets include *preaggregated* data. In many cases, aggregates are reported in total local currency units, so we can only hope to deflate such measures to arrive at implicit volumes. In other cases, the volumes that are reported are unweighted totals of heterogeneous commodities or factors of production. Secondary data may be available to adjust such volumes, but comparable volumes may only be derived indirectly.

5.2.1 Value Aggregates

When confronted with value aggregates measured in a variety of local currency units, each must usually be deflated to reflect changes over time in each country's average price level and converted to arrive at aggregates in comparable real values or volumes.

The choice of an appropriate local price index entails some conceptual difficulties. Readily available price indices are typically general indices that may not reflect price developments in specific sectors of an economy such as agriculture. World Bank (1989) statistics indicate that implicit deflators of GDP and AgGDP are systematically different. Broadly speaking, AgGDP deflators indicate lower average rates of inflation in more-developed countries than do deflators defined over all sectors of the economy. The opposite holds true in most less-developed countries. Thus, using AgGDP deflators instead of GDP deflators will yield lower estimates of implied growth in real agricultural output for less-developed countries, as indicated in table 5.1.

Another problem is that price indices are commonly constructed using fixed quantity weights, as in a Laspeyres price index. The advantage of these measures is their ease of interpretation; they tell us how much the cost of purchasing exactly the same basket of goods has changed over time. Their disadvantage lies in the fact that they tend to overstate

Table 5.1: *Difference in Growth in Real Agricultural Output Using Alternative Deflators*

Region	1961-70	1971-80	1981-85	1961-85
	%	%	%	%
Sub-Saharan Africa (20)[a]	0.8	-0.8	-3.1	-0.7
China	-1.1	-1.2	-0.5	-1.7
Asia & Pacific, excl. China (12)	-0.7	1.0	-0.3	0.5
Latin America & Caribbean (21)	0.9	0.3	-2.1	0.1
West Asia & North Africa (7)	-1.1	1.3	-1.3	-0.1
Less-Developed Countries (61)	-0.3	0.3	-1.0	-0.2
More-Developed Countries (13)	0.5	0.7	2.9	0.9
Sample Total (74)	0.0	0.4	-0.2	0.1

Source: Implicit GDP and AgGDP deflators and AgGDP data primarily taken from World Bank (1989), and PPPs taken from Summers and Heston (1988).

Note: Percentages indicate absolute differences in compound annual rate of growth of "real" AgGDP deflated with implicit AgGDP less that deflated with implicit GDP deflators. A positive difference in growth means that agricultural prices have grown more slowly than the average price level.

[a] Bracketed figures indicate the number of countries in regional totals.

changes in the general price level by failing to allow for changes in the composition of the basket of goods produced or consumed that are likely to occur if there are changes in relative prices over the period being considered. The longer the time horizon of the study, the more likely are fixed-weight indices to understate the volume of economic activity by deflating with an index that fails to account for substitution. As argued in the index-number literature, the use of Divisia price indices would alleviate this last problem.[7] However, in an international context, these indices are so rarely constructed that they are currently not an option for international comparative analysis.

The problems of currency conversion have already been touched on, so the only new question is the order in which one employs deflators and converters. From the various algorithms available for translating values into comparable volumes, a practical alternative is to select a two-step procedure. One can first convert local currency values into a numeraire currency, such as US dollars, then apply an appropriate price index to account for price-level variability in the numeraire currency. The other option is to first deflate local currency values using local price indices then convert local prices into a numeraire currency using some base-year measure of relative currency values.

There are numerous deflators and currency converters that can be incorporated into either algorithm. Unfortunately, the choices matter. Since we have no independent measure of the truth, we are forced to proceed using some rules of thumb.

In choosing a price deflator, one should use the price index that most nearly reflects

---

[7] See Diewert (1978). He demonstrates the quantitative differences between fixed weight, chained, implicit, and explicit quantity indices using time series data on Canadian consumption expenditures.

the composition of the aggregate value to be deflated. In multicountry studies, this rule of thumb will argue for an algorithm in which aggregates are deflated first with a local price index whenever adequate price indices are available for each country in the sample. The basket of goods covered in a local price index may be quite different from that of a numeraire country's index when living standards and local relative prices vary substantially across the countries in a sample. This cross-sectional variability would lead to biases in measurement whose direction and magnitude would be difficult to predict.

A more subtle problem is the combined choice of deflator and converter. If the values to be compared are the total values of a single uniform good, the two algorithms (deflation then conversion or conversion then deflation) yield the same result if and only if the deflator and converter are defined over the specific good. If the values to be compared are aggregates, the deflator and the converter must be defined over the specific basket of goods represented by the aggregate. General price indices, market and/or official exchange rates, and nonspecific PPPs all introduce biases to the extent that they reflect aggregates whose composition may differ from the aggregate of interest.

Even with properly defined deflators and converters, the problems of aggregation cannot be escaped. As demonstrated in Pardey, Roseboom, and Craig (forthcoming) the two algorithms will yield different volume series unless it is the scale and not the composition of the aggregates that varies over time and across countries. Both algorithms diverge from the desired volume measure as the composition of the aggregate changes across the sample. So, when using the convert-first procedure, the volume measure will be biased unless the composition of the numeraire country's aggregate is representative of all other countries in all years of the sample. The deflate-first procedure will generate biases in the volume measure whenever the base-year basket within each country is not representative of that country for the period being considered.

So, in a particular application, the choice of algorithm must be made on the basis of whether it is the temporal or cross-sectional composition of the aggregate that is likely to vary most. Researchers have shown a preference for converting local currencies to dollars first and then deflating using a US price index. However, in a data set that includes countries at diverse stages of development, it is quite likely that cross-country differences in the composition of the aggregates will dominate the temporal variability unless the data span several decades; hence, a deflate-first procedure would demand far less of the data.

Pardey, Roseboom, and Craig (forthcoming) contrast the results of applying the two procedures to data on agricultural research expenditures in a sample of 90 countries. Volume measures were constructed using the convert-first algorithm with annual average exchange rates and PPPs as alternative currency converters; both series were then deflated using the US implicit GDP deflator. These were contrasted with volume measures produced by deflating first with country-specific implicit GDP deflators and converting with each of the two base-year currency converters.

For this application, no price index covering the specific mix of labor, materials, and equipment peculiar to agricultural research was available in each country, so the GDP deflator was a practical compromise.[8] The annual average exchange rate used was the yearly official market rate, which generally corresponds to the IMF's *rf* or inverted *rh* rate. The PPP series, which was defined over GDP, represented another compromise. Published PPPs either cover too few countries or a basket of goods that is not particularly representative of agricultural research.[9] The commodity coverage of PPPs obtained from Summers and Heston (1988) did, at least, correspond closely to that of the implicit GDP deflators being used.

Table 5.2 reports the 1981-85 average annual volume of resources committed to agricultural research implied by each of four measures.[10] In each column the regional total is indexed on the total sample volume implied by the particular conversion method.

For the 1981-85 period, the regional shares exhibited nontrivial sensitivity to the choice of translation. Converting the series first with annual average exchange rates lowered the measured total research commitment by at least one billion dollars. In general terms, the differences between the estimates were more dramatic for the less-developed than for the more-developed countries. A difference of approximately 55% in the less-developed countries' share of global research expenditures arose simply from the choice of converters. In particular, using PPPs rather than exchange rates approximately doubled the Asia & Pacific region's share of total research resources from around 6.3% to more than 13%. This pattern can be traced to the fact that relative price levels in less-developed countries reflected in Summers and Heston's (1988) PPPs are much lower on average than those implied by market exchange rates.

The volume measure was somewhat sensitive to the order of deflation and conversion near the base year but over longer time periods the two algorithms produced more obviously divergent results — particulary when volumes were obtained using annual average exchange rates as converters. Figure 5.2a presents the percentage deviation of the deflate-first

---

[8] A long-run agricultural research deflator for the US which takes account of annual variations in the mix of labor, capital, and materials used in agricultural research is given by Pardey, Craig, and Hallaway (1989). For additional discussion relating to R&D deflators, see also NSF (1970), Jaffe (1972), Mansfield, Romeo, and Switzer (1983), Mansfield (1987), and Bengston (1989).

[9] MacDonald (1973) and OECD (1981) discuss the concept of a PPP for R&D at some length. However, MacDonald provides such series for only a very small set of more-developed countries. PPPs defined over subsectors of the economy differ substantially, as described in Kravis, Heston, and Summers (1982). They point out that, on average, currencies for less-developed countries have substantially less purchasing power over a basket of investment goods and services than over a more general basket of goods and services. For government goods and services, the converse is true. We chose to use the broadly based Geary-Khamis PPPs of Heston and Summers (1988) calculated over GDP rather than any of its subaggregates because Heston and Summers themselves were concerned about the robustness of these more specific PPPs.

[10] These four measures all involved deflating with implicit GDP deflators. This contrasts with the method used by Evenson and Kislev (1975a), Judd, Boyce, and Evenson (1986), and Mergen et al. (1988). The clearest description of the translation procedure used in these studies appears to be in Judd, Boyce, and Evenson (1983, p. 3) where it is stated that "[research] expenditures were converted to US dollars using official exchange rates and were then inflated to 1980 dollars using a general wholesale price index."

Table 5.2:   *Alternative Measures of the Volume of Agricultural Research Resources,
1981-85 Average*

Region	Convert-first		Deflate-first	
	AAER	PPP	AAER	PPP
	%	%	%	%
Sub-Saharan Africa (31)[a]	4.1	4.7	4.6	4.7
Asia & Pacific, excl. China (11)	6.3	13.3	6.3	13.4
Latin America & Caribbean (17)	6.7	9.0	5.8	8.9
West Asia & North Africa (8)	2.1	2.9	2.1	2.8
Less-Developed Countries (68)	*19.2*	*29.8*	*18.8*	*29.8*
MDCs other than US (21)	55.4	49.2	59.7	49.7
United States (1)	25.4	21.0	21.5	20.5
More-Developed Countries (22)	*80.8*	*70.2*	*81.2*	*70.2*
Total Sample (90)	100.0	100.0	100.0	100.0
Total Sample Volume [b]	*5491*	*6646*	*6493*	*6821*

Source: Annual average exchange rates and implicit GDP deflators are primarily taken from World Bank (1989), PPPs from Summers and Heston (1988), and agricultural research expenditure data from Pardey and Roseboom (1989a).

Note: Translation procedures involved deflating with either US or local implicit GDP deflators and converting with either annual average exchange rates (AAER) or purchasing power parities (PPP) over GDP. Figures represent regional shares of a 90-country total. Data may not add up exactly because of rounding.

[a] Figures in brackets indicate the number of countries in regional totals.
[b] Millions of 1980 US Dollars.

versus the convert-first volume measures when annual average exchange rates and implicit GDP deflators are used to derive the respective volume measures. In figure 5.2b the same graph is presented for the volume series which used PPP exchange rates and GDP deflators.

When annual average exchange rates are used, the deflate-first algorithm led to a consistently larger volume measure than that obtained when expenditures were converted first. This suggests that, ceteris paribus, either the US dollar was undervalued with respect to virtually every country's currency in 1980, or that movements in local price levels were imperfectly translated by changes in the official annual exchange rates. The difference between these two volume measures is most pronounced in the Bretton Woods years when all exchange rates were essentially fixed. This gives further credence to the idea that official exchange rates may carry little or no information about changes in the relative purchasing power of different currencies, and so will be inappropriate converters for the purposes of international comparisons of long time series.

The temporal pattern of deviations of the PPP-converted measures in figure 5.2b is far less dramatic than those in figure 5.2a. By construction, changes in PPPs over time should do a better job of capturing changes in relative price levels between countries. In contrast

Figure 5.2a: *Percentage deviation of convert-first from deflate-first formula using annual average exchange rate convertors and implicit GDP deflators (Base-year = 1980)*

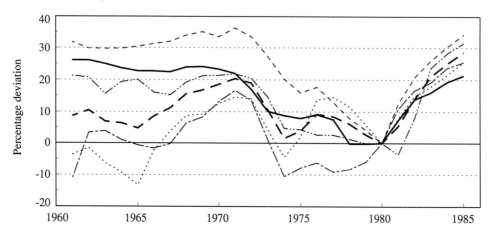

Figure 5.2b: *Percentage deviation of convert-first from deflate-first formula using PPP convertors and implicit GDP deflators (Base-year = 1980)*

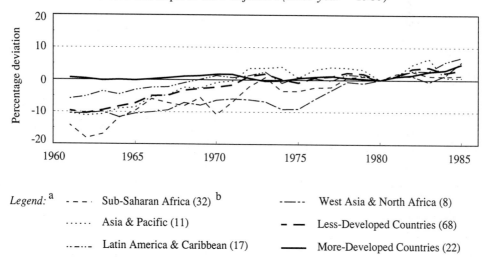

Legend:[a] - - - - Sub-Saharan Africa (32)[b]   -·—·· West Asia & North Africa (8)

······ Asia & Pacific (11)   – – Less-Developed Countries (68)

··—··· Latin America & Caribbean (17)   ——— More-Developed Countries (22)

Source: See table 5.2.

[a]Legend applies to both figures. Regional averages weighted by proportion of the 1981-85 average of agricultural research expenditures (expressed in 1980 PPPs) for each group accounted for by each country.
[b]Figures in brackets indicate the number of countries.

to the measures with exchange rate conversions, there appear to be no systematic differences between the convert- and deflate-first methods for the more-developed countries in any particular subperiod and for most less-developed country regions in the post Bretton Woods years. With these data, the convert-first procedure generates a larger volume measure than the deflate-first method for many less-developed country groupings during the Bretton Woods years. It is difficult to make too much of this trend as pre-1975 PPPs for many of the less-developed countries were derived using so-called short-cut extrapolation methods based, among other things, on market exchange rates without the benefit of local price measures based on benchmark survey data (Summers and Heston 1988).

5.2.2 Noncomparable Volumes

Statistics on many important agricultural inputs such as land, labor, tractors, and fertilizer are published as real totals or volumes. This would appear to make the job of international comparisons easier; however, the totals do not always count strictly comparable units.

The hectares of land in agriculture are far from homogeneous even when broken down into categories of cropland, pastureland, and rangeland. In an aggregate counting of just cropland hectares, one hectare of cropland that receives ten centimeters of rain per year may well have been added to another hectare of cropland that receives ten centimeters of rain per month. Moreover, some croplands are irrigated while others are not. While the aggregation of heterogeneous cropland hectares is less likely within a small region, it will almost invariably be the case when forming totals within a country, and it is certainly a problem for comparisons of cropland totals in a large international sample.

The problem of aggregating heterogeneous cropland hectares is compounded by the fact that we often want a measure of total land in agriculture that combines hectares of cropland with even more dissimilar hectares of pastureland, rangeland, and so on. If local values or rents for different land types can be observed, we have some direct way of measuring the relative productivity of different hectares. The use of local relative rents to reflect actual quality differences rests on the implicit assumption that local rents reflect the marginal value of a hectare of land in agriculture.

With a series of local rents, we could construct a direct "quality-adjusted" Divisia index for land volumes using methods described in section 5.1. The Divisia index would then provide a measure of real changes in land in agriculture over time that could then be scaled to get a volume aggregate measured in any type of numeraire hectare. For example, if we designate cropland as the numeraire or representative hectare, then when one hectare of pastureland rents for a fraction of a hectare of cropland, its relative rent would lead us to count it as that same fraction of a hectare of cropland. To arrive at internationally comparable cropland totals, cross-sectional scaling of the resulting time series indices for individual countries would be necessary. This would require the most difficult calculation; namely, constructing real estate converters that translate local cropland into an international cropland hectare of constant quality.

An additional problem with using unweighted total hectares of land, especially in international comparisons, is the fact that it is difficult to find measures on the intensity of land use. If a hectare of cropland is used for several crops in one year in Asia, but is rarely used for more than one planting in other parts of the world, simple counts of total hectares will give a distorted view of the cross-sectional variation in the flow of land services. This could be solved if land were measured in service flow units such as hectare-plantings rather than as a stock measure of hectares.[11]

As with the land input, the ideal measure of labor inputs in agriculture would reflect the service flow from labor and not merely the stock of workers available for the sector. Moreover, a human-capital perspective on labor gives rise to an analagous aggregation issue for labor variables. An hour worked by a farmer with no experience and a primary education is quite likely to be less effective than an hour worked by an experienced farmer with the same or higher level of education. In addition to differences in the human capital characteristics of farmers, there are differences in the effectiveness of hours worked by farmers, family members, and hired workers.

Once again, indices of the quantity of a quality-adjusted labor input could be constructed using the Divisia methods outlined in section 5.1 if wages and hours of different worker types were available. Scaling the index to get local aggregates measured in consistent units over time would require the use of base-year local relative wages. To make international comparisons, one also needs converters to capture the cross-sectional differences in the human-capital characteristics of the representative or numeraire worker.

5.3 CONSTRUCTING AGRICULTURAL DEVELOPMENT MEASURES

The ideal aggregation methods discussed above can hardly ever be implemented when using international data sets because detailed information on prices and quantities is typically not available. In addition, a distinction is rarely made between stocks of inputs and the service flows from those inputs. In this section, the compromises required to analyze agricultural development or productivity measures are discussed along with the likely biases inherent in using preaggregated data.

5.3.1 Agricultural Output Measures

In measuring agricultural output, one would often like an output measure that represents gross production less *inside* inputs, i.e., those inputs produced and reused within agriculture (Star 1974). In other words, products such as seeds or eggs which are required as inputs in their own production or feeds such as hay or milk which will be used as inputs in livestock should be deducted from gross production to avoid double counting. It is possible to start from FAO statistics[12] on "gross-gross output" and deduct inside inputs to get a final output aggregate, but comparable and reliable data on these inside inputs are difficult to come by for all countries and all time periods. FAO also publishes agricultural production indices

---

[11] A comparative study of Asian agriculture factored in multiple cropping levels to distinguish total cropped area versus cultivated land area. It gave land utilization rates ranging from 189% for Taiwan in 1966 to 93% for Thailand during 1980-81 (APO 1987, p. 17).

[12] See FAO (1974) for a discussion on output concepts used by FAO.

which are simply Laspeyres indices of final agricultural output. But, using this index for international comparisons of aggregates requires a great deal of data in order to calculate cross-sectional scaling factors.[13]

An alternative is to use value-added figures of AgGDP from national accounts data. By construction, these have the advantage of comparability with broader measures of economic activity such as GDP. However, they may present some problems for international comparisons if their calculation does not result in strictly comparable output aggregates. The problem does not arise because there is something intrinsically wrong with netting out intermediate inputs whether purchased from *outside* or produced *inside* the agricultural sector. Rather, the problem stems from asymmetric treatment of these two types of inputs. For example, in more-developed economies, fertilizer inputs are more likely to be purchased inputs than in less-developed economies where the same services may be provided largely from inside inputs such as manure. If the inside and outside inputs are not treated symmetrically in arriving at value-added figures, this will introduce some biases when using value-added measures in comparisons of levels and growth rates of agricultural output in cross-country studies.

The ratio of value-added to final agricultural output differs across countries. As one would expect, this ratio is much higher in less- than in more-developed countries. In 1975, this ratio ranged from 39% in Switzerland to 96% in Thailand (FAO 1986b). Within most countries, changes in this ratio over the past two decades are much less pronounced than cross-country differences (ECE and FAO 1981, 1989; FAO 1986b). However, there are shifts in this ratio. Changes in the structure of agriculture — in particular degrees of specialization — or changes in relative prices which lead to substitution between inside and outside inputs are likely to affect the ratio of value-added to gross output even in the absence of changes in technology and productivity.

The measures of agricultural output that are used throughout this volume are time series of AgGDP in current local currency units extracted primarily from World Bank (1989b). AgGDP measures were chosen over the alternative of scaling the FAO's agricultural production index (*FAO Production Yearbook*). The FAO production index excludes forestry and fishery outputs, which introduces problems of mismatched coverage in research and in most of our conventional input variables. Moreover, direct comparisons of AgGDP and non-AgGDP were deemed important for contrasting the development of agriculture with the rest of the economy.

To get comparable volume measures, these nominal output aggregates were first deflated using implicit AgGDP deflators based in 1980 (World Bank 1989). They were then

---

[13] Hayami et al. (1971) and Hayami and Ruttan (1971, 1985) first constructed country-specific estimates of final agricultural output, averaged over 1957-62 and measured in wheat-equivalent units. They then extrapolated these country-level estimates using FAO's agricultural production indices. Potential biases in using wheat-equivalent measures are discussed in section 5.1, while biases from inferring output growth rates when using fixed weight indices such as the FAO production indices are discussed in section 5.2 and also in Craig and Pardey (1990a).

converted to US dollars in the base year using PPPs defined over gross agricultural output (FAO 1986). We were able to use the preferred order of translation as well as representative deflators and converters for the bulk of the sample. An even more representative converter would have been PPPs defined over value added in agriculture, but no such converter was available for the whole sample.[14]

5.3.2 Agricultural Input Measures

Data on total hectares in agriculture used in this volume are unfortunately aggregates of stocks of heterogeneous land types. Information is sparse on multiple cropping, so one is also forced to use stocks of land instead of service flows. An annual breakdown of land types was available but not the local rent or value data that would have allowed calculation of economically meaningful aggregates.

Table A5.1 gives some idea of the heterogeneity of agricultural lands by indicating the percentage of total agricultural land accounted for by permanent pasture, or by arable and permanently cropped land. The latter is further disaggregated to indicate the percentage of land under irrigation. The differences across countries in the types of land and the extent of irrigation are quite dramatic. In China, for example, a low percentage of total agricultural land is either arable or permanently cropped, but almost half of that land is irrigated. No country except Japan irrigates as large a percentage.

The percentage of agricultural land in permanent pasture is dictated more by agroecological characteristics than by stage of development. Countries in Asia have a significantly lower percentage of land in pastures than do countries anywhere else in the world, and they irrigate their arable and permanent cropland more intensively than do countries in any other region. At the other end of the spectrum, agricultural land in sub-Saharan Africa is predominantly pastureland, and a very small percentage of arable land is irrigated.

An international index of land quality has been calculated by Peterson (1984) using an hedonic approach to valuing the cross-sectional differences in agricultural land characteristics. First, value weights for different land characteristics are derived by regressing a cross section of US land values on the differing characteristics of land in agriculture in the US. These weights are then used to place a relative value — and therefore a measure of relative quality — on hectares of agricultural land in different countries. The indices of relative quality of total agricultural land range from a regional average of 67 for West Asia & North Africa to an average of 161 for Asia & Pacific. A group of 83 less-developed countries had an average index of 101, while the average for the 21 more-developed countries in his study was 81.

These land-quality indices could be used to scale up or down the unadjusted total hectares in agriculture instead of implementing the more demanding Divisia input indices,

---

[14] Terluin (1990) provides a recent attempt at constructing such a converter for 10 EEC countries.

but they have some shortcomings. First of all, the index fails to account for changes over time in the quality of the average hectare in agriculture. Given the increased irrigation usage and the changing mix of land types evident in table A5.1, it is clear that the index will not fully account for changing quality differentials. More subtle problems of the correct weights and the relevant land characteristics lead us to think that these indices are useful but not completely satisfactory indicators of cross-sectional land quality.

Labor aggregates available for international studies of agriculture are as inadequate as those available for land. There are no broadly based studies that allow one to distinguish between stocks of labor available to agriculture and actual hours worked in agriculture. Instead, the available aggregate counts the economically active agricultural population, whether actually engaged or seeking employment in agriculture, forestry, hunting, and fishing.

In addition, the information on labor force characteristics is so difficult to come by that it is virtually impossible to construct Divisia indices using current international data sets. Even hedonic procedures analogous to Peterson's land-quality index are difficult to implement for a very large set of countries in the absence of country-specific information on age, education, and income profiles of agricultural workers.

What information is available indicates that the educational attainment of workers has varied dramatically, both across countries and over time (table 5.3). In the past decade, the secondary school enrollment ratio in less-developed countries has ranged from a low of 19% in sub-Saharan Africa to 45% in Latin America & Caribbean. For the same period, this ratio averaged 89% across more-developed countries. Over the past two decades, the ratio of primary enrollment has doubled in sub-Saharan Africa, and secondary school enrollment has more than quadrupled. These relatively recent changes in human capital investments may not have shown up yet in the labor input to agriculture in Africa but they are suggestive of the impact of development on the quality of the labor force.

5.3.3   Productivity Measures

To assess the development of the agricultural sector and, in particular, the sectoral rate of productivity change, one needs detailed price and quantity information on outputs and the whole range of inputs such as land, labor, energy, fertilizer, pesticides, and capital. With such data, it would be possible to construct total factor productivity indices (TFPs) which seek to separate out that part of output growth that can be attributed to increased or altered input usage from that which is interpreted as a pure productivity change (Capalbo and Antle 1988). The construction of such TFPs for a large international sample makes such demands on the currently available data that it is difficult to interpret the unexplained changes in output as productivity changes in the face of so many potential measurement errors.

While meaningful TFPs may be difficult to construct, we can learn a great deal about the patterns of development in agriculture with a judicious use of partial productivity indices. The interpretation of all productivity measures requires some care. Some, if not all, of the change in a particular factor's productivity may be attributable to increased usage of

Table 5.3: *Primary and Secondary School Enrollment Ratios*

Region	1961-65	1966-70	1971-75	1976-80	1981-85
	Primary school enrollment ratio				
	%	%	%	%	%
Sub-Saharan Africa (38)[a]	38	43	50	67	74
China	95	89	95	100	100
Asia & Pacific, excl. China (15)	67	73	76	80	85
Latin America & Caribbean (25)	91	95	92	95	98
West Asia & North Africa (18)	66	75	80	86	91
Less-Developed Countries (97)	76	78	81	87	90
More-Developed Countries (21)	99	99	99	99	100
Total (118)	81	82	85	89	92
	Secondary school enrollment ratio				
	%	%	%	%	%
Sub-Saharan Africa (38)[a]	4	5	8	12	19
China	21	24	36	47	38
Asia & Pacific, excl. China (15)	20	24	25	29	34
Latin America & Caribbean (25)	17	24	30	39	45
West Asia & North Africa (18)	16	23	30	38	44
Less-Developed Countries (97)	18	22	28	34	36
More-Developed Countries (21)	66	74	82	85	89
Total (118)	29	33	39	44	45

Source: World Bank (1989) and *UNESCO Statistical Yearbook 1983 and 1987*.

Note: Primary school enrollment ratio represents enrollment of students of all ages at primary level as a percentage of primary age students. Secondary school enrollment ratio is calculated in the same way. Definitional inconsistancies make it possible to obtain school enrollment ratios greater than 100%. Where this has occurred we have rounded to 100%.

[a] Bracketed figures indicate the number of countries in regional totals.

nonmeasured inputs. Nevertheless, the measured changes (especially when used in conjunction with data on other inputs) still provide useful information on development patterns.

Measurement errors are of concern in both total and partial productivity measures. In assessing changes in output per hectare, it is as important to have a consistently defined denominator in the fraction as it is to define output in a uniform way across countries and over time. Accurate representations of output per worker in agriculture likewise require one to use labor inputs in units that are comparable over time and space. Because the input aggregates one is forced to use do not have the same composition or average characteristics across countries, they make the comparison of productivity measures problematic. However, it is far easier to anticipate the magnitude and direction of biases from mismeasuring individual inputs in partial productivity indices than it is to disentangle multiple sources of measurement error in TFPs.

Using state-level data on the continental US, the implications of systematically adjusting land and labor totals for cross-sectional quality differences are clear (Craig and Pardey 1990b). When land rents are used to account for differences in the quality of pastureland, nonirrigated cropland, and irrigated cropland, measured levels and growth rates of output per acre are systematically changed. The implied growth rates of land productivity are reduced somewhat when one accounts for the fact that there has been some improvement over time in the average quality of land in US agriculture. As arid western cropland is turned into irrigated cropland, total unadjusted acres do not increase even though the average quality of land in agriculture does. Once this increase in input quality is accounted for, part of the likely increase in output will be attributed to increased quality of land and not to increased productivity of a quality-constant acre of land.

The most dramatic effect of accounting for quality differentials in land comes when rescaling implied *levels* of output per acre. The spread in measured regional differences in land productivity is greatly reduced with quality adjustment of land. Since there is a wide range of land quality in the US, these results can be used to anticipate the problems of using unweighted total hectares in international productivity studies.

Craig and Pardey (1990b) have also constructed a quality-adjusted index of labor for the US using actual hours worked by hired and family workers, human capital characteristics of farm operators, and data reflecting the shift from full- to part-time farming by operators. Not surprisingly, quality adjustment has a significant effect on measured labor productivity. Measured growth rates of output per hour are reduced when the increases in the average age and educational attainment of farm workers are taken into account. Regional differences in measured levels of output per hour are also reduced when the labor input is quality-adjusted. While regional differences in agricultural labor markets in the US have become less pronounced, differences across states in the mix of labor types and the average quality of labor have been historically important in accounting for part of the measured cross-sectional differences in levels and growth rates of agricultural output per hour.

Figures 5.3a and 5.3b illustrate the combined effect of quality adjustments on land and labor productivity measures for the US. Each path represents the average level of agricultural output per acre and output per hour in a different region in the US.[15] The more

---

[15] The regions depicted in figures 5.3a and 5.3b correspond to the 10 USDA production regions with one exception. The states in the USDA's Northeast region were split into two groups to distinguish between states whose agricultural growth has slowed dramatically or stopped in recent years and those that have continued to grow at rates more typical of the Corn Belt and Lake States. The regions comprise Northeast 1 (Maine, New Hampshire, Vermont, Connecticut, Rhode Island, Massachusetts, New Jersey), Northeast 2 (Delaware, Maryland, New York, Pennsylvania), Corn Belt (Illinois, Indiana, Iowa, Missouri, Ohio), Lake States (Michigan, Minnesota, Wisconsin), Northern Plains (Kansas, Nebraska, North Dakota, South Dakota), Appalachian (Kentucky, North Carolina, Tennessee, Virginia, West Virginia), Southeast (Alabama, Florida, Georgia, South Carolina), Delta States (Arkansas, Louisiana, Mississippi), Southern Plains (Oklahoma, Texas), Mountain (Arizona, Nevada, New Mexico, Idaho, Colorado, Montana, Utah, Wyoming), and Pacific (California, Oregon, Washington).

Figure 5.3a: *Quality-unadjusted land and labor productivity paths for the US, 1949-85*

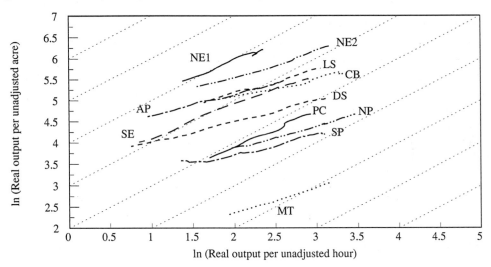

Figure 5.3b: *Quality-adjusted land and labor productivity paths for the US, 1949-85*

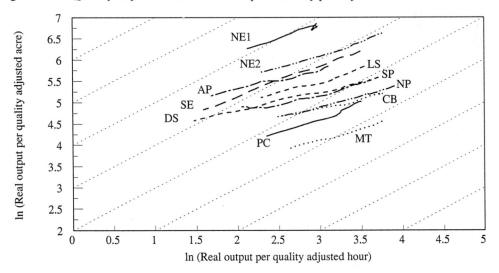

Legend: NE1=Northeast 1; NE2=Northeast 2; CB=Corn Belt; LS=Lake States; NP=Northern Plains; AP=Appalachian; SE=Southeast; DS=Delta States; SP=Southern Plains; MT=Mountain; PC=Pacific.

Source: Based on data reported in Craig and Pardey (1990b).

Note: All partial productivities expressed in natural log terms. For the sake of clarity, productivity paths represent five-year moving averages. Regional groupings of states correspond with USDA's farm production regions except for NE1 and NE2 which together form the USDA's Northeast region. See footnote 15 for details. Diagonal lines represent constant land/labor ratios.

careful quality-adjusted aggregation procedures broadly reduce measured differences in land and labor productivity. Since land is completely immobile across states and is not nearly as mobile intersectorally as labor, it is not surprising that land productivity still remains variable even within one country. If the nonagricultural uses of arid lands in the Mountain states are not pressing for its removal from agriculture, there is no economic necessity for its marginal product to match that of the Corn Belt or Lake States. Similarly, given the economic alternatives to agriculture in the Northeast, only land with a high marginal product is likely to remain in agriculture.

Because the regional labor markets are increasingly integrated within the US, we expect that, over time, significant differences in the marginal product of labor will be eroded. Workers in all regions will leave agriculture in response to opportunities outside agriculture (Kislev and Peterson 1982). With an increasingly national market for most labor skills, this should tend to eliminate spatial differences in returns to agricultural workers. As is evident from these figures, once we account for the difference in the labor mix across states, the spatial dispersion in the average productivity of labor appears to narrow more rapidly over time. Because there have been substantial cross-sectional differences in labor inputs to US agriculture over the post-war period, the effects described here of quality adjustment on US partial labor productivity measures can help us anticipate problems in comparing international labor productivity measures.

5.4   CONSTRUCTING AGRICULTURAL RESEARCH INDICATORS

The primary source for the public-sector agricultural research personnel and expenditure data used in this volume is the Indicator Series country files reported in Pardey and Roseboom (1989a). The Indicator Series represents a fully documented and sourced compilation of benchmark survey data plus information from approximately 1000 additional data sources for NARSs in 154 more- and less-developed countries, where possible, for the 27 years 1960 through 1986.[16] Extensive efforts went into achieving completeness and commensurability in the series. However, the unavoidably disparate nature of the data sources, plus the subject of the data series itself, means that these statistics should be considered indicative rather than definitive. Nevertheless, the series represents a major effort to consolidate and restructure previously available data compilations.

The following three sections briefly describe the statistical concept of a NARS used to compile the series, as well as some measurement issues that are germane to constructing comparative measures of agricultural research activity. While the scope of the series, in terms of country and time-period coverage plus number of indicators, constitutes a substan-

---

[16] The Indicator Series reports data on a calendar-year basis whenever possible. However, in numerous instances data were recorded on a fiscal- or academic-year basis. The procedure adopted in such cases was to place the observation in the calander year that overlaps most with the respective fiscal or academic year. Consequently, a fiscal year running from April 1, 1980 to March 31, 1981 was placed in calendar year 1980. A fiscal year running from July 1, 1980 to June 30, 1981 was placed in the calendar year 1981.

tial extension of or addition to global compilations, there remains a problem with missing observations that limits the use of these data for purposes of policy analysis. The procedures used to tackle this problem are detailed in section 5.4.4, while section 5.4.5 compares prior data compilations with those presented in this volume.

5.4.1 Defining a NARS

There is no universal agreement as to what constitutes a NARS, and while the concept of a NARS has value as an analytical tool, it is difficult to operationalize for statistical purposes. A useful beginning was to split the concept of a NARS into three dimensions, namely, (a) national, (b) agricultural, and (c) research, and to consider each of these dimensions separately.

National

The notion of what constitutes a "national" set of statistics on agricultural research is open to many interpretations. One option is to adopt a geographic interpretation and include all agricultural research performed within the boundaries of a country. Another possibility is to pursue a sectoral approach and include domestically targeted research activities funded and/or executed by the public sector of a particular country.[17] This latter approach was adopted for the Indicator Series, which attempts to include all agricultural research activities that are financed and/or executed by the public sector, including private, nonprofit, agricultural research. It explicitly excludies private, for-profit, agricultural research. This sectoral coverage corresponds to that adopted by the OECD (1981, pp. 83-91) and includes the government, private nonprofit, and higher-education sectors but excludes the business-enterprise sector.

The government sector was taken to include those federal or central government agencies, as well as provincial or state and local government agencies, that undertake agricultural research and development (R&D). One must be careful to avoid double-counting federal resources that fund agricultural research at the state or provincial level and to ensure that nonresearch activities are excluded. This is a particular problem for research performed by government agencies at the state and local level, which, in many instances, also deliver nonresearch services such as rural extension.

The private nonprofit sector generally includes only a small number of institutions,

---

[17] Classifying by source of funds is known as a "funder-based" system of classification as opposed to a "performer-based" system, which classifies according to the nature of those institutions that actually execute the research. Clearly, these classification systems can give rise to different measures of research capacity, and a preferred approach would be to classify research activity by one or the other method. However, at a practical level, when attempting to construct a global database of agricultural research statistics, we were forced to adopt an eclectic approach and use an ad hoc combination of both procedures to arrive at a set of statistics.

154 Craig, Pardey, and Roseboom

which are nevertheless very important for some countries. Some commodity research in less-developed countries, particularly that concerned with export-oriented estate crops such as tea, coffee, and rubber, is often financed wholly or in part by (industry-enforced) export or production levies and performed by private or semiprivate nonprofit research institutions. These institutions often operate as pseudo-public-sector research agencies or, at the very least, substitute directly for such agencies, so it was appropriate to include them in our measure of public agricultural research.

The higher-education sector is fairly readily identified but does present special problems when agricultural research statistics are compiled. Care was taken in constructing the Indicator Series to isolate research from nonresearch activities (e.g., teaching and extension) and to prorate personnel and expenditure data accordingly.

The national agricultural research statistics reported in the Indicator Series exclude the activities of research institutions with an international or regional mandate, such as CIMMYT, IRRI, and WARDA, along with bilateral institutions such as ORSTOM and CIRAD. The research operations for many of these multilateral agencies are quantified and discussed in chapter 9. While their research output may often have substantial impact on the agricultural sectors of their host countries, their mandates direct their research activities towards international and regional, rather than national, applications of their findings. However, all foreign research activities (including those associated with organizations such as those noted above) that are either funded or executed in collaboration with the national research agencies (or administered by them) were included in the series.

Agricultural

When measuring science indicators by socioeconomic objective, the OECD (1981, p. 113) recognizes that two approaches to classification are possible. They can be classified

(a) according to the *purpose* of an R&D program or project;
(b) according to the general *content* of the R&D program or project.

For example, a research project to improve the fuel efficiency of farm machinery could be placed under "agriculture" if classified by purpose, but "energy" if classifies by R&D content. The Indicator Series adopted the procedure used by the OECD and classifies research by purpose rather than content, as it is generally the purpose for which research is undertaken that has the greatest relevance for policy.

The notion of agricultural research used for the Indicator Series includes research in primary agriculture (crops, livestock, plus factor-oriented topics) as well as forestry and fisheries.[18] In general terms, this corresponds with the coverage used by both OECD (1981)

---

[18] Prior compilations of NARS indicators, e.g., Evenson and Kislev (1971), Boyce and Evenson (1975), and Judd, Boyce, and Evenson (1983, 1986), have sought to limit their coverage by excluding forestry, fisheries, and sometimes veterinary research. A substantive argument in favor of adopting the wider

and UNESCO (1984). For policy and analytical purposes, it would be desirable to differentiate agricultural research among commodities, but for many systems this is practically impossible, particularly on a time-series basis going back to 1960. Quite a few countries only report data on national research expenditures that are not differentiated according to socioeconomic objectives, even at this rather aggregate level. For these systems, it was simply not possible, given currently available data, to generate plausible time series at this level. Nevertheless, for a sample of 83 less-developed countries, we constructed preliminary estimates of research personnel stratified by four commodity aggregates (namely, crops, livestock, forestry, and fisheries) for the post-1980 period. Our findings are reported in chapter 8.

A further difficulty is that a significant amount of agricultural research has an effect at the postharvest stage, while the technology is embodied in inputs that are applied at the farm level. Take, for example, the efforts of plant breeders to improve the storage life of horticultural crops or to alter the baking quality of cereals. These characteristics are embodied in new crop varieties that are adopted by farmers. Furthermore, there is a lack of uniformity in the way research that is applied directly at the postharvest stage is currently reported. The OECD (1981, p. 115) classification omits "... R&D in favor of the food processing and packaging industries" from their socioeconomic objective of *agriculture, forestry and fisheries*,[19] while UNESCO (1984, p. 64) includes "... R&D on the processing of food and beverages, their storage and distribution." The Indicator Series sought to implement a variant of these approaches, excluding, where possible, research applied directly at the postharvest stage. Omitting research on food processing and packaging improves the compatibility of these statistics with value-added measures such as agricultural GDP and the like. Nevertheless, public-sector research targeted directly to food and beverage storage (and in some cases, processing) may in practice be included in this series, although this is more likely to be true of advanced systems in the more-developed countries.

A final difficulty was to obtain statistics for the higher-education sector, classified by purpose or "socioeconomic objective." The more general case is to find personnel and, possibly, expenditure data classified by field of science, where the basis of classification is the nature rather than the purpose or objective of the research activity itself.[20] In those cases where it was necessary to rely on field-of-science data, the series attempted to follow the UNESCO (1984, p. 77) procedure and consider agronomy, animal husbandry, fisheries, forestry, horticulture, veterinary medicine, and other allied sciences, such as agricultural

---

definition of agricultural research, as reflected in the statistics reported in this volume, is that the resulting series is then consistent with the agricultural aggregates of GDP, population, and so on, as published by the World Bank and United Nations organizations

[19] OECD (1981) includes it instead under the socioeconomic objective of "promotion of industrial development."

[20] Classifying research on the basis of the nature of the R&D activity itself, rather than its principal economic objective, is called a "functional" approach (OECD 1981, p. 53).

sciences, thereby excluding fields such as bacteriology, biochemistry, biology, botany, chemistry, entomology, geology, meteorology, and zoology. These latter fields are more appropriately classified as natural sciences, although in some cases the classification is a little hazy. It was therefore necessary to apply a "purpose or objective test" to some of these so-called natural science disciplines and to include in the series research undertaken in these areas when the ultimate purpose or objective of that research could have a direct impact on the agricultural sector.

Research

It is possible to identify a continuum of basic, or upstream, research to applied, or downstream, research. Much agricultural research has been characterized as mission-oriented in the sense that it is problem-solving in orientation, whether or not the solution to the problem requires basic or applied research. OECD (1981, p. 28) states that "... the basic criterion for distinguishing R&D from related activities is the presence in R&D of an appreciable element of novelty." For instance, simply monitoring the incidence of plant and animal diseases in and of itself is not considered research and may only be undertaken to enforce quarantine regulations or the like. But, using this information to study the causes or control mechanisms associated with a particular disease is considered research. Of course, some screening of the literature, newly available plant and animal material, and alternative production practices should be included as part of measured research activity, given its importance in the many countries that are undertaking substantial efforts to adapt existing agricultural technology to their local conditions.

   Agricultural research also includes a significant amount of maintenance research that attempts to renovate or replace any deterioration in gains from previous research.[21] Gains in output are often subject to biological degradation as pests and pathogens adapt to research-conferred resistance and control mechanisms. The role of maintenance research is substantial not only in many more-developed countries where current production practices employ technologies that are biologically intensive, but also in many less-developed countries, particularly those situated in the tropics where relatively rapid rates of pest and pathogen adaptation tend to shorten the life of research-induced gains.

   The difficulties of differentiating research from nonresearch activities is especially pertinent in the case of agricultural research, given the dual role of many public-sector agencies charged with agricultural research responsibilities. It is common to find such agencies involved in additional nonresearch activities such as teaching; extension services; certification, multiplication, and distribution of seeds; monitoring and eradication of plant and animal diseases; health maintenance, including veterinary medicine; and analysis and certification of fertilizers. In general, it is separating the research component from the joint

---

21 For additional discussions on maintenance research see Ruttan (1982), Miranowski and Carlson (1986), Plucknett and Smith (1986), and Adusei and Norton (1990).

teaching-research activities (in the case of universities) and the joint extension-research activities (of ministerial or department-based agencies) that is most difficult. If direct measures of expenditure and personnel data were not available at the functional level, then secondary data were often used to estimate the appropriate breakdown of aggregate figures into their research versus nonresearch components.

Even in the case of those institutions whose mandate is ostensibly limited to research, there were problems in obtaining consistent coverage of research-related activities. For example, general overhead services, including administrative personnel or expenditures required to support research, can be excluded from reported figures for a variety of reasons. In some instances, the institutional relationship between a national research agency and the ministry within which it is located means that overhead services and the like are charged against the ministry and not the research agency. Alternatively, some research agencies report total personnel and expenditure statistics based on an aggregation of project-level rather than institution-level data. In such cases, administrative overheads cannot be allocated across projects and thus may be omitted entirely or in part from the agency-level statistics.

A further issue involved identifying the research component of the farm operations that are usually undertaken in support of agricultural research. To the extent that such farm operations are necessary to execute a program of research, it seems appropriate that they be included in a measure of the commitment of national resources to agricultural research. However, some systems undertake farm operations at levels well above those required to support research, with the surplus earnings from farm sales being siphoned off to support research and even various nonresearch activities. In some instances, including all the resources devoted to the farm operations of a NARS substantially overstates the level of support to agricultural research within the system.

There was also the need to make a clear distinction between economic development and experimental development. According to OECD (1981, p. 25), "... experimental development is systematic work, drawing on existing knowledge gained from research and/or practical experience that is directed to producing new materials, products, or devices, to installing new processes, systems, and services, or to improving substantially those already produced or installed." Experimental development is therefore concerned with applying new findings from formal and informal research activities. This contrasts with the notion of economic development, which in general terms, is concerned with improving the well-being or standard of living of members of a society in a particular country or region.

Clearly, while improvements in agricultural productivity that follow from experimental development contribute to the process of economic development, they represent only part of the story. Improvements in rural infrastructure, via investments in irrigation, transportation and communication facilities plus improved rural health and education services, also contribute to the economic development of the agricultural sector and, ultimately, to society as a whole (Antle 1983).

A problem arises when one attempts to compile statistics on agricultural research and experimental development activities in less-developed countries. This occurs when a substantial portion of R&D activity is financed and/or executed as part of an economic-development aid package. It is often difficult to identify the experimental versus economic-development component of an aid package, particularly given the project orientation of much development aid. For instance, development assistance to establish, upgrade, or rehabilitate irrigation facilities can often incorporate research to evaluate water quality and identify preferred crop varieties as well as agronomic and irrigation practices. However, including all of the project's resources in a measure of NARS capacity could seriously overestimate the level of resource commitment to agricultural research.

Another less obvious difficulty concerns the somewhat transient nature of some of the agricultural research funded through development projects, which tends to be of relatively short duration, often between one to five years. In some cases, it is undertaken largely by expatriates and is seldom a part of the existing national research infrastructure. This type of research presumably contributes to the overall level of national research activity and should be captured in a NARS indicator, particularly if one is concerned with measuring sources of growth or technical change within a country. However, to the extent that such research is not integrated into the existing national research infrastructure, it is not a good measure of the "institutionalized research capacity" of a national system. The strategy pursued in this case was to include such development-financed research only when the research component could be isolated from the nonresearch component with an acceptable level of precision, and when it appeared to be integrated into the existing agricultural research infrastructure within a country.

5.4.2 Research Personnel Indicators

One possibility for measuring the human resource commitment to a NARS is simply to report the total number of personnel employed within a research system. This personnel aggregate would not only sum together scientific staff regardless of their qualifications and skills, but would also include, in an unweighted fashion, research technicians and other support staff. Because support staff often substitute directly for other capital and operating expenses in the research process, such a series may be driven largely by differences in the relative cost of research labor and nonlabor inputs, resulting, for example, in quite volatile fluctuations in the ratio of researchers to nonresearchers. As a consequence, all-inclusive research personnel aggregates would not accurately reflect differences in the underlying scientific capacity that is relevant for many purposes and is our measurement objective here. Thus, the Indicator Series sought to include only research personnel, i.e., researchers engaged directly in the conception or creation of new knowledge, products, processes, methods, and systems.[22] The series attempted to exclude technicians as well as support and clerical staff who normally perform research and technical tasks under the supervision of a

_____

[22] This corresponds to the OECD (1981, p. 67) definition of a researcher.

researcher.

A practical procedure for differentiating research from nonresearch staff was to rely principally on educational levels rather than occupational classes. While there are clearly substantial difficulties in standardizing educational levels on a global basis, an international standard classification of education (ISCED) has been developed and is in general use (UNESCO 1976). The Indicator Series sought to include only NARS personnel who held at least a third-level university degree (ISCED-level categories 6 and 7) as researchers.[23] This included holders of first and postgraduate degrees (or their equivalent) earned at bona fide universities or at specialized institutes of university status.

The series further attempted to classify national research personnel by degree status — PhD, MSc, BSc, or equivalent. This substantially improved our ability to use the personnel data as an indicator of the human capital or "quality-adjusted" research commitment to national agricultural research, as reported in chapter 8. There was also an attempt to differentiate between local and expatriate scientific staff in order to enhance the information contained in the personnel series. As discussed earlier, the series sought to include only those expatriates who were working directly on domestic issues in an integrated fashion with the national research system.

Personnel who were classified as research managers or administrators presented special problems. To the extent that they are engaged in the planning and management of the scientific and technical aspects of a researcher's work, they should be classified as researchers and included, at least on a prorated basis, in the series. They are usually of a rank equal to or above that of persons directly employed as researchers and are often former or part-time researchers (OECD 1981, p. 67). However, in many cases, it is not at all clear if research managers or administrators maintain any direct involvement with the scientific process itself.

The problems of dealing with data on research administrators are analogous to those of dealing with data on other NARS personnel who may hold dual research and nonresearch appointments. This is particularly important when including personnel from institutions in the ministry or department of agriculture, who perform, for example, a dual research-extension function, or from universities where personnel often hold joint research-teaching appointments. In all cases, an attempt was made to measure researchers in full-time equivalent (FTE) units. If direct measures in FTE units were not available, secondary data, which enabled total researcher figures to be plausibly prorated to FTE units, were used.

---

[23] An alternative procedure (see OECD 1981, pp. 67-69) is first to classify researchers, technicians, and other supporting staff on the basis of the ILO (1986) classification scheme and then use ISCED procedures to classify researchers by educational level. Given the rather heavy reliance on secondary data in the Indicator Series, it was not possible to operationalize the ILO classification scheme.

5.4.3   Research Expenditure Indicators

There are several commonly accepted methods of measuring the (annual) commitment of financial resources to R&D (OECD 1981, pp. 72-82):

(a) *performer-based* reporting of the sum of funds received by all relevant R&D agencies for the performance of intramural R&D;

(b) *source-based* reporting of the funds supplied by all relevant agencies for the performance of extramural R&D;

(c) *total intramural* expenditures for R&D performed within a statistical unit or sector of the economy, whatever the source of funds.

The Indicator Series sought to report actual research expenditures, not simply budgeted funds, appropriations, or funds available, and so was based on method *c* wherever possible. A substantial number of the major discrepancies in prior compilations were due to large variations — sometimes upward of 30% to 50% — between funds budgeted or appropriated and funds actually spent. Some funds allocated to research at the beginning of a fiscal year, for example, may never materialize, especially if governments are forced to trim proposed outlays over the course of the year because of unforeseen budgetary shortfalls. Conversely, some research systems may actually receive more funds than are spent, and thus carry funds over to future budgetary periods. This is particularly true for systems experiencing substantial capital investments where funds are allocated initially in a lump-sum fashion and then drawn on over a period of time as needed.

The expenditures reported in the Indicator Series are total, inclusive of salary, operating, and capital expenses. While the series reports actual expenses, for some purposes it may be more appropriate to measure resources used rather than funds spent. This would involve explicitly separating capital from noncapital expenditures. Capital expenditures, which measure (gross) additions to the stock of capital invested in agricultural research, could then be converted into flow terms by estimating the future service flows derived from them.[24] These capital service flows could then be added back to noncapital expenditures to derive an overall measure of the resources actually used for research over time.

One of the major undertakings in compiling the Indicator Series was to collect all expenditure data in current local currency units. This allowed the standardized translation procedures described in section 5.2.1 to be applied to all countries. All expenditures were deflated using local implicit GDP deflators based in 1980. The series was then converted using PPPs defined over GDP.

---

[24] See Pardey, Craig, and Hallaway (1989) for details. Unfortunately, there are simply not enough data available at the international level to construct a time series that differentiates research expenditures by factor type. However, in chapter 8 we do report our preliminary attempts to differentiate research expenditures by factor type in a sample of 43 less-developed countries for the post-1980 period.

5.4.4 Shortcut Estimation Methods

The data base that underpins the analysis reported in this volume substantially upgrades and extends previously available indicators on research personnel and expenditures. But the nontrivial number of missing observations that remain impedes our ability to form aggregates or undertake comparative analyses.

The practical and, by necessity, ad hoc approach of dealing with this problem has been to implement a hierarchical series of shortcut estimation procedures. The preferred and most data-demanding approach uses econometric procedures to estimate a series of reduced form equations first. These are then used to construct ordinary least squares (OLS) predictions of missing research personnel and expenditure observations. In the absence of suitable regressors, various non-econometric extrapolation and interpolation procedures have been applied directly to the country-level data.

Naturally the appropriate level of precision for any data set is a function of the uses to which it is to be put. In this instance, the objective was to construct aggregate measures that represent broad trends at the regional or subregional level over the 1961-85 period. With an underlying unit of analysis consisting of annual, national-level data, we opted first to construct simple, quinquennial averages of the country-level observations beginning with the 1961-65 period.[25] While this aggregation procedure may artificially dampen variability in data where there are strong trends, we argue that five-year averages offer a basis for more realistic global comparisons than the point estimates used by previous analysts.[26] Such aggregation also serves to minimize the influence of spurious variability as well as substantially reducing the number of observations to be estimated by shortcut methods. Specifically, the primary data matrix includes 151 countries over a 25-year period for a total of 3775 entries per indicator. Averaging over a five-year period reduces the size of this matrix to 755 entries, which in turn, given the available data, reduces the number of personnel data points to be estimated by short-cut methods from 67% to 30% and the number of expenditure data points from 65% to 45%.

The regression procedures used to derive shortcut research personnel and expenditure estimates do not presume any causal relationship between the set of right-hand-side (RHS) "explanatory" variables and the research indicators for which estimates are being sought. But, as Ahmad (1980) and Clague (1986) point out when using analogous shortcut procedures in a different context, an informed choice of candidate RHS variables should draw on some understanding of the likely partial correlations these variables may exhibit with research expenditure and personnel indicators.

On the presumption that research personnel and expenditure aggregates at the national level exhibit relatively stable trends over time, then one-period lags or leads of these

---

[25] The 1961-65 period averages were centered on 1963 and so on for later periods.

[26] For instance Boyce and Evenson (1975), Evenson and Kislev (1971, 1975a), Judd, Boyce, and Evenson (1983, 1986), and Oram and Bindlish (1981).

variables are credible regressors for our purposes.[27] With personnel expenditures accounting for an average of 56% of total research expenditures in less-developed countries and probably an even higher percentage for more-developed countries as US data suggest (section 8.2), current research expenditures were also deemed suitable for inclusion in the set of variables regressed against research personnel. Similarly, research personnel were used as a regressor in the equation used to predict research expenditures. Measures of the absolute (AgGDP) and relative (AgGDP/GDP) size of the agricultural sector also appear to be systematically related to the level of agricultural research activity and were added to the set of RHS variables used to generate shortcut estimates.[28] To capture the effects of a myriad of complex socioeconomic influences that would otherwise be difficult if not impossible to quantify, regional dummies were also incorporated into the estimating procedure. Finally, a series of Chow tests rejected the hypothesis that a set of time dummies added significant "explanatory" power to the regressions, and these dummies were omitted from the final specifications summarized in table 5.4.

Table 5.4: *Specification of Shortcut Estimating Equations*

	Research personnel (RP_t)						Research expenditures (RE_t)					
	specification number						specification number					
Regression	1	2	3	4	5	6	1	2	3	4	5	6
Constant	*	*	*	*	*	*	*	*	*	*	*	*
$AgGDP_t$	*	*	*	*	*	*	*	*	*	*	*	*
$(AgGDP/GDP)_t$	*	*	*	*	*	*	*	*	*	*	*	*
RP_{t-1}	*		*									
RP_t							*	*			*	
RP_{t+1}		*		*								
RE_{t-1}							*		*			
RE_t	*	*			*							
RE_{t+1}								*		*		
Regional dummies	*	*	*	*	*	*	*	*	*	*	*	*

Note: To minimize the influence of spurious observations these equations were estimated in double-log form.

Before fitting the six research personnel and six research expenditure specifications used to generate shortcut estimates, the data were stratified across five income classes, so that a total of 60 empirical relationships were estimated. This enabled the predictive influence of each RHS variable to be conditioned by stage-of-development considerations.

27 In some cases, data permitting, the research personnel and/or expenditure figures were predicted by averaging the estimates obtained from equations that involved regressors consisting of leads and lags of the respective RHS variable.

28 Evenson and Kislev (1975b) and Pardey, Kang, and Elliott (1989) present empirical support for this notion, while in chapter 1 some of the conceptual underpinnings of such a relationship are discussed.

Adjusted R^2s for the specifications reported in table 5.4 averaged 0.94 across the total of 60 personnel prediction equations and 0.90 across the 60 expenditure prediction equations that were estimated by OLS procedures, with the best fits being obtained for the high-income countries. In all but one instance, the current and/or lagged RHS research personnel and expenditure variables entered with positive signs, as, in general, did the coefficient on the AgGDP variable. The sign on the AgGDP/GDP variable was somewhat more volatile but was uniformly negative and significant for the low-income countries.

Prior to incorporating these OLS estimates into the respective research personnel and expenditure series, a variety of screening procedures were used to ensure that they gave rise to plausible time series. The actual and first-differenced series for research expenditures, research personnel, and the implied expenditures per researcher for each country were jointly scrutinized for evidence of outliers that could not be accounted for by secondary information contained in various sources, including the country-level documentation of the Indicator Series. OLS estimates that were identified as outliers were then estimated by straightforward extrapolation and interpolation procedures.[29]

Figure 5.4 gives the percentage of observations and the share of research personnel and expenditure totals that were derived by various methods. Around 55% of the expenditure observations were accounted for by direct estimates, while a significantly higher proportion of research personnel observations (70%) were directly estimated. Regression-based procedures generated 26% of the research expenditure estimates and 19% of the research personnel figures. The remaining observations were derived using various non-econometric extrapolation and interpolation procedures.

For our purposes, it is significant that a substantially higher share of research expenditure and personnel totals were accounted for by direct observations because the problem of missing observations was concentrated in the group of small, less-developed NARSs. It must be emphasized, however, that a nontrivial share of the personnel and expenditure totals of the less-developed countries were constructed by shortcut procedures. One should bear this in mind when interpreting the various indicators presented elsewhere in this book.

---

[29] A variety of procedures were employed that sought to make maximum use of the available data. Various country-specific ratios of expenditures per researcher were used in conjunction with expenditure-only or researcher-only figures for a particular year to infer corresponding researcher and expenditure data. In some instances ratios of expenditures per researcher for the region in which the country is located and/or expenditure ratios for countries of comparable size, stage of development, or time period were used in a similar manner. For a few, often small, countries where only recent observations were available, the series was backcast by assuming that the country's rate of growth for the indicator in question was approximated by the regional average.

Figure 5.4:   *Proportion of estimates derived by direct and shortcut methods*

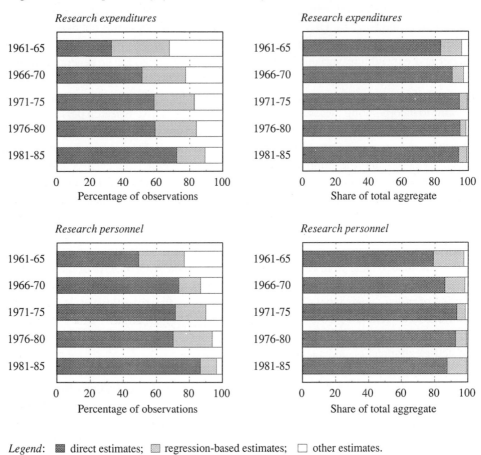

Legend: ▨ direct estimates; ▨ regression-based estimates; ☐ other estimates.

5.4.5   Comparison with Prior Compilations

Internationally comparable data sets of investments in NARSs are sparse and up to now have been based largely on the series constructed by Evenson and colleagues[30] supplemented by the efforts of Oram and Bindlish (1981). The data sources, NARS and variable definitions, data processing, and aggregation procedures used by Pardey and Roseboom to construct the compilation described in this chapter and detailed in the appendix to this book differ in substantive ways from those used in prior compilations. Rather than simply extending both the country and temporal coverage of these existing data sets, Pardey and Roseboom endeavored to rework the recent historical record concerning the global pattern of invest-

---

[30] See footnote 26 for list of references.

ments in public agricultural research. To permit a more informed use of these new data, the nontrivial and systematic differences between the Pardey and Roseboom series and the most recently published series by Judd, Boyce, and Evenson (1986) will be quantified in this section and placed in context.

The principal data sources for the Pardey and Roseboom series is the ISNAR Indicator Series (Pardey and Roseboom 1989a) supplemented by recently updated data on the US (Pardey, Eveleens, and Hallaway 1991) and China (Fan 1991b). In compiling the Indicator Series, Evenson-related sources (including the Judd, Boyce, and Evenson series) were directly used for only 11% of the personnel data and 6% of the expenditure data. In quite a few instances, however, they drew from the same primary sources used by Evenson and his colleagues, although these were often supplemented with information from additional primary sources that were considered more complete for our purposes. The multiplicity of sources that were used to compile the primary data underlying both the Pardey and Roseboom and the Judd, Boyce, and Evenson series compounds the difficulties of maintaining consistent coverage in several dimensions: namely, (a) over time within a country, (b) among countries, and (c) across the personnel and expenditure series. A key to improving consistency in the data underlying the Pardey and Roseboom series was to identify and track the institutional coverage of the available data, paying particular attention to dates of creation, organizational mergers or divisions, details of name or mandate changes, and the like. This involved gathering quantitative and qualitative data from as many documentable sources as possible, including several ISNAR benchmark surveys, then reconciling and synthesizing these multiple observations into a data series that represented as closely as possible the NARS concept identified earlier in this chapter. Boyce and Evenson (1975, p. iv) note that their "... data [may] appear to be in error simply because no attempt to determine how many agencies are involved in the support of research in the country has been made." Improvements over time in the institutional coverage of available data sources means that spurious, and in some cases substantial, growth in national research capacity can be introduced if issues related to institutional coverage are ignored.

The concepts of a NARS underpinning the two series differ substantially. The notion of agricultural research used in the Pardey and Roseboom series includes primary agricultural research (crops, livestock, plus factor-oriented areas) as well as forestry and fisheries, while the Judd, Boyce, and Evenson series tries to exclude forestry and fisheries research (Judd, Boyce, and Evenson 1986, pp. 79-80). There are several practical and conceptual issues which on balance, at least from our perspective, favor the broader definition. First, such a definition is consistent, in general, with the coverage used by both OECD (1981) and UNESCO (1984). Second, while for policy and analytical purposes it is desirable to differentiate agricultural research among commodities, for many systems this is practically impossible, particularly in time-series back to 1961. Certainly for a significant number of systems, disparate agencies are charged with the responsibility of funding and/or executing these various areas of research and care needs to be taken to ensure that they are included, where appropriate, in the aggregate measures. On the other hand, attempts to exclude forestry and fisheries research are confounded by the fact that quite a few countries only report preaggregated national research data that fail to differentiate among socioeconomic objectives. Finally, a substantive argument in favor of adopting the wider definition of agricultural research is that the resulting series is then generally consistent with the definitions of agricultural GDP, population, etc., commonly published by the World Bank, United Nations, and FAO.[31]

There are also significant and quantitatively important differences between the Pardey and Roseboom and the Judd, Boyce, and Evenson series concerning the manner in which research expenditures are translated into commensurable units. Research expenditures for the Pardey and Roseboom series were all compiled in current local currency units and then translated, in a standardized fashion, to constant 1980 US dollar aggregates. Our two-step translation procedure first deflates to base year 1980 using local implicit GDP deflators and then converts to US dollars using Summers and Heston's (1988) PPP measures. This is far from ideal, but in our judgment, it is likely to introduce less aggregation bias[32] than the approach used by Judd, Boyce, and Evenson that first converts using annual exchange rates and then deflates using a US wholesale price index.[33]

A final source of discrepancy between the Pardey and Roseboom and the Judd, Boyce, and Evenson compilations lies in the different regional and subregional aggregations they use. In brief, the principal differences are as follows: Judd, Boyce, and Evenson include Japan and China in their Asian aggregates while Pardey and Roseboom include Japan in the more-developed country grouping and, given the system's overwhelming size coupled with the preliminary nature of Chinese agricultural research statistics, list China as a stand-alone figure. Judd, Boyce, and Evenson include South Africa in their Southern Africa totals while the Pardey and Roseboom series omits South Africa from its sub-Saharan Africa aggregates in order to maintain consistency with existing World Bank and UN practice. Judd, Boyce, and Evenson do not report a West Asia & North Africa grouping. Judd, Boyce, and Evenson group Australia and New Zealand along with the Pacific islands into an Oceania aggregate while Pardey and Roseboom include Australia and New Zealand in their more-developed country totals and merge the Pacific islands into an Asia & Pacific aggregate. And, finally, Judd, Boyce, and Evenson include Ireland and the United Kingdom in a Northern Europe total, while Pardey and Roseboom group them under a Western Europe aggregate.

The cumulative effects of these different measurement and compilation procedures are captured in the comparative data on level and rate of growth presented in tables 5.5 and 5.6, where the Pardey and Roseboom data have been reaggregated to match the regional

---

[31] An inconsistency that remains involves agricultural research or agricultural output (inclusive of forestry and, of particular concern here, fisheries) indexed over "agricultural" land.

[32] See section 5.1.

[33] In numerous cases Boyce and Evenson (1975) and Judd, Boyce, and Evenson (1983) recorded research expenditures directly in current or constant US dollars rather than current local currency units. This leaves their figures subject to the possibly capricious conversion methods of their source authors.

Table 5.5:   *Level of Investment in Agricultural Research — Comparison of Pardey and Roseboom with Judd, Boyce, and Evenson Estimates, Percentage Differences*

Region[a]	Researchers[b]				Expenditures[b]				Expenditure per researcher[b]			
	1961-65	1966-70	1971-75	1976-80	1961-65	1966-70	1971-75	1976-80	1961-65	1966-70	1971-75	1976-80
	%	%	%	%	%	%	%	%	%	%	%	%
North Africa	14	44	49	40	36	41	48	46	26	-5	-2	10
West Africa	6	8	-13	-6	18	23	29	1	13	16	37	7
East Africa	31	17	-5	-34	55	50	35	29	35	40	38	47
Southern Africa	-8	-20	-29	-51	74	60	56	41	75	66	66	61
Africa	*14*	*28*	*21*	*12*	*36*	*38*	*37*	*20*	*25*	*14*	*21*	*9*
West Asia	37	16	11	18	52	65	59	53	25	59	54	44
South Asia	61	62	63	52	73	71	73	67	31	23	26	30
Southeast Asia	34	14	13	27	83	83	83	68	74	80	80	56
East Asia	29	14	2	-3	41	29	25	29	18	17	24	31
Asia (excl. China)	*36*	*27*	*21*	*21*	*55*	*50*	*48*	*47*	*29*	*32*	*34*	*33*
Caribbean and Central America	-2	-41	-25	-10	53	40	46	49	54	58	57	54
Temperate South America	9	4	1	-6	27	31	25	28	20	28	25	32
Tropical South America	8	2	-2	-20	50	48	44	46	46	48	45	55
Latin America	*7*	*-4*	*-5*	*-15*	*44*	*44*	*40*	*44*	*40*	*46*	*43*	*51*
Northern Europe	35	13	3	-25	45	41	39	29	15	32	37	43
Central Europe	18	7	0	-28	-5	-8	10	-7	-28	-16	10	16
Southern Europe	3	-10	-4	18	36	-6	-12	8	34	4	-8	-13
Western Europe	*21*	*6*	*1*	*-17*	*20*	*10*	*18*	*7*	*-2*	*4*	*17*	*20*
Oceania	23	19	16	42	16	-2	-10	-30	-8	-25	-31	-123
North America	47	45	42	38	17	15	8	17	-57	-53	-59	-33
North America & Oceania	*43*	*40*	*36*	*39*	*17*	*13*	*5*	*10*	*-46*	*-45*	*-50*	*-47*
Total	33	25	19	15	31	27	26	25	-3	3	8	12

Source: Judd, Boyce, and Evenson (1983) and the appendix to this book.

[a] Regional groupings correspond to aggregations presented in Judd, Boyce, and Evenson (1983, 1986).

[b] Pardey and Roseboom minus Judd, Boyce, and Evenson estimates expressed as a percent of Pardey and Roseboom.

Table 5.6:   *Growth of Investment in Agricultural Research — Comparison of Pardey and Roseboom with Judd, Boyce, and Evenson Estimates*

	Researchers		Expenditures	
	PR[a]	JBE[b]	PR[a]	JBE[b]
	%	%	%	%
North Africa	9.4	6.8	4.7	3.6
West Africa	8.1	9.0	6.8	8.1
East Africa	5.2	9.9	4.5	7.7
Southern Africa	2.5	4.9	6.4	12.2
Africa	*7.9*	*8.1*	*5.6*	*7.2*
West Asia	4.9	6.8	7.8	7.7
South Asia	6.8	8.2	8.0	9.4
Southeast Asia	8.8	9.5	6.4	10.8
East Asia	1.1	3.6	5.5	7.0
Asia (excl. China)	*3.6*	*5.1*	*6.5*	*7.6*
Caribbean and Central America	7.7	8.3	9.7	10.3
Temperate South America	4.2	5.3	3.7	3.6
Tropical South America	7.8	9.7	8.7	9.3
Latin America	*6.9*	*8.4*	*7.7*	*7.8*
Northern Europe	3.0	7.6	4.7	6.4
Central Europe	1.5	4.5	6.1	6.3
Southern Europe	3.4	2.3	5.2	7.8
Western Europe	*2.5*	*5.2*	*5.5*	*6.5*
Oceania	4.9	2.9	3.6	6.6
North America	1.0	2.1	3.3	3.3
North America & Oceania	*1.8*	*2.3*	*3.3*	*3.9*
Total	3.3	5.0	5.2	5.8

Source: See table 5.5.

[a] PR = Pardey and Roseboom; figures given here are compound annual rate of growth between 1961-65 and 1976-80 period averages.

[b] JBE = Judd, Boyce, and Evenson; figures given here are compound annual rate of growth between 1959 and 1980 estimates.

groupings used by Judd, Boyce, and Evenson. The Pardey and Roseboom series estimates an overall level of full-time equivalent researchers some 15% to 33% higher than the Judd, Boyce, and Evenson series, with the downward bias in the Judd, Boyce, and Evenson series being magnified in the earlier years, particularly for East Africa, Asia, Northern Europe, and North America. The systematic attempt to exclude forestry and fisheries research from the Judd, Boyce, and Evenson series, combined with the more extensive, but still not ideal, coverage of research performed by "secondary" public agencies, including universities, in the Pardey and Roseboom series, could account in large measure for these differences. Contrary to this general pattern, the Judd, Boyce, and Evenson series appears to substantially overestimate the number of research personnel for the 1976-80 period throughout Africa (excluding North Africa) and Latin America, plus Northern and Central Europe. A comparative examination of the underlying data for this period suggests that the Judd, Boyce, and Evenson series for these regions relies more heavily than the Pardey and Roseboom series on extrapolated rather than directly observed data, thereby (presumably erroneously) carrying forward the somewhat higher growth rates of earlier periods.

The Judd, Boyce, and Evenson series also reports a much lower level of real research expenditures than the Pardey and Roseboom series. The downward bias in the Judd, Boyce, and Evenson figures is much more dramatic for the less-developed countries than for the more-developed countries, presumably due to differences in coverage, compounded by substantial differences in currency translation procedures.

Although the Pardey and Roseboom series suggests that overall there are significantly more public resources devoted to agricultural research at the national level than the Judd, Boyce, and Evenson figures reveal, the rates of growth in research personnel and expenditures appear somewhat lower than has hitherto been reported.[34] It is to be hoped that the substantial efforts invested by Pardey and Roseboom (1989a) in tracking institutional coverage within a country over time has resulted in a series that is, in some senses, more consistent than the Judd, Boyce, and Evenson series. There is a tendency for more readily available, often preaggregated, data to improve implicitly in coverage over time, and it is particularly such aggregate data that were used by Judd, Boyce, and Evenson. Therefore, we believe the Pardey and Roseboom growth rates are more realistic than those reported by Judd, Boyce, and Evenson. The Pardey and Roseboom data form the basis for most of the analyses reported in subsequent chapters, particularly in chapters 7 and 8.

---

[34] Notwithstanding the fact that neither series adjusts for quality differences over time or across countries.

Table A5.1:   *Agricultural Land*

Region	Total agricultural land[a]					Arable land and land under permanent crops									
						Percentage of total area					Percentage irrigated[b]				
	61-65	66-70	71-75	76-80	81-85	61-65	66-70	71-75	76-80	81-85	61-65	66-70	71-75	76-80	81-85
	(millions of hectares)					%	%	%	%	%	%	%	%	%	%
Nigeria	48.5	49.3	50.5	51.1	51.6	60.6	60.3	59.3	59.2	59.4	2.7	2.7	2.7	2.7	2.7
Western Africa (17)	191.2	194.1	196.3	198.3	198.9	14.8	16.2	17.2	18.4	19.0	0.7	0.8	1.1	1.3	1.5
Central Africa (7)	38.5	39.0	39.5	39.9	40.2	25.3	26.5	27.7	28.9	29.7	0.1	0.2	0.4	0.5	0.7
South Africa (10)	206.1	212.4	216.0	217.2	217.6	8.2	9.0	9.6	9.8	10.0	2.4	2.4	2.8	4.1	5.2
East Africa (8)	210.3	211.9	213.5	214.4	215.0	16.8	17.6	18.3	18.8	19.1	5.1	5.1	5.1	5.3	5.5
Sub-Saharan Africa (43)	*694.6*	*706.9*	*715.8*	*720.9*	*723.4*	*17.2*	*18.1*	*18.8*	*19.4*	*19.8*	*2.7*	*2.7*	*2.8*	*3.1*	*3.4*
China	*422.7*	*421.1*	*419.7*	*418.7*	*419.1*	*24.5*	*24.2*	*24.0*	*23.8*	*23.9*	*30.2*	*35.1*	*40.1*	*44.6*	*44.3*
South Asia (8)	264.3	267.2	269.8	272.4	271.3	79.9	80.4	80.8	81.1	81.2	19.2	21.1	23.3	25.7	28.6
South-East Asia (9)	54.4	58.1	62.4	65.7	68.3	74.6	76.4	78.0	79.2	79.9	20.0	19.0	19.7	20.9	23.1
Pacific (10)	1.4	1.4	1.5	1.5	1.6	65.4	67.4	65.9	66.5	67.5	0.1	0.1	0.1	0.1	0.1
Asia & Pacific, excl. China (27)	*320.1*	*326.7*	*333.7*	*339.7*	*341.2*	*78.9*	*79.6*	*80.2*	*80.7*	*80.8*	*19.4*	*20.8*	*22.6*	*24.8*	*27.5*
Caribbean (17)	4.9	5.3	6.1	6.2	6.3	53.6	51.8	45.7	48.0	48.7	8.7	9.3	10.4	10.8	11.6
Central America (8)	114.5	114.7	115.9	117.5	119.4	25.9	25.7	25.4	25.8	26.3	11.1	12.3	14.9	17.4	17.4
South America (12)	502.2	531.6	560.5	582.6	599.6	16.1	19.3	20.9	22.2	23.2	6.0	5.3	5.3	5.5	5.7
Latin America & Caribbean (37)	*621.5*	*651.5*	*682.6*	*706.3*	*725.2*	*18.2*	*20.7*	*21.9*	*23.0*	*23.9*	*7.4*	*6.9*	*7.3*	*7.8*	*8.0*
North Africa (5)	86.5	87.2	88.9	89.2	85.8	26.4	26.9	27.5	27.9	29.4	16.9	17.9	18.0	17.3	17.6
West Asia (15)	228.6	229.6	230.9	229.1	227.3	24.6	25.5	26.0	25.6	25.3	15.7	16.3	18.0	18.4	19.5
West Asia & North Africa (20)	*315.1*	*316.8*	*319.7*	*318.3*	*313.1*	*25.1*	*25.9*	*26.4*	*26.2*	*26.4*	*16.1*	*16.8*	*18.0*	*18.1*	*18.9*
Less-Developed Countries (108)	*2374.1*	*2423.0*	*2471.5*	*2503.9*	*2522.0*	*28.1*	*29.2*	*29.8*	*30.3*	*30.7*	*15.6*	*16.5*	*17.8*	*19.0*	*19.9*

Table A5.1:   *Agricultural Land (Contd.)*

	Total agricultural land[a]					Arable land and land under permanent crops									
						Percentage of total area					Percentage irrigated[b]				
Region	61-65	66-70	71-75	76-80	81-85	61-65	66-70	71-75	76-80	81-85	61-65	66-70	71-75	76-80	81-85
	(millions of hectares)					%	%	%	%	%	%	%	%	%	%
Japan	6.0	5.8	5.7	5.5	5.4	97.9	95.8	92.6	90.1	88.8	49.9	61.1	62.4	62.7	62.1
Australia & New Zealand (2)	493.0	503.2	511.8	509.3	504.9	6.9	7.9	8.2	8.5	9.3	3.5	3.7	3.8	3.8	4.0
Northern Europe (5)	13.1	12.8	12.5	12.4	12.3	72.7	72.1	71.8	71.5	72.2	1.0	1.6	2.8	5.6	7.1
Western Europe (8)	84.0	81.3	79.1	77.8	77.0	50.5	49.0	48.5	49.0	49.2	3.3	3.7	4.2	4.9	5.7
Southern Europe (4)	66.1	65.0	62.4	61.4	60.8	65.2	65.0	64.3	64.5	65.0	13.1	14.6	17.0	18.5	19.6
North America (2)	503.3	500.1	497.5	497.8	507.1	43.9	45.8	46.6	46.9	46.6	6.8	7.0	7.3	8.4	8.4
More-Developed Countries (22)	*1165.6*	*1168.2*	*1169.1*	*1164.3*	*1167.5*	*30.6*	*31.3*	*31.3*	*31.6*	*32.1*	*7.4*	*7.9*	*8.3*	*9.2*	*9.4*
Total (130)	3539.7	3591.2	3640.5	3668.1	3689.5	28.9	29.9	30.3	30.7	31.2	12.8	13.5	14.6	15.8	16.5

Source: Compiled from FAO Production Yearbooks.

[a] Agricultural land includes arable land, land under permanent crops and permanent pastures. It excludes forest and woodland areas. Variable levels and intensities of commercial exploitation of forest and woodland areas both over time within countries and between countries suggests that less distortion in the agricultural input-output relationship is induced by excluding rather than including forest and woodland areas in a measure of agricultural land.

[b] Percent of arable land and permanently cropped land under irrigation.

Table A5.2: *Economically Active Agricultural Population*

Region	1961-65	1966-70	1971-75	1976-80	1981-85
	(millions)				
Nigeria	14.2	16.0	18.1	20.7	23.3
Western Africa (17)	20.0	21.4	22.7	24.2	25.9
Central Africa (7)	11.5	12.0	12.6	13.4	14.4
Southern Africa (10)	13.1	14.2	15.6	17.6	19.5
Eastern Africa (8)	27.4	30.5	33.8	37.5	41.2
Sub-Saharan Africa (43)	*86.1*	*94.0*	*102.9*	*113.4*	*124.3*
China	*294.8*	*318.7*	*348.3*	*383.5*	*417.7*
South Asia (8)	190.3	202.0	215.9	231.9	250.2
South-East Asia (9)	59.7	63.8	67.8	71.4	75.2
Pacific (11)	1.2	1.2	1.3	1.3	1.4
Asia & Pacific, excl. China (28)	*251.1*	*267.0*	*285.0*	*304.7*	*326.8*
Caribbean (18)	2.9	2.9	2.9	3.0	3.1
Central America (8)	8.8	9.2	9.9	11.0	11.9
South America (12)	22.2	23.3	24.0	24.0	24.1
Latin America & Caribbean (38)	*33.9*	*35.4*	*36.8*	*38.0*	*39.1*
North Africa (5)	9.2	9.2	9.3	9.7	10.1
West Asia (15)	18.9	19.4	19.8	20.0	20.3
West Asia & North Africa (20)	*28.1*	*28.6*	*29.1*	*29.6*	*30.5*
Less-Developed Countries (130)	*694.0*	*743.6*	*802.2*	*869.2*	*938.5*
Japan	13.6	11.4	9.3	7.2	5.6
Australia & New Zealand (2)	0.6	0.6	0.6	0.6	0.6
Northern Europe (5)	1.5	1.3	1.1	1.0	0.8
Western Europe (8)	10.1	8.0	6.7	5.8	5.0
Southern Europe (4)	13.1	10.6	8.8	7.5	6.4
North America (2)	5.3	4.7	4.4	4.4	4.1
More-Developed Countries (22)	*44.3*	*36.5*	*30.9*	*26.5*	*22.4*
Total (152)	738.2	780.1	833.1	895.7	960.9

Source: Compiled from FAO (1987b).

Note: Economically active agricultural population includes all persons engaged or seeking employment in an economic activity related principally to agriculture, forestry, hunting, or fishing, whether as employees, own-account workers, salaried employees, or unpaid workers assisting in the operation of a family farm or business *FAO Production Yearbook 1987* (p. 4).

Chapter 6

Patterns of Agricultural Growth and Economic Development

Barbara J. Craig, Philip G. Pardey, and Johannes Roseboom

Perspectives on the process of economic growth and development have been far from static. Analysis has at various times focused on industrialization as the key to both growth and development. Intermittent swings in fashion have emphasized the importance of a dynamic or leading agricultural sector. Still other approaches have attempted to synthesize sectoral views of development with notions of dual economies on the one hand and balanced or unbalanced growth on the other. Layered on top of these more structural views of economies are notions that comparative advantage and thus the linkages between the domestic and the world economy condition development patterns.[1] Since we will not be able to settle the question about the ultimate theory of economic development, we present information about observed patterns of change in agriculture alongside information on growth in nonagricultural output and factor movements between the two sectors.

6.1 AGGREGATE AGRICULTURAL GROWTH IN PERSPECTIVE

An informed discussion of developments and sources of growth in agriculture requires some background in broader measures of economic change. The past two decades provide a richer set of information on international economic aggregates than was previously available, although, as always, it covers an all-too-brief time series on a subset of the world. On a global scale, the sustained growth in population has been more than matched by unprecedented increases in output of both the agricultural and nonagricultural sectors of the world's economy. The net effect is an increase of approximately 50% in real GDP per capita between

---

[1] Agricultural-led development strategies have been reviewed recently by Adelman, Bournieux, and Waelbroeck (1986). Building on the earlier work of Lewis (1954), dual economy models were formalized by Jorgenson (1961) and Fei and Ranis (1964) among others. For a perspective on unbalanced growth see Baumol (1967). Hayami and Ruttan (1985, ch. 2) review the treatment of agriculture in economic development models.

1961-65 and 1981-85 (table 6.1). When GDP is split into agricultural and nonagricultural components, as in table 6.2, it is evident that per capita increases in output have come from both sectors. The patterns do differ, however, with substantially larger increases in nonagricultural output per capita and somewhat more erratic increases in agricultural output per capita.

Table 6.1:   *Development of GDP, Total Population, and GDP per Capita, 1961-65 to 1981-85*

Region	1961-65	1966-70	1971-75	1976-80	1981-85
	GDP [a]	*GDP indexed on 1961-65 = 100*			
Sub-Saharan Africa (36)[b]	107	115	157	182	191
China	167	151	221	278	419
Asia & Pacific, excl. China (15)	461	125	158	208	269
Latin America & Caribbean (23)	434	128	177	231	261
West Asia & North Africa (13)	184	150	229	299	320
Less-Developed Countries (88)	1,354	132	181	234	286
More-Developed Countries (18)	3,532	128	155	180	201
Total sample (106)	4,886	129	163	195	224
	Population [c]	*Total population indexed on 1961-65 = 100*			
Sub-Saharan Africa (36)	205	114	131	152	178
China	688	113	127	138	148
Asia & Pacific, excl. China (15)	834	113	127	142	158
Latin America & Caribbean (23)	224	114	130	147	166
West Asia & North Africa (13)	123	115	132	151	173
Less-Developed Countries (88)	2,074	113	128	143	158
More-Developed Countries (18)	615	105	110	114	118
Total sample (106)	2,689	111	124	136	149
	GDP per capita [d]	*GDP per capita indexed on 1961-65 = 100*			
Sub-Saharan Africa (36)	524	101	120	120	107
China	243	134	175	201	284
Asia & Pacific, excl. China (15)	553	111	125	147	170
Latin America & Caribbean (23)	1,936	112	136	157	157
West Asia & North Africa (13)	1,492	131	174	198	185
Less-Developed Countries (88)	653	116	142	164	181
More-Developed Countries (18)	5,744	121	141	158	170
Total sample (106)	1,817	116	131	143	150

Source: GDP data primarily taken from World Bank (1989) and total population data from FAO (1987b).

[a] In billions of 1980 PPP dollars.
[b] Bracketed figures indicate the number of countries in regional totals.
[c] In millions.
[d] In 1980 PPP dollars.

Table 6.2: *Development of Nonagricultural and Agricultural per Capita GDP, 1961-65 to 1981-85*

Region	1961-65	1966-70	1971-75	1976-80	1981-85
	NonAgGDP per capita[a]	*Indexed on 1961-65 = 100*			
Sub-Saharan Africa (36)[b]	270	113	149	153	136
China	147	137	192	227	310
Asia & Pacific, excl. China (15)	309	115	139	181	220
Latin America & Caribbean (23)	1,618	116	141	165	168
West Asia & North Africa (13)	1,187	136	190	220	202
Less-Developed Countries (88)	445	122	156	188	207
More-Developed Countries (18)	5,391	123	144	162	175
Total sample (106)	1,576	118	135	149	156
	AgGDP per capita[a]	*Indexed on 1961-65 = 100*			
Sub-Saharan Africa (36)	255	87	88	84	76
China	96	129	147	161	243
Asia & Pacific, excl. China (15)	245	105	106	104	105
Latin America & Caribbean (23)	318	95	109	113	101
West Asia & North Africa (13)	324	104	106	105	111
Less-Developed Countries (88)	209	105	111	113	123
More-Developed Countries (18)	355	99	105	100	88
Total sample (106)	242	102	108	107	110

Source: GDP and AgGDP data primarily taken from World Bank (1989) and population data from FAO (1987b).

[a] In constant 1980 PPP dollars.

[b] Bracketed figures indicate the number of countries in regional totals.

Disaggregated figures on GDP growth provide a different picture of growth and development. By 1981-85, sub-Saharan Africa had by far the lowest average per capita GDP. It represents one extreme case in that it is the region with the highest average population growth rate and the lowest real GDP growth rate. Both in total GDP and in the sectoral components of GDP, output per capita stagnated in sub-Saharan Africa and then declined in real terms in the past decade.

China represents the other extreme in that it experienced the lowest population growth rate of the less-developed regions in this sample and its fourfold increase in real output was the largest of any region. Growth in real agricultural output per capita dominated that of the nonagricultural sector; however, both more than doubled over the past two decades.

To provide a snapshot of the distribution of key economic aggregates, the regional shares of the sample output and population are presented in table 6.3 for 1981-85. In this period, less-developed countries account for just 36% of world output although they have

Table 6.3: *Regional Shares of Total GDP and Population and of Agricultural GDP, Population, and Land, 1981-85 Average*

Region	Total		Agricultural		
	GDP	Population	GDP	Population	Land[b]
	%	%	%	%	%
Sub-Saharan Africa (42)[a]	2	10	7	14	20
China	7	25	22	35	12
Asia & Pacific, excl. China (21)	11	32	31	39	8
Latin America & Caribbean (32)	10	9	11	6	20
West Asia & North Africa (18)	6	6	7	4	8
Less-Developed Countries (114)	36	82	78	98	68
More-Developed Countries (22)	64	18	22	2	32
Total Sample (136)	100	100	100	100	100

Source: GDP and AgGDP data primarily taken from World Bank (1989), population data from FAO (1987b), and land data from *FAO Production Yearbooks*.

[a] Bracketed figures indicate the number of countries in regional totals.
[b] Agricultural land comprises arable land, permanently cropped land, and permanent pastures.

82% of the world's population. Only 18% of the world's population reside in more-developed countries but these countries account for nearly two-thirds of output.

The difference between distribution of people and output is even more dramatic for agriculture. Although more-developed countries have only 2% of the world's agricultural population, they generate 22% of the world's agricultural output using 32% of the world's agricultural land. It takes the remaining 98% of the world's agricultural population to generate 78% of global agricultural GDP.

The disparities among less-developed countries are quite striking as well. Asia & Pacific plus China account for 53% of agricultural GDP, with 20% of the agricultural land and 74% of the agricultural population. On the same share of the world's agricultural land, sub-Saharan Africa produces less than one-fifth of the output of Asia. The differences in factor usage and endowments that explain such disparities are examined in section 6.3.

Looking at the regional averages in per capita GDP, it is evident that the less-developed countries have made bigger strides over the past two decades than have the more-developed countries; however, the absolute per capita GDP gap between less- and more-developed countries has widened considerably. Given the declining but still high proportion of GDP accounted for by agriculture in less-developed countries (figure 6.1), it seems that the comparative lack of progress in their agricultural sectors is an important factor in the widening per capita GDP gap.

These figures drive home the point that world economic growth, in and of itself, need not generate a more egalitarian income distribution across countries. Evidence with respect to the income distribution effects of the green revolution technologies has been widely and hotly debated (Lipton with Longhurst 1989). Few have questioned that absolute levels of poverty have decreased as a result of the introduction of these technologies. But, notwithstanding that in principle such technologies are thought to be scale neutral, a number of the initial studies concluded that the benefits arising from these technologies accrued, in an "unacceptably" large measure, to large- not small-scale farmers.

Figure 6.1: *AgGDP as a percentage of GDP, 1961-85*

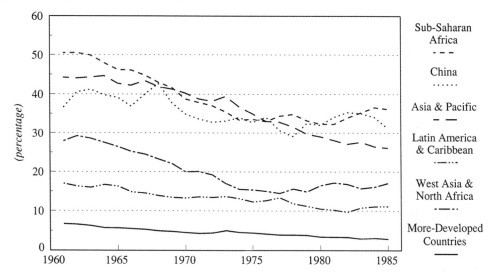

Source: Data primarily taken from World Bank (1989).

Note: Weighted averages. Sample includes 94 countries.

There is, however, some later evidence that offers a contrary perspective. Pinstrup-Andersen and Hazell (1985) gave four reasons why earlier studies erred, namely, (a) the studies were conducted too soon after the release of the green revolution technologies, (b) benefits to the poor, as consumers of rice and wheat, through lower prices were largely overlooked, (c) they gave little or no attention to the linkages between these green revolution technologies and the (rural) nonfarm economy, and (d) there are major difficulties associated with partitioning observed changes into those resulting from green revolution technologies per se and those arising from population growth, institutional arrangements, plus agricultural and broader, economy-wide policies and the like. A recent micro-level investigation of the North Arcot region in India (Hazell and Ramasamy 1991) — the site of an earlier study by Cambridge and Madras universities that was critical of the (income) distributional consequences of the green revolution technologies — provides longitudinal evidence that in many respects is at odds with the negative findings of the earlier studies. It underscores the dangers of reading too much into the few available studies of the distributional consequences of the green revolution technologies that took a partial perspective and were undertaken before the ensuing adjustment processes had more fully run their course.

Cross-country evidence from Fields (1989) is in keeping with the findings from these more recent micro-level studies and suggests that it is the nature and not the rate of economic growth that is decisive in determining whether inequality increases or decreases. Development, as it is commonly defined, often includes the notion of a more-equal distribution of income, but the structure of the world economy evidently provides no guarantees that a convergence of incomes across countries necessarily follows from economic growth.[2]

6.2   AGRICULTURE'S SHARE IN GROWTH

Broadly speaking, economic development is associated with a decline in the relative importance of agriculture as a larger proportion of an economy's resources is directed to nonagricultural production. This pattern can be attributed, in large measure, to the nature of demand for agricultural products. As a society becomes wealthier, a smaller fraction of its income is devoted to food and fiber consumption. This has two effects that contribute to the decline in the relative importance of (production) agriculture. The prices of agricultural relative to nonagricultural goods typically fall. Furthermore, the increasing sophistication of manufactured goods or processed agricultural products increases the value-added margins to a greater degree in the nonagricultural sector.[3] There are subtle shifts within agriculture toward more highly valued products — a move from grain production to livestock production, a shift from root crops to small grains and feeds — but these shifts are unlikely to be more important quantitatively than the intersectoral price and value-added effects mentioned above.

The declining importance of agriculture need not imply that the absolute level of resources devoted to agriculture declines. In fact, the economically active population in agriculture increased by 35% over the past two decades in less-developed countries — with 81% of that increase accounted for by Asia & Pacific and China (table A5.2). While more-developed countries have experienced a 49% decrease in agricultural labor, the total land devoted to agriculture has remained basically unchanged while the use of purchased inputs over this same period has increased dramatically. There have been some modest increases in the total land devoted to agriculture in less-developed countries, particularly those in South America (table A5.1).

Following Kuznets (1961), it is possible to decompose changes in GDP per capita into three elements, namely, (a) the increment in output per worker in the agricultural sector, (b) the increment in output per worker in the nonagricultural sector, and (c) the shift in the labor

---

[2] Earlier work on the "convergence" hypothesis by Gerschenkron (1952) has recently been revisited and reanalysed by Baumol (1986), de Long (1988), and Baumol and Wolff (1988).

[3] A good treatment of marketing margins in the context of agricultural prices is presented by Gardner (1975). The analytics of marketing margins are further elaborated by Fisher (1981) and, with particular reference to less-developed countries, by Yotopolous (1985).

force from agriculture to the nonagricultural sector of the economy. The first two changes are weighted by the respective end-of-period labor force shares of the sector, and the shift in the labor force is weighted by the initial difference in output per worker in the two sectors. The decomposition of changes in GDP in this manner highlights the fact that different sectors of an economy are not independent. The agricultural and nonagricultural sectors are connected through factor markets.

More formally, Kuznets' identity is as follows:

$$q_t - q_{t-1} = (q_{at} - q_{at-1}) L_{at} + (q_{nt} - q_{nt-1}) L_{nt} + (q_{nt-1} - q_{at-1})(L_{nt} - L_{nt-1}) \quad (6.1)$$

where q_{it} is the output per worker and L_{it} is the share of the total labor force in sector i in period t where $i = a$ denotes the agricultural sector and $i = n$ the nonagricultural sector.

This formula is implemented to quantify agriculture's contribution to overall economic growth in different regions of the world (figure 6.2). The contribution of agriculture to the total increase in GDP per capita has been relatively small for all regions except China, and, in the case of sub-Saharan Africa, it has at times been negative.

The contribution of the nonagricultural sector has also not always been positive either. The recent stagnation or fall in per capita GDP across all regions except China is mostly accounted for by a slowdown or absolute decline in nonagricultural output. For West Asia & North Africa the most recent drop in nonagricultural production reflects the sustained decline in the price of oil during the 1980s. The generally poor performance of the nonagricultural sector in the most recent period is partly a result of the global recession but may also be indicative of the tighter world credit market. The nonagricultural sector of highly indebted less-developed countries is more susceptible than the agricultural sector to a shortage of foreign exchange, given its far greater reliance on imported inputs.

The shift of the labor force out of agriculture has made a positive contribution to per capita GDP growth in all regions of the world.[4] The movement of workers out of agriculture has a positive effect on total GDP when the marginal product of a worker is higher outside of agriculture than in it. This positive effect is magnified if a relatively large share of the labor force remains active in agriculture.

The available data on average value-added per worker in the two sectors indicate that value-added per worker in agriculture has clearly been lower than for nonagriculture for every region (figure 6.3). It may be tempting to regard this statistic as a measure of the income earned per worker in a particular sector. This need not be the case. Those employed part-time in agriculture earn a significant portion and, in some cases, a majority of their incomes off-farm, while substantial earnings are also repatriated to agricultural households by family members working in non-agricultural sectors. The income of rural households in

---

[4] The movement of labor from agriculture into other sectors does not always imply a shift from rural locations to urban ones. Those working off-farm in rural towns are counted as nonagricultural workers in international labor statistics.

Figure 6.2:   *Changes over five-year intervals in GDP per capita decomposed into contributions by the agricultural and non-agricultural sector and the intersectoral shift in labor, 1961-65 to 1981-85*

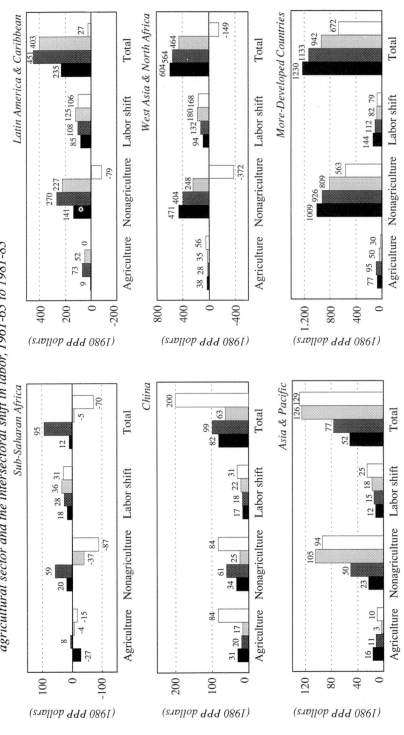

Legend: The changes in GDP per capita have been calculated between the following period averages: ■ 1961-65 and 1966-70;

■ 1966-70 and 1971-75;   ▨ 1971-75 and 1976-80;   ☐ 1976-80 and 1981-85.

Source: Calculations based on data primarily taken from World Bank (1989b) and FAO (1987b).

both less- and more-developed countries thus includes returns from agricultural and nonagricultural employment. Total value-added per worker in any industry will almost surely overstate actual wages since part of it constitutes returns to capital and land.[5] Although the figures we do have are not perfect indicators of the relative marginal product or income of agricultural workers, they do provide indicators of the relative productivity of workers in the two sectors.

Figure 6.3: *Value-added per worker in the agricultural relative to the nonagricultural sector, 1961-85*

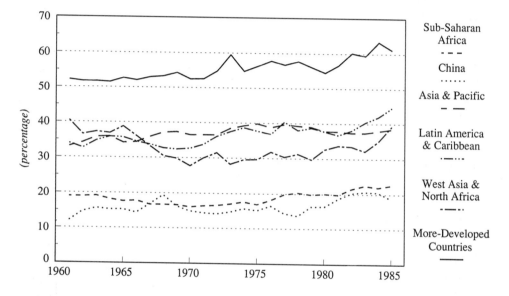

Source: Calculated ratio based on data primarily taken from World Bank (1989) and FAO (1987b).

Note: Ratio represents AgGDP per economically active population in the agricultural sector relative to non-AgGDP per economically active population in the nonagricultural sector. Regional averages represent simple averages. Sample includes 93 countries.

In more-developed countries, the gap between average value-added by an agricultural and nonagricultural worker is narrowing, and the rate at which labor has been moving between the two sectors has slowed over the past two decades. In less-developed countries, where the gap remains wide and the majority of the labor force is still engaged in agriculture, movements of labor out of agriculture still make important contributions to growth in total GDP per capita. In the most recent period, the shift of labor and its contribution to total GDP

---

5 Real GDP per capita may also be a poor comparative indicator of well-being. For a discussion of the notion of well-being from an economic perspective, see McLean (1987).

per capita shrank somewhat in regions that experienced sharp declines in nonagricultural output.

6.3   AGRICULTURAL PRODUCTIVITY

To understand the various regional trends in per capita output in the agricultural sector, it is helpful to analyze the productivity of the most widely measured factors of production. Hayami and Ruttan (1971, 1985) described the historical development of agriculture in 44 countries, illustrating their development paths with changes in labor and land productivity. Their method is adapted here to a larger set of countries using annual observations over a longer period.[6]

In figure 6.4 the logged ratios of output per hectare and output per worker for the years 1961 to 1985 are graphed for 73 countries that have been grouped into six regions and three countries. The real agricultural output aggregate used here represents AgGDP denominated in 1980 US dollars. The deflate-first then convert procedure used to translate nominal AgGDP aggregates, measured in local currency units, into internationally comparable aggregates is discussed fully in chapter 5 along with the efficacy of using value-added aggregates as measures of agricultural output. The land measure is a stock of total hectares of land in agriculture whether they be arable, permanently cropped, or pasture lands. The number of workers used is the total population economically active in agriculture.[7] The dark arrows indicate the path of these two productivity measures, and the diagonals indicate constant factor ratios. A productivity path that crosses such a diagonal from left to right indicates an increase in the number of hectares per worker. The longer a productivity path, the greater the *percentage* change in productivity.

As is evident from figure 6.4, there are considerable differences across regions both in the levels of these partial productivity measures and their paths over time.[8] The highest measured output per hectare occurs in Asia and Europe, and the lowest in Australia. Output per worker is highest in more-developed countries and is lowest in China. The paths of these partial productivity measures over the past two decades display informative differences. In some regions, such as Europe, North America, and especially Japan, increases in output per

---

6 See also Hayami and Ruttan (1969); FAO (1974); Yudelman, Butler, and Banerji (1976); Nguyen (1979); Yamada and Ruttan (1980); Scandizzo (1984); Kawagoe and Hayami (1985); APO (1987); Capalbo and Antle (1988); Peterson (1988); and Schmitt (1988) for additional insights concerning the measurement of agricultural productivity in an international context.

7 The labor variable used by Hayami et al. (1971) and Hayami and Ruttan (1971, 1985) included only male workers economically active in agriculture. We included females also became they are a significant component of the agricultural work force and statistics on female participation have been improved considerably in the more recent data series.

8 The regional differences mask some even greater disparities in factor productivity at the country level, but the groupings provide a reasonable way to summarize the information without unduly distorting the relative position of the different regions and their development over time.

Figure 6.4: *International comparison of land and labor productivities, 1961-85*

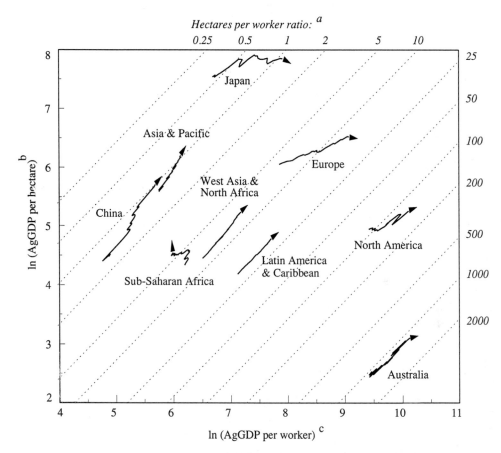

Source: AgGDP and implicit AgGDP deflator data primarily taken from from World Bank (1989), final agricultural output PPPs from FAO (1986b), agricultural labor from FAO (1987b), and agricultural land from *FAO Production Yearbooks*.

Note: AgGDP in nominal local currency units was first deflated to base year 1980 using country-specific AgGDP deflators and then converted to US dollars using agricultural output PPPs. The number of countries on which the regional (weighted) averages are based is as follows: sub-Saharan Africa (17), Asia & Pacific (11), Latin America & Caribbean (18), West Asia & North Africa (9), Europe (13), and North America (2).

[a] Hectares of agricultural land per economically active member of the agricultural population.
[b] Hectares of agricultural land includes arable plus permanently cropped and permanently pastured land.
[c] Agricultural workers is here defined as economically active agricultural population.

worker have exceeded increases in output per hectare, which has allowed increased output with fewer workers per hectare of land (table 6.4). In China and Asia & Pacific, increases in land productivity have been dominant, and these regions now employ the most workers per hectare of all the regions sampled. In Latin America & Caribbean as well as West Asia

Table 6.4:   *Hectares of Agricultural Land per Agricultural Worker*

Region	1961-65	1966-70	1971-75	1976-80	1981-85
Sub-Saharan Africa (17)[a]	6.2	5.7	5.3	4.8	4.4
China	1.4	1.3	1.2	1.1	1.0
Asia & Pacific, excl. China (11)	1.2	1.1	1.1	1.0	1.0
Latin America & Caribbean (18)	18.7	18.8	18.8	18.9	18.8
West Asia & North Africa (9)	7.8	7.6	7.6	7.4	7.0
Less-Developed Countries (56)	*2.9*	*2.8*	*2.6*	*2.5*	*2.3*
Japan	0.4	0.5	0.6	0.8	1.0
Europe (13)	6.4	7.8	9.1	10.3	12.1
North America (2)	94.4	107.3	112.2	112.5	124.3
Australia	1,046.0	1,118.4	1,134.2	1,084.5	1,105.3
More-Developed Countries (17)	*26.5*	*32.4*	*38.2*	*44.4*	*52.8*
Sample Total (73)	4.4	4.2	4.0	3.7	3.5

Source: Labor data extracted from FAO (1987b) and land data from *FAO Production Yearbooks*.

[a] Bracketed figures indicate the number of countries in regional totals.

& North Africa, productivity increases in both factors have been roughly equal, and their land/labor ratios have remained fairly static.

Sub-Saharan Africa's productivity path is clearly an outlier. Although there were some small increases in productivity in both labor and land in the immediate post-colonial period, this was followed by a noticeable deterioration in output per worker and stagnation in output per hectare. Without more detailed data, it is difficult to diagnose precisely what has happened, but the decline in productivity can variously be attributed to deterioration in infrastructure (including national agricultural research systems), disturbances caused by wars in several of the countries in the region, government economic policies that systematically discriminated against agriculture, and increased population pressure on marginal lands.

In their original work, Hayami and Ruttan (1971) noted cross-sectional patterns in productivity changes that reflected underlying differences in relative factor endowments. Here, with a sample covering a wider range of countries, the data reinforce the notion that countries with relatively low land-to-labor endowments — such as China and those in Asia & Pacific — tend to follow a path of development in agriculture that economizes on the use of land. By contrast, in countries where labor is relatively scarce — such as North America — the development path is biased towards changes that are labor- rather than land-saving.[9]

---

[9] There are various notions of biased (or factor-saving) versus neutral technical change, particularly in the context of several as opposed to two (say land and labor) factors of production. These are reviewed in Binswanger and Ruttan (1978).

The results here indicate that land and labor endowments cannot tell the whole story. Initial factor endowments encouraged land-saving technological change in Japan and labor-saving technological change in North America and Australia — although for the last two regions, much of this change happened prior to the start of the sample reported here. However, in densely populated regions such as Japan and Europe, the most recent partial productivity changes indicate the use of labor-saving rather than land-saving technologies. The fact that labor, and not land, has been induced to leave agriculture by the higher returns available in other sectors means that these regions have looked to other factors to substitute for labor that has left agriculture and to augment the productivity of workers remaining in the sector. Without information on agricultural inputs other than land and labor, however, economic interpretation of these partial productivity measures is difficult.

6.3.1 Other Inputs to Agriculture

Careful interpretation of these partial-productivity measures requires some recognition of the possibility of substituting other inputs for land and labor. While it is difficult to imagine a perfect substitute for land, there are many ways to alter the productivity of any given unit of land through complementary inputs such as fertilizers, pesticides, irrigation, and both physical and human capital. The same purchased inputs can also augment the productivity of labor.

Figures on fertilizer consumption in table A6.1 indicate some dramatic changes in fertilizer usage over the past two decades. It is not surprising that China, the region with the relatively largest increases in output per hectare, is also the region with the largest increase — over 1000% — in fertilizer usage per hectare. There is a remarkable correspondence between exceptionally high rates of fertilizer usage in Japan and Europe and high levels of output per hectare in these regions. Among less-developed countries, the highest regional rates of output per hectare occur in China and Southeast Asia, where fertilizer inputs per hectare are far above the less-developed country average. At the other end of the spectrum, Australia has the lowest regional output per hectare and has fertilizer application rates that are lower than those for most less-developed regions except sub-Saharan Africa.

Over the period 1961-65 to 1981-85, fertilizer usage has increased, on average, almost nine-fold in the less-developed countries. This contrasts sharply with nearly twofold increase on average in the more-developed countries. As with other inputs, deminishing returns to additional applications of fertilizer suggests that the rate of increase in fertilizer usage will eventually trend downward. The growth rates in table 6.5 are consistent with this expectation. The more-developed countries are characterized by relatively high application rates (table A6.1) and, as just noted, experienced much slower rates of increase in fertilizer usage over the 1961-85 period than the less-developed countries. Indeed, the use of fertilizer in the more-developed countries actually declined in absolute terms during the early 1980s. Correspondingly, the growth in fertilizer usage declined to quite low levels in sub-Saharan Africa while in Latin America & Caribbean growth virtually ceased — two regions that

Table 6.5:   *Average Annual Growth of Fertilizer Use*

Region	1961-70	1970-80	1980-85	1961-85
	%	%	%	%
Sub-Saharan Africa (41)[a]	12.5	5.5	2.9	8.7
China	22.6	13.4	4.2	14.2
Asia & Pacific, excl. China (22)	15.7	9.5	7.9	11.0
Latin America & Caribbean (33)	11.9	10.5	0.7	9.6
West Asia & North Africa (20)	11.8	10.3	6.6	10.6
Less-Developed Countries (117)	15.9	11.0	5.0	11.6
More-Developed Countries (22)	6.2	2.7	-0.2	3.0
Total (139)	8.0	5.5	2.3	5.6

Source: Fertilizer data extracted from FAO (1990a).

Note: Fertilizer use based on the unweighted sum of mass of nitrogen, phosphorus (as oxide), and potassium (as oxide) used in agricultural production. Period growth rates represent log-linear compound growth rates.

[a] Bracketed figures indicate the number of countries in regional totals.

experienced the slowest growth in land productivity. This slow growth in fertilizer usage was in spite of the fact that world fertilizer prices were lower throughout the 1980s compared with earlier periods.[10] In Bumb's (1989) view this recent pattern of fertilizer use follows largely because (a) agricultural surpluses in the more-developed countries led to lower agricultural prices[11] and stimulated efforts to control agricultural production through, for example, acreage reduction programs and production quotas, and (b) a shortage of foreign exchange to buy fertilizers has been most limiting for the highly indebted countries of Latin America & Caribbean and sub-Saharan Africa, while the economic readjustment programs on which these countries embarked severly affected fertilizer subsidies.

The use of capital services in agriculture over the past two decades is virtually impossible to document. Even information on the agricultural capital stock is spotty. Complete information on tractors, animal traction, combines, harvesters, threshers, milking machines, irrigation equipment, storage facilities, and public infrastructure is available for very few countries. Even if the data were available, aggregating such stocks over a region and converting them to a useful measure of the service flow from capital requires detailed information on capital prices, utilization rates, economic depreciation rates, and the lifespan

---

[10] According to Bumb (1989), in earlier years the impact of changing world market prices for fertilizer on global fertilizer use was muted because in many countries farmers were insulated from world market prices. High fertilizer prices during the 1970s, for example, were often offset by government subsidies. More recently, declining global fertilizer prices combined with tighter government budgets meant that, in an increasing number of cases, subsidies have been removed or progressively lowered.

[11] During the first half of the 1980s agricultural prices in the more-developed countries grew considerably slower than the average price level, while in the less-developed countries the opposite was the case (table 5.1, chapter 5).

of different capital types.

In table A6.2, total tractors in use in agriculture are reported. These figures are available for a wide range of countries, but they provide — at best — a crude indicator of total services from capital. Changes in the stock of tractors have been used as a proxy for the change of capital use in agriculture. The danger in doing this lies in forgetting the change over time in the quality of tractors and the probably more significant cross-sectional differences in average tractor quality. A new 1961 tractor is not identical to a new 1985 tractor any more than the average tractor in use in Thailand is identical to the average tractor in use on the Great Plains of North America.

Over the past two decades, the inventories of tractors have increased at a much higher rate in the less-developed countries than in the more-developed. Clearly, less-developed countries started with a much lower stock of tractors and are far from catching up. The number of tractors in use per hectare is still far lower in less-developed countries, and it is most probable that the gap in capital services per hectare is even larger. So, it is no surprise that the levels of output per worker for the group of less-developed countries fall short of those of all the more-developed regions, excluding Japan.

Europe and Japan are the regions that employ the largest numbers of tractors and largest input of fertilizer per hectare. The big surge in Japan's labor productivity and its switch to a more land-intensive agriculture corresponds more with the dramatic increase in the number of tractors per hectare than with fertilizer use. Without due accounting for the use of capital and other purchased inputs, the shift in both Europe and Japan towards labor-saving technologies would be hard to rationalize.

6.3.2 Mismeasurement of Agricultural Outputs and Inputs

Any international comparison of partial-productivity measures requires some attention to the conceptual and technical problems involved in using such broad aggregates as AgGDP, total hectares of land in agriculture, and the total economically active agricultural population.

As discussed in chapter 5, the value-added measures may introduce biases in international comparisons of agricultural output. If there is an asymmetric treatment of inside and outside inputs, value-added will not bear the same relationship to final output for all countries and all points of time if inside and outside inputs constitute different shares of final output.[12] Value-added is a much smaller fraction of final output in more-developed economies than in less-developed. Part of that reflects the more intensive use of fertilizers, machinery, pesticides, and other purchased inputs. Since outside inputs are more likely to be accurately subtracted from final output than inside inputs, the value-added measures used

---

[12] Inside inputs refer to intermediate inputs such as seed, feed, and manure produced and then reused within the sector. Outside inputs refer to purchased materials such as commercial fertilizers, pesticides, and energy which are produced outside of the agricultural sector.

above probably understate the output of more-developed countries relative to that of less-developed countries. Correct subtraction of both inside and outside inputs would act to increase the separation between productivity paths for less- and more-developed countries in figure 6.4.

The figures on total hectares of land in agriculture include total hectares of arable, permanently cropped, and pasture land. These land types are heterogeneous, and the average quality of any one of these land types differs internationally, so the levels of measured output per hectare must be interpreted with some care. As was demonstrated in section 5.3.3, quality adjustment of heterogeneous land aggregates in the US tends to reduce, but not eliminate, measured disparities in the regional levels of output per unit of land. For international comparisons, regional differences in land quality are probably even more extreme than in the US, so accounting for differences in land quality would tend to collapse the vertical displacement of the regional productivity paths in figure 6.4.

Quality adjustment of land will affect more than just the relative position of regions in figure 6.4. As hectares of unimproved cropland are replaced with improved land, at least part of the likely increase in output should be attributed to the use of higher-quality land instead of being counted as an increase in yield per hectare of constant quality. So, for regions such as China and Asia that have made substantial permanent improvements in land quality through irrigation and terracing, quality adjustment of land would reduce the measured rates of increase in output per hectare and thus shorten their partial-productivity paths. If one could also account for multiple cropping of hectares and measure instead hectare-plantings, this would no doubt serve to reduce the measured output per hectare in China, Asia, and Japan still further.

Measured worker input in agriculture may be even less satisfactory than measures of land. The figure is a stock of economically active workers in agriculture. There are no international data available from which to infer part- or full-time status of workers, let alone a measure of actual hours worked.

The cross-sectional quality differentials in the average worker may not be as great as those in land. However, differences in levels of educational attainment of the general population do suggest that the average worker in a less-developed country is far less likely to have enrolled in secondary school than the average worker in a more-developed country. The fact that regional disparities in secondary and, especially, primary school enrollments have been disappearing, would lead one to expect that the gaps in human capital endowments per worker in agriculture are narrowing but have by no means been eliminated.

Schultz (1964), among others, has highlighted the importance of differences in human capital characteristics in explaining gaps in labor productivity. The quality adjustment of labor inputs in US agriculture had a significant effect on measured output per hour, which suggests that accounting for international differences in labor input would also change the placement and lengths of the productivity paths in figure 6.4. Accounting for differences in cross-sectional human capital characteristics would lead one to discount the services of less-educated and less- experienced workers. If one worker is replaced by another with more experience or education, the likely increase in output per worker would be discounted to reflect the fact that some of the output change is properly attributed to changes in the quality of the inputs.

Broadly speaking, quality adjustment should reduce the measured output per worker in more-developed regions relative to that in less-developed regions, thus reducing the horizontal dispersion of the productivity paths in figure 6.4. Accounting for increased educational attainment in all regions will tend to shrink any measured increases and magnify any measured decreases in output per worker.

6.4   CONCLUDING COMMENTS

The change in international patterns of growth and productivity in agriculture described in this chapter reflect distinctly different regional development paths. The paths were not unexpected; they tend to reflect the underlying and changing patterns of factor endowments and reinforce the notion that production techniques are adjusted to economize on the use of relatively scarce or expensive factors. The measured changes in factor productivity reflect substantial regional variation in output per hectare, output per worker, and in the factor mix of land and labor. However, much of the cross-sectional differences in the factor productivity levels can probably be explained by quality differentials in land and labor inputs, as well as the employment of capital and other purchased inputs.

A thorough understanding of agricultural development and the design of agricultural research policy requires an appreciation of historical development patterns. Since the nature and level of research investment are driven by the productivity of conventional inputs to agriculture and will, in turn, affect their future productivity, an understanding of the linkages between agriculture and other sectors and the dynamics of factor substitution within agriculture is critical for analysis and choice in research policy.

Table A6.1: *Fertilizer Consumption*

Region	Total consumption[a] (thousand tons)					Average consumption per hectare arable & permanently cropped land									
						Total[a] (kg per hectare)					Nitrogen (kg per hectare)				
	61-65	66-70	71-75	76-80	81-85	61-65	66-70	71-75	76-80	81-85	61-65	66-70	71-75	76-80	81-85
Nigeria	2	8	25	101	250	0.1	0.3	0.8	3.3	8.1	0.0	0.1	0.4	1.7	3.8
Western Africa (17)[b]	30	61	110	172	183	1.0	1.9	3.2	4.7	4.9	0.3	0.8	1.2	1.9	1.9
Central Africa (6)	3	10	15	13	17	0.3	0.9	1.4	1.1	1.4	0.1	0.5	0.6	0.6	0.6
Southern Africa (10)	109	169	284	304	358	6.4	8.8	13.8	14.3	16.4	3.1	4.4	7.1	7.6	8.9
Eastern Africa (7)	49	97	159	164	230	1.4	2.6	4.1	4.1	5.6	1.0	1.7	2.6	2.5	3.5
Sub-Saharan Africa (41)	*192*	*344*	*593*	*753*	*1,038*	*1.6*	*2.7*	*4.4*	*5.4*	*7.2*	*0.8*	*1.4*	*2.3*	*2.8*	*3.7*
China	*1,258*	*3,143*	*5,400*	*10,557*	*16,800*	*12.1*	*30.6*	*53.2*	*105.1*	*166.5*	*9.2*	*23.5*	*38.8*	*85.6*	*129.7*
South Asia (8)	779	2,214	3,657	6,221	9,598	3.7	10.3	16.8	28.1	43.6	2.6	7.1	11.9	19.4	29.5
South-East Asia (8)	851	1,333	2,137	2,937	4,060	21.0	30.0	43.7	56.3	74.4	12.5	17.5	25.1	32.7	43.9
Pacific (6)	5	7	16	23	24	6.6	8.7	18.7	26.6	27.8	5.6	6.6	12.2	17.9	17.3
Asia & Pacific, excl. China (22)	*1,635*	*3,554*	*5,810*	*9,181*	*13,682*	*6.5*	*13.6*	*21.7*	*33.5*	*49.6*	*4.2*	*8.8*	*14.3*	*21.9*	*32.3*
Caribbean (13)	62	92	148	125	119	26.8	37.3	56.7	44.6	40.9	12.6	16.6	25.6	21.1	20.3
Central America (8)	368	679	1,090	1,479	2,001	12.4	23.0	37.0	48.8	63.7	9.1	16.0	24.9	34.0	44.7
South America (12)	666	1,151	2,425	4,360	3,968	8.2	11.1	20.6	33.6	28.5	2.9	3.8	6.4	9.5	9.1
Latin America & Caribbean (33)	*1,095*	*1,922*	*3,663*	*5,964*	*6,088*	*9.7*	*14.2*	*24.5*	*36.7*	*35.1*	*4.7*	*6.7*	*10.4*	*14.3*	*15.8*
North Africa (5)	394	515	799	1,067	1,428	17.2	21.9	32.7	42.9	56.7	11.7	15.0	21.2	27.3	35.7
West Asia (15)	222	569	1,156	2,122	3,034	3.9	9.7	19.2	36.2	52.7	2.1	5.3	11.1	20.2	31.1
West Asia & North Africa (20)	*616*	*1,084*	*1,955*	*3,189*	*4462*	*7.8*	*13.2*	*23.1*	*38.2*	*53.9*	*4.9*	*8.1*	*14.0*	*22.3*	*32.5*
Less-Developed Countires (117)	*4,795*	*10,048*	*17,420*	*29,644*	*42,070*	*7.2*	*14.2*	*23.6*	*39.0*	*54.2*	*4.5*	*9.1*	*14.6*	*25.1*	*35.9*

Table A6.1: *Fertilizer Consumption (Contd.)*

Region	Total consumption[a] (thousand tons)					Average consumption per hectare arable & permanently cropped land									
						Total[a] (kg per hectare)					Nitrogen (kg per hectare)				
	61-65	66-70	71-75	76-80	81-85	61-65	66-70	71-75	76-80	81-85	61-65	66-70	71-75	76-80	81-85
Japan	1,730	2,060	2,015	2,118	2,020	221.6	259.0	250.5	263.2	250.8	113.4	133.3	136.6	142.6	141.7
Australia & New Zealand (2)	1,195	1,491	1,628	1,627	1,650	34.6	37.5	38.9	37.8	35.0	1.8	3.2	4.3	5.7	6.8
Northern Europe (5)	1,257	1,630	1,955	1,927	1,902	131.9	177.2	217.4	216.9	214.2	45.9	71.6	97.2	104.8	109.1
Western Europe (8)	8,557	10,696	12,230	13,288	13,678	202.2	268.8	318.3	349.1	361.0	62.8	94.5	120.2	149.7	170.8
Southern Europe (4)	2,058	2,701	3,493	4,398	4,477	47.8	64.0	86.9	110.9	113.3	21.8	31.5	43.4	57.0	61.1
North America (2)	9,913	14,892	17,985	22,025	20,805	44.8	65.0	77.6	94.3	88.0	18.4	29.2	36.9	45.8	45.5
More-Developed Countries (22)	*24,710*	*33,470*	*39,306*	*45,384*	*44,532*	*69.3*	*91.4*	*107.3*	*123.2*	*118.9*	*24.8*	*36.4*	*45.5*	*55.8*	*57.7*
Total (139)	29,505	43,518	56,726	75,028	86,602	28.8	40.5	51.4	66.5	75.3	11.6	18.4	24.9	35.1	43.0

Source: Fertilizer data extracted from *FAO* (1990a) and agricultural land data from *FAO Production Yearbooks.*

[a] Unweighted sum of mass of nitrogen, phosphorus (as oxide), and potassium (as oxide).
[b] Bracketed figures indicate the number of countries in regional totals.

Table A6.2: *Tractors in Use in Agriculture*

Region	Total number of tractors in use in agriculture[a] (thousands)						Tractors per 1000 hectares of agricultural land[b]				
	1961-65	1966-70	1971-75	1976-80	1981-85	Growth rate[c] (%)	61-65	66-70	71-75	76-80	81-85
Nigeria	0.7	1.9	6.4	8.1	9.5	14.3	0.01	0.04	0.13	0.16	0.18
Western Africa (17)[d]	2.9	5.1	7.2	9.3	11.1	6.9	0.02	0.03	0.04	0.05	0.06
Central Africa (7)	1.2	2.3	3.1	3.9	4.5	6.8	0.03	0.06	0.08	0.10	0.11
Southern Africa (10)	23.6	33.3	43.7	59.8	52.5	4.1	0.11	0.16	0.20	0.23	0.24
Eastern Africa (8)	27.1	31.5	37.9	42.5	51.3	3.3	0.13	0.15	0.18	0.20	0.24
Sub-Saharan Africa (43)	*55.4*	*74.2*	*98.2*	*113.6*	*129.0*	*4.3*	*0.08*	*0.10*	*0.14*	*0.16*	*0.18*
China	*60.4*	*154.0*	*291.0*	*573.8*	*830.3*	*14.0*	*0.14*	*0.37*	*0.69*	*1.37*	*1.98*
South Asia (8)	55.5	110.70	243.4	436.6	693.4	13.5	0.21	0.41	0.90	1.60	2.56
South-East Asia (9)	13.9	23.9	41.8	84.1	164.3	13.1	0.26	0.41	0.67	1.28	2.41
Pacific (11)	2.7	3.9	5.2	6.4	7.2	5.0	1.98	2.77	3.42	4.16	4.55
Asia & Pacific, excl. China (28)	*72.1*	*138.5*	*290.3*	*527.2*	*864.9*	*13.2*	*0.23*	*0.42*	*0.87*	*1.55*	*2.53*
Caribbean (18)	12.0	13.3	14.5	14.9	14.6	1.0	2.47	2.54	2.36	2.40	2.33
Central America (8)	73.9	97.9	114.5	131.2	176.7	4.5	0.65	0.85	0.99	1.12	1.48
South America (12)	340.5	435.2	547.5	762.2	996.2	5.5	0.68	0.82	0.98	1.31	1.66
Latin America & Caribbean (38)	*426.4*	*546.4*	*676.5*	*908.3*	*1,187.5*	*5.3*	*0.69*	*0.84*	*0.99*	*1.29*	*1.64*
North Africa (5)	63.9	86.4	109.5	140.8	179.7	5.3	0.74	0.99	1.23	1.58	2.10
West Asia (15)	87.1	147.6	267.7	516.0	740.7	11.3	0.38	0.64	1.16	2.25	3.26
West Asia & North Africa (20)	*151.0*	*234.0*	*377.2*	*656.8*	*920.4*	*9.5*	*0.48*	*0.74*	*1.18*	*2.06*	*2.94*
Less-Developed Countries (130)	*765.3*	*1,147.1*	*1,733.3*	*2,779.7*	*3,932.0*	*8.5*	*0.32*	*0.47*	*0.70*	*1.11*	*1.56*

Table A6.2: *Tractors in Use in Agriculture (Contd.)*

Region	Total number of tractors in use in agriculture[a] (thousands)					Growth rate[c] %	Tractors per 1000 hectares of agricultural land[b]				
	1961-65	1966-70	1971-75	1976-80	1981-85		61-65	66-70	71-75	76-80	81-85
Japan	23.9	194.1	436.3	1,089.4	1,605.4	23.4	3.98	33.25	77.19	198.16	296.73
Australia & New Zealand (2)	370.2	418.4	428.0	422.0	421.9	0.7	0.75	0.83	0.84	0.83	0.84
Northern Europe (5)	485.3	581.9	643.8	709.4	751.8	2.2	36.98	45.56	51.37	57.09	61.07
Western Europe (8)	2,808.0	3,484.9	3,914.0	4,182.3	4,412.7	2.3	33.44	42.87	49.47	53.79	57.31
Southern Europe (4)	496.6	828.9	1,187.3	1,601.8	2,007.3	7.2	7.51	12.76	19.01	26.07	33.01
North America (2)	5,319.7	5,974.0	5,791.1	5,597.3	5,362.5	0.0	10.57	11.95	11.64	11.24	10.57
More-Developed Countries (22)	*9,504.2*	*11,482.1*	*12,401.2*	*13,602.3*	*14,560.7*	2.2	8.15	9.83	10.61	11.68	12.47
Total (152)	10,269.5	12,629.3	14,134.5	16,382.0	18,492.7	3.0	2.90	3.52	3.88	4.47	5.01

Source: Tractor data extracted from FAO (1990b) and agricultural land data from *FAO Production Yearbooks.*

[a] Unweighted sum of total wheel and crawler tractors (excluding garden tractors) in use.
[b] Data on agricultural land given in Table A5.1.
[c] Compound annual average over the period 1961-65 to 1981-85.
[d] Bracketed figures indicate the number of countries in regional totals.

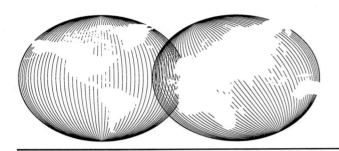

PART III

PUBLIC–SECTOR
AGRICULTURAL RESEARCH

Chapter 7

Regional Perspectives on National Agricultural Research

Philip G. Pardey, Johannes Roseboom, and Jock R. Anderson

Significant regional differences in the patterns of growth in agricultural output were identified in chapter 6. These differences are driven by and, in turn, have a substantial impact upon the policy environment within which the agricultural production and supporting science sectors operate. While it is possible to identify the general nature and scope of these development trends, more specific details — many of which have crucial policy implications — are difficult to quantify. Moreover, such patterns that have been identified may be more apparent than real, because they are based on data and concepts that involve difficulties when comparable measures of agricultural research and of agricultural output and inputs are constructed (chapter 5). Nevertheless, recognizing and understanding these measurement problems greatly enhances the insights to be had from using quantitative evidence on the nature and causes of productivity growth in the agricultural sector when evaluating alternative research policy options.

Of particular interest here is the role of formal agricultural R&D conducted in the public arena in achieving various national development objectives. To facilitate a comparative assessment of national agricultural research systems (NARSs) at the regional and subregional levels, a completely revised and updated set of agricultural research indicators has been juxtaposed against relevant measures of agricultural output and inputs. Careful consideration of these indicators is no substitute for quantitative attempts to "explain" regional patterns of partial (or, ideally, total) factor productivity growth and the development consequences thereof. But such indicators can usefully inform national, regional, and international research policy choices that will not wait for the more formal analysis that may eventually be undertaken.

The inherent and, in some cases, substantial lags in the agricultural research process itself as well as institutional precedents play an important role in shaping the nature and impact of agricultural research endeavors. Placing our contemporary agricultural research indicators in an historical context serves to anchor them to these institutional precedents.

7.1   GLOBAL AGRICULTURAL RESEARCH CAPACITY

A useful point of departure for the intraregional analyses that follow in sections 7.2 to 7.7 is to portray first the global development of the capacity of public-sector NARSs between 1961-65 and 1981-85. For this purpose, the data have been grouped into quinquennial averages for six broad "regions," including four regional aggregates of less-developed countries covering sub-Saharan Africa, Asia & Pacific, Latin America & Caribbean, and West Asia & North Africa, a separate entry for China, and a "regional aggregate" of the more-developed countries.[1] Although much useful detail is necessarily lost when data are aggregated, the present intention is to examine broad trends rather than country-specific details.

7.1.1   Personnel and Expenditure Aggregates

The pattern of global investment in public agricultural research has undergone a dramatic change over the past two decades. Global agricultural research capacity has grown substantially while, at the same time, the less-developed countries have significantly increased their share (at least in quality-unadjusted terms). Recent trends, however, indicate a marked departure from this pattern of growth. There are signs that new investment is slowing, particularly with regard to financial support for agricultural research in sub-Saharan Africa and Latin America & Caribbean, which may bring possibly unanticipated and untoward changes in the future.

Averaged over the 1981-85 period, the global total of agricultural researchers working in the public sector stood at slightly more than 134,000 full-time equivalents. This represents a 2.2-fold increase from the 1961-65 period, which translates into an average annual growth rate of 4.1%. The number of researchers in less-developed countries grew at just over four times the annual rate of the more-developed countries (7.1% against 1.7%). As a result, the global share of researchers in less-developed countries increased from 33% in 1961-65 to 58% in 1981-85 (figure 7.1a). The Asia & Pacific region plus China accounted for about 70% of the less-developed country total in 1981-85 (table 7.1). Latin America & Caribbean and West Asia & North Africa each accounted for about 12%, while the remaining less-developed country researchers (6%) worked in sub-Saharan Africa. Significantly, the total number of researchers in sub-Saharan Africa would increase by around 39% if the region were redefined to include the Republic of South Africa's public research system.

Global spending on public agricultural research averaged $8.4 billion per annum in 1981-85, up by a factor of 2.6 on the level of real expenditures two decades earlier. The

---

[1]   The assignment of countries to regions is detailed in the appendix to this volume. Because of data limitations the world totals used throughout this volume exclude USSR, Eastern Europe, Mongolia, North Korea, Vietnam, Cambodia, Djibouti, Bhutan, South Africa, and Cuba.

Figure 7.1a: *Agricultural researchers, regional shares*

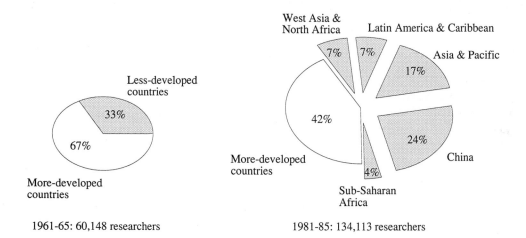

1961-65: 60,148 researchers   1981-85: 134,113 researchers

Figure 7.1b: *Agricultural research expenditures, regional shares*

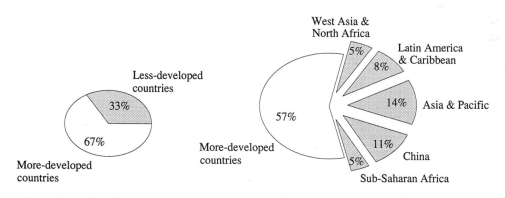

1961-65: 3284 million 1980 PPP $ 1981-85: 8443 million 1980 PPP $

Note: Percentages do not sum exactly because of rounding.

expenditure share of the less-developed countries grew from 33% in 1961-65 to only 43% in 1981-85. This is considerably less than the corresponding fraction of agricultural researchers (58%) who work in the public-sector NARSs of these less-developed countries (figure 7.1b).

Comparative patterns of growth in research personnel and expenditures are given in table 7.2. The 6.2% rate of increase in real spending for the less-developed countries was approximately 50% larger than the increase for the more-developed countries over the period from 1961-65 to 1981-85. However, it fell short of the 7.1% increase in research

Table 7.1: *Agricultural Research Personnel and Expenditures, Regional Totals*

Region	1961-65	1966-70	1971-75	1976-80	1981-85
	Agricultural research personnel *(full-time equivalents)*				
Sub-Saharan Africa (43)[a]	1,323	1,841	2,416	3,526	4,941
China	6,966	9,900	11,563	20,048	32,224
Asia & Pacific, excl. China (28)	6,641	9,480	12,439	18,559	22,576
Latin America & Caribbean (38)	2,666	4,122	5,840	6,991	9,000
West Asia & North Africa (20)	2,157	3,485	4,746	6,019	8,995
Less-Developed Countries (130)	19,753	28,829	37,004	55,143	77,737
More-Developed Countries (22)	40,395	44,039	48,123	51,602	56,376
Total (152)	60,148	72,868	85,126	106,745	134,113
	Agricultural research expenditures *(millions 1980 PPP dollars per year)*				
Sub-Saharan Africa (43)[a]	149.5	227.2	276.9	359.1	372.3
China	271.4	296.2	485.4	689.3	933.7
Asia & Pacific, excl. China (28)	316.7	475.4	651.5	928.3	1,159.6
Latin America & Caribbean (38)	229.1	355.1	486.6	679.3	708.8
West Asia & North Africa (20)	126.9	249.7	300.8	341.2	455.4
Less-Developed Countries (130)	1,093.6	1,603.7	2,201.0	2,997.3	3,629.8
More-Developed Countries (22)	2,190.7	3,057.2	3,726.3	4,171.4	4,812.9
Total (152)	3,284.3	4,660.9	5,927.3	7,168.7	8,442.7

Note: (Sub) totals may not add up exactly because of rounding.

[a] Bracketed figures indicate the number of countries in the regional totals. The appendix to this volume indicates which specific countries were included in the regional aggregates.

personnel experienced by the less-developed countries over the corresponding period. By contrast, the more-developed countries increased their real research expenditures at more than double the rate of increase of research personnel. Asia & Pacific is the only less-developed region for which the overall annual rate of growth in real expenditures (6.7%) exceeded the rate of growth in number of researchers (6.3%). In fact, this region exhibited the largest rate of increase in real expenditures but the slowest growth in research personnel over the past two decades, when compared with other less-developed regions. The sub-Saharan Africa region experienced the slowest rate of growth in real spending levels of any of the less-developed country regions, despite (or perhaps, to a degree, in response to) substantial donor support, while research personnel growth was about average. The causes and consequences of these divergent patterns of research spending and personnel growth are explored below.

The period-to-period averages in table 7.2 reveal a general contraction in the growth of agricultural research expenditures in the less-developed countries during the final period

Table 7.2: *Growth of Agricultural Research Personnel and Expenditures, Compound Annual Averages*

	Annual average growth rate				
Region	1961-65 to 1981-85	1961-65 to 1966-70	1966-70 to 1971-75	1971-75 to 1976-80	1976-80 to 1981-85
	Agricultural researchers				
	%	%	%	%	%
Sub-Saharan Africa (43)[a]	6.8	6.8	5.6	7.9	7.0
China	8.0	7.3	3.2	11.6	10.0
Asia & Pacific, excl. China (28)	6.3	7.4	5.6	8.3	4.0
Latin America & Caribbean (38)	6.3	9.1	7.2	3.7	5.2
West Asia & North Africa (20)	7.4	10.1	6.4	4.9	8.4
Less-Developed Countries (130)	7.1	7.9	5.1	8.3	7.1
More-Developed Countries (22)	1.7	1.7	1.8	1.4	1.8
Total (152)	4.1	3.9	3.2	4.6	4.7
	Agricultural research expenditures				
	%	%	%	%	%
Sub-Saharan Africa (43)	4.7	8.7	4.0	5.3	0.7
China	6.4	1.8	10.4	7.3	6.3
Asia & Pacific, excl. China (28)	6.7	8.5	6.5	7.3	4.5
Latin America & Caribbean (38)	5.8	9.2	6.5	6.9	0.9
West Asia & North Africa (20)	6.6	14.5	3.8	2.6	5.9
Less-Developed Countries (130)	6.2	8.0	6.5	6.4	3.9
More-Developed Countries (22)	4.0	6.9	4.0	2.3	2.9
Total (152)	4.8	7.3	4.9	3.9	3.3

[a] Bracketed figures indicate the number of countries in the regional totals.

of the sample, except in West Asia & North Africa. The precipitous decline in the rate of growth in real spending for sub-Saharan Africa over this same period reflects a widespread slowdown throughout the region. This was compounded by a 23% decline in total spending by the Nigerian system, which alone accounts for approximately one-quarter of public spending on agricultural research in sub-Saharan Africa. Anecdotal evidence suggests that this contractionary pattern of support has continued or even accelerated over the more recent past for many less-developed countries and may have been matched in some of the more-developed countries as well. Latin America & Caribbean also witnessed a widespread slowdown in total agricultural research spending between 1976-80 and 1981-85, with 26 of the 38 countries in the region experiencing declines in absolute terms.

Agricultural research represents but one investment option for the public sector and it must compete for funds against alternative claims on expenditures including those for other rural public goods such as roads, communications, and education (chapter 1). Contractions in the level of investment in agricultural research within sub-Saharan Africa and Latin America & Caribbean during the early 1980s mirror a deteriorating pattern of capital accumulation that is widespread throughout the economies of these two regions, economies that are severely affected by burgeoning levels of debt. Figure 7.2 shows that throughout the early 1970s all less-developed regions significantly increased the share of their GDP being invested rather than consumed. But, the rates of overall public- and private-sector investment, particularly in Latin America & Caribbean and sub-Saharan Africa, peaked in the mid-1970s and declined rapidly thereafter. Ossa (1990) points to the combined influence of reduced export earnings, a slowdown and reversal in capital inflows, and increased public- and private-sector interest obligations as placing severe constraints on the availability of resources for domestic use in these two regions.

Figure 7.2:   *Total investment as a percentage of GDP, 1971-85*

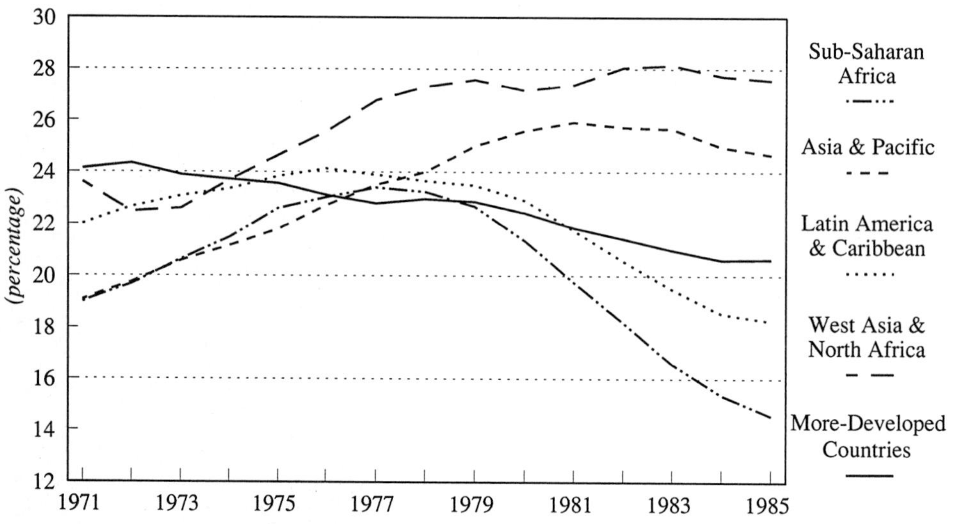

Source: Data compiled from IMF (1989b).

Note: Total investment includes both private and public investment. The 1971 to 1983 observations were centered five-year moving averages. The latter two years are averaged over four and three observations, respectively.

7.1.2   Real Expenditures per Researcher

In table 7.3, indicators of real expenditures per researcher are compared. This expenditure ratio exhibits a substantial degree of variability, both within a region over time and among regions during any given period. With real expenditures measured in 1980 purchasing power parity (PPP) terms, the overall ratio of spending per researcher for more-developed

Table 7.3: *Real Expenditures per Researcher per Year*

Region	Expenditures per researcher per year				
	1961-65	1966-70	1971-75	1976-80	1981-85
	(1980 PPP dollars)				
Sub-Saharan Africa (43)[a]	113,000	123,400	114,600	101,800	75,300
China	39,000	29,900	42,000	34,400	29,000
Asia & Pacific, excl. China (28)	47,700	50,100	52,400	50,000	51,400
Latin America & Caribbean (38)	85,900	86,200	83,300	97,200	78,800
West Asia & North Africa (20)	58,800	71,700	63,400	56,700	50,600
Less-Developed Countries (130)	55,400	55,600	59,500	54,400	46,700
More-Developed Countries (22)	54,200	69,400	77,400	80,800	85,400
Total (152)	54,600	64,000	69,600	67,200	63,000

Note: Data rounded to nearest hundred dollars.

[a] Bracketed figures indicate the number of countries in the regional totals.

countries increased steadily from $54,200 in 1961-65 to $85,400 in 1981-85.[2] The more-developed countries have continued to move toward more capital-intensive — in both human and physical terms — research systems over the past two decades. Evidence based on detailed data on the changing factor mix of the US state agricultural experiment stations points to a significant increase in human capital relative to physical capital over the long run (see section 8.2). By contrast, a mixed pattern of capital deepening appears to characterize the national research systems of the less-developed countries since the early 1960s. On average the less-developed countries spent $55,400 per researcher in 1961-65. This amount peaked during the early 1970s, followed by a steady decline, and reached $46,700 by 1981-85.

One widely observed factor that has contributed to the overall decline in spending per researcher among less-developed countries can be traced to the substantial growth in university graduates resulting from an expansion in local university capacity. Governments in numerous less-developed countries often oblige public-sector, including research, agencies to offer employment to these graduates, but in many instances they fail to provide sufficient matching funds to preserve spending-per-researcher ratios.

Both China and Asia & Pacific have historically displayed low levels of real support per researcher when compared with other regions of the world. When translating research expenditures that are measured in nominal local currency units into real aggregates, we attempted to account for the relatively low average price levels that have prevailed in Asia. Compared with the alternative translation procedures used by others in the past (see sections 5.2 and 5.4.5), this substantially increased (in fact, it doubled) the region's share of the

---

[2] This increase is driven, in part, by Japan's exceptionally rapid increase in spending per researcher from a relatively low $32,300 in 1961-65 to $69,100 by 1981-85.

global volume of resources committed to agricultural research. As a consequence, although our translation procedures did not eliminate regional differences in the volume of resources expended per researcher, they did narrow them.

Economies of size and scope accruing to the large (and, in some ways, less fragmented) research systems that dominate Asia would tend to lower average costs per unit of research output. They could also account, to some extent, for the region's lower spending per researcher. In addition, relatively lower labor costs, resulting from a comparative abundance of labor, would induce a substitution of labor for capital and other inputs in the knowledge-production process. This would also tend to reduce the region's ratio of spending per scientist.

A striking feature of table 7.3 is the historically high level of expenditures per researcher in sub-Saharan Africa. This peaked in the late 1960s at $123,400 and has declined steadily since. During the 1960s, recently decolonized NARSs in the region were still staffed by a high proportion of relatively expensive expatriate researchers (ex-colonial initially but now increasingly American and other). The region's infrastructure was poorly developed at that time, which raised the cost of basic communication, transport, and electrical services. Research hardware and instrumentation often had to be imported. Further, the region includes numerous small NARSs, many of which are attempting to address production issues arising from diverse agroecological and socioeconomic environments, which could give rise to diseconomies of size and scope that further force up average research costs. These and other factors discussed in more detail in chapter 8 may account in large measure for the region's historically high expenditure ratios.

While infrastructural constraints surely remain, and in some instances have probably intensified, the substantial decline in the levels of support per researcher may in part reflect the Africanization of the research system that has occurred during this period. Several forces are at work here. There has clearly been a trend to replace more expensive expatriate researchers with less expensive (but on average, possibly less skilled) local researchers. Anecdotal evidence suggests that, in the late 1950s prior to independence, about 90% of the agricultural researchers in the region were expatriates. By the late 1960s, the share of expatriates had declined to around 60%, according to data provided by Cooper (1970) for some 30 sub-Saharan African countries. Our data (section 8.3.2) suggest that the decline has continued, so that by 1981-85, the share of expatriates had fallen to less than 30%. There have also been substantial changes in the financial support for research in many countries of the region. Donor funds continue to account for a major share of total funding for public agricultural research, but with independence there was a shift from institution-based support through various colonial administrations to largely project-based, bilateral support mechanisms (section 9.3). As a consequence, the policy forces that shape staffing decisions have, to an apparently increasing degree, been decoupled from at least some of the forces that determine funding levels.

7.1.3 Size of NARSs

When measured in terms of full-time equivalent researchers, the average size of public-sector NARSs has more than doubled over the two decades since 1961-65 from approximately 400 to 880 full-time equivalent researchers. The average size of less-developed country systems had increased from 150 to 600, while more-developed country systems grew, on average, from 1840 to 2560 researchers.[3] Average research expenditures across all systems, expressed in constant 1980 PPP dollars, increased from around $22 million per system to $56 million. There were 74 NARSs in 1961-65 with fewer than 25 researchers but by 1981-85 only 39 were that small (figure 7.3). All of the micro NARSs (i.e., those with fewer than 25 researchers) in this 152-country sample are located in less-developed regions, in particular the Caribbean (12), Pacific (9), and sub-Saharan Africa (10). Correspondingly, the number of larger NARSs employing more than 1000 researchers increased from nine to 26, of which six now employ more than 4000 researchers.[4]

Figure 7.3: *Size distribution of NARSs*

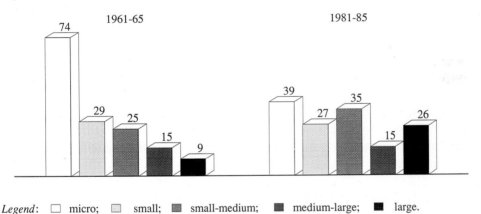

Legend: □ micro; ▨ small; ▥ small-medium; ▦ medium-large; ■ large.

Note: Size ranges are defined as follows: Micro: < 25 researchers; Small: 25-99 researchers; Small-medium: 100-399 researchers; Medium-large: 400-999 researchers; Large: > 1000 researchers.

In spite of the increasing number of medium- to large-sized NARSs, there remain a substantial number of small NARSs with little capacity to undertake anything but highly focused adaptive research on a few commodities and/or to maintain search and screening capabilities on a slightly broader front in an endeavor to capture potential research spillovers. Cross-country research spillovers arise through various channels, ranging from technology transfers by private seed, machine, and chemical companies to formal and informal

---

[3] Excluding China from these totals reduces average system size in 1981-85 from 880 to 790 researchers and the average size of research systems in less-developed countries from 600 to 350 researchers.

[4] Related figures for less-developed NARSs only are given in section 8.4.

networking structures among public-sector NARSs. Success in capturing these potential spillovers in a timely manner continues to elude many of these small systems. Evidence of this fact was provided in a recent study of agricultural knowledge transfers among Argentina, Bolivia, Brazil, Chile, Paraguay, and Uruguay (Thorpe and Pardey 1990). As might be expected, the smaller NARSs within the region tapped regional and international sources of scientific knowledge at a somewhat higher rate than the larger systems, but it was especially revealing to observe that the currency of their knowledge sources declined dramatically over time. By the early to mid-1980s, the average "age" of their cited material had increased by 126% when compared with the average citation lags of a decade earlier.

This contemporaneity indicator measures an important quality dimension of local agricultural research programs. In a dynamic environment, the comparative advantage of domestic agriculture vis-à-vis international competitors is, in part, a function of the speed with which agricultural research systems can adapt new knowledge into technological packages appropriate for their particular circumstances. If local agricultural research systems in smaller countries draw on an ever-aging knowledge base, it will lead to further erosion of their indigenous technological capacity relative to international competitors. This, in turn, will raise domestic production costs relative to competitors, thereby choking off export markets and increasing still further the social opportunity costs of any programs aimed at import substitution or domestic food self-sufficiency.

7.1.4 Agricultural Research Productivity and Intensity Ratios

An analysis of research personnel numbers and expenditure levels tells only so much. Juxtaposing these research-input indicators against various measures of agricultural outputs and conventional inputs brings these data closer to the issues of agricultural growth and development broached in chapter 6.

Agricultural Research Intensity Ratios

Output intensity ratios, which express agricultural research expenditures as a percentage of agricultural output, are subject to a variety of interpretations. From a demand-side perspective, they can, with appropriate caveats, be used in conjunction with other indicators to gain insights into the political economy forces that shape support for public agricultural research (chapter 1).

Figure 7.4 maps agricultural research intensity ratios averaged over the 1981-85 period.[5] It is clear that in general a large number, in fact 49, of the less-developed countries continue to invest in agricultural research at levels that fall below 0.5% of their AgGDP,

---

[5] For the sake of clarity a considerable number of small island states were excluded from figure 7.4. In general, these small island states have intensity ratios that are considerably higher than the less-developed country average (see section 8.4).

Figure 7.4: *Agricultural research intensity ratios, 1981-85 average*

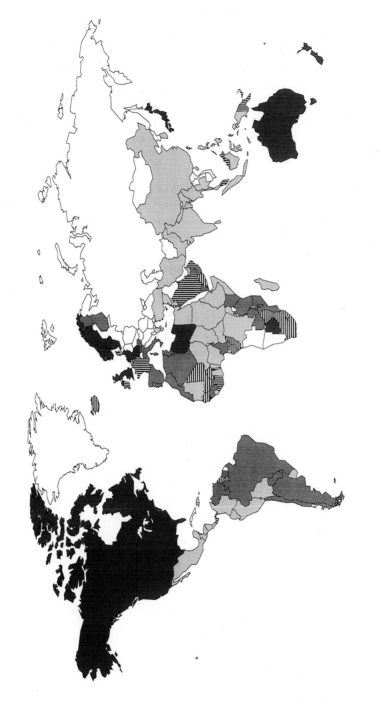

Legend: ▦ 0.5% <;   ▨ 0.5 - 1.0%;   ▥ 1 - 1.5%;   ▥ 1.5 - 2%;   ■ > 2%;   □ na

although there are noticable regional disparities. Asian countries typically have intensity ratios that are low by international standards. Intensity ratios throughout Central America and much of the Andean region (other than Chile) are uniformily lower than for the rest of Latin America, which still spends only at around the 0.8% level. In marked contrast with the other less-developed regions of the world, there is a great deal of heterogeneity in agricultural research intensity levels across sub-Saharan Africa. The SADCC countries,[6] and several of the former French colonies in Western and Central Africa, have intensity ratios that are above the regional average of 0.5%. There is also a marked dichotomy of spending levels within West Asia & North Africa. Some of the rich oil-exporting countries spend in excess of 2% of their AgGDP on agricultural research, while the most important agricultural economies of the region, such as, Egypt, Turkey, and Iran, spend less than 0.5%. Intensity ratios among the more-developed countries are generally higher than 2% except for the Southern European countries whose intensity ratios are less than half those of their northern neighbours.

Between 1961-65 and 1981-85 agricultural research intensity ratios increased on average by 3.5% per annum across the less-developed world and 4.6% across more-developed countries. Contrary to this general longer-run pattern, intensity ratios for Central Africa and China were actually lower in 1981-85 than they were two decades earlier (table 1.6, chapter 1). Equally worrisome is that, over the more recent 1976-80 to 1981-85 period, relative spending levels declined in 37 of the 92 less-developed countries for which data were available; with 10 of the countries experiencing a decline located in Latin America & Caribbean and 18 in sub-Saharan Africa.

Agricultural Research Productivity and Factor-Intensity Ratios

On the supply side, the *inverse* of an output-intensity ratio takes on some of the characteristics of a partial-productivity measure. As discussed in chapter 5, measures of aggregate agricultural product indexed over a single input, in this case research, vary over time and across countries or regions in response to a variety of factors. These include changes in the level and composition of the output basket, changes in the quality and level of utilization of the input itself, and changes in the level of utilization of other inputs that are either substitutes or complements in production. These measurement and interpretation issues aside, the notion that ratios of agricultural output to research inputs can be interpreted as measures of partial productivity, in a manner analogous to that commonly used with regard to output per unit of agricultural land and labor, should not be taken too far.

Agricultural research is appropriately seen as an investment activity. Research usually leads to an eventual increase in the stock of knowledge or an improvement in technology,

---

[6] The Southern African Development Coordination Conference (SADCC) member countries are Angola, Botswana, Lesotho, Malawi, Mozambique, Swaziland, Tanzania, Zambia, and Zimbabwe. Angola has not been included in the figure because of a lack of data.

which in turn generates a stream of benefits that continues until the new technology or knowledge is superseded or becomes obsolete. But, for research to realize its growth-promoting impacts takes some time. There are lags in the research process itself (Pardey 1989) and further lags in the uptake of new technologies and new ideas (Lindner 1981; Tsur, Sternberg, and Hochman 1990). As a result, the productivity effects of research can persist for up to 30 years (Pardey and Craig 1989). Thus, interpreting agricultural output relative to *contemporaneous* agricultural research expenditures as a partial research productivity measure raises some thorny issues. And these are confounded by these lag effects to the extent that current research expenditures are not proportional to the service flows derived from the existing stock of useful knowledge.

Bearing such caveats in mind, the inverse output-intensity ratios presented in table 7.4 measure the average level of output, specifically AgGDP, per unit of contemporaneous research expenditure and per full-time equivalent researcher. These "average product" ratios declined over the 1961-85 period for both the more- and less-developed countries. Ratios of AgGDP per dollar of research expenditure dropped by roughly one-half during this period, although the rate of decline across the less-developed countries was somewhat lower than in the more-developed. This is because it was buoyed up by a sustained rather than

Table 7.4: *Agricultural Research Partial Productivity Ratios*

Region	1961-65	1966-70	1971-75	1976-80	1981-85
	AgGDP per dollar of research expenditures				
	(1980 PPP dollars)				
Sub-Saharan Africa (37)[a]	390	257	238	198	202
China	243	325	254	213	254
Asia & Pacific, excl. China (15)	701	550	452	350	314
Latin America & Caribbean (26)	328	228	215	179	173
West Asia & North Africa (13)	351	205	199	209	192
Less-Developed Countries (92)	422	341	293	244	244
More-Developed Countries (18)	104	78	71	62	49
Total (110)	205	156	144	130	117
	AgGDP per researcher				
	(millions 1980 PPP dollars)				
Sub-Saharan Africa (37)[a]	43.5	30.6	26.6	19.8	15.2
China	9.5	9.7	10.6	7.3	7.3
Asia & Pacific, excl. China (15)	33.1	27.5	23.7	17.5	16.3
Latin America & Caribbean (26)	28.6	19.7	17.8	17.4	13.6
West Asia & North Africa (13)	20.3	14.8	12.7	11.6	9.4
Less-Developed Countries (92)	23.1	18.7	17.3	13.2	11.3
More-Developed Countries (18)	5.6	5.4	5.5	5.1	4.3
Total (110)	11.3	10.6	10.6	9.3	8.4

[a] Bracketed figures indicate the number of countries in regional totals.

declining ratio for China.

Just as higher ratios of output per unit of labor or land are considered (if carefully measured) to be indicative of more "productive" labor or land inputs, a higher ratio of output per research dollar may also be interpreted in this way. If research were subject to diminishing returns, however, increasing the research intensity of a production regime would lead to a decline in the average product of research inputs. The factor ratios presented in table 7.5, measuring research expenditures and personnel per unit of agricultural land and labor, show that all of the more- and less-developed regions have indeed been steadily increasing their research intensities since 1961-65. In 1981-85, the less-developed countries spent nearly $4 on agricultural research per agricultural worker while the more-developed countries spent over $210. These factor intensities represent a 2.5- and 4.4-fold increase for the less- and more-developed countries, respectively, over the corresponding ratios that prevailed in the 1961-65 period.

Among the less-developed countries, Asia & Pacific records the highest output per unit of research expenditure and personnel for the 1981-85 period. By contrast, Latin America & Caribbean and China show the lowest output per research dollar and per researcher, respectively (table 7.4). On the presumption that diminishing returns to agricultural research eventually prevails, these partial-productivity patterns would be expected if the factor-intensity ratios for agricultural research were comparatively low for Asia & Pacific and high for Latin America & Caribbean. At first glance, the factor-intensity ratios in table 7.5 are consistent with this premise in the case of intensity ratios per unit of labor but are at odds with it if related to land. As is to be expected, the level of research spending per unit of agricultural labor for Asia & Pacific is among the lowest in the world. The region's research expenditures per unit of agricultural land are, however, by far the highest of those for the less-developed regions.

As noted, a problem with taking these factor-intensity ratios at face value is that they fail to account for significant interregional differences in input quality. In particular, if land aggregates were formed in terms of quality-adjusted or "effective" land units, these interregional relativities between research expenditure and land input would be markedly changed. In the Asia & Pacific region, 28% of arable land and permanently cropped land is under irrigation, compared with 19% in West Asia & North Africa, 8% in Latin America & Caribbean, and 3% in sub-Saharan Africa. Also, Asia & Pacific has a markedly higher proportion of agricultural land that is either arable or under permanent crops — 81%, compared to a less-developed country average of 31% during 1981-85 (see chapter 5, table A5.1). Factoring in these differences would substantially lower the Asia & Pacific ratio of research expenditures per unit of quality-adjusted land vis-à-vis the other less-developed regions of the world.[7]

A variety of additional factors further cloud the relationship between these factor-in-

---

[7] This ratio would be lowered even further for Asia & Pacific relative to other regions if multiple cropping rates were also factored in.

Table 7.5: *Agricultural Research Factor Intensity Ratios*

Region	Agricultural research expenditures[a]					Agricultural researchers				
	Per economically active person in agriculture					Per million economically active persons in agriculture				
	1961-65	1966-70	1971-75	1976-80	1981-85	1961-65	1966-70	1971-75	1976-80	1981-85
Sub-Saharan Africa (37)[b]	1.71	2.37	2.71	3.27	3.12	15	20	24	33	42
China	0.92	0.93	1.39	1.80	2.24	24	31	33	52	77
Asia & Pacific, excl. China (15)	1.20	1.70	2.22	2.93	3.42	25	34	42	59	66
Latin America & Caribbean (26)	6.47	9.66	12.79	17.50	17.74	74	112	154	180	266
West Asia & North Africa (13)	4.48	8.99	10.51	11.60	14.31	77	125	165	201	292
Less-Developed Countries (92)	1.52	2.09	2.69	3.37	3.78	28	38	45	63	82
More-Developed Countries (18)	48.64	82.44	119.07	156.08	213.50	902	1191	1537	1905	2458
Total (110)	4.35	5.85	6.99	7.88	8.65	80	92	101	117	137
	Per hectare of agricultural land					Per million hectares of agricultural land				
	1961-65	1966-70	1971-75	1976-80	1981-85	1961-65	1966-70	1971-75	1976-80	1981-85
Sub-Saharan Africa (37)	0.22	0.33	0.41	0.54	0.56	2	3	4	5	7
China	0.64	0.70	1.16	1.65	2.23	16	24	28	48	77
Asia & Pacific, excl. China (15)	1.04	1.54	2.08	2.88	3.59	22	31	40	58	69
Latin America & Caribbean (26)	0.35	0.52	0.69	0.94	0.95	4	6	8	10	12
West Asia & North Africa (13)	0.40	0.81	0.96	1.05	1.41	7	11	15	19	29
Less-Developed Countries (92)	0.46	0.67	0.91	1.22	1.46	8	12	15	23	32
More-Developed Countries (18)	1.84	2.55	3.11	3.52	4.02	34	37	40	43	47
Total (110)	0.93	1.30	1.64	1.97	2.30	17	20	24	29	36

[a] Constant 1980 PPP dollars.
[b] Bracketed figures indicate the number of countries in regional totals.

tensity and partial-productivity measures. It is not only the economic quality of land but also agroecologies that vary markedly between and within regions. All else being equal, in a region with higher-quality land, labor, and other inputs, plus benign and relatively homogeneous agroecological zones, one would expect research expenditures to have a greater impact on aggregate agricultural productivity and output than if these same resources were directed toward a region less fortunate in these aspects.

Many less-developed countries also have heterogeneous socioeconomic environments, with a wide range of agricultural commodities grown, stored, and marketed under highly variable, often localized, conditions, ranging from subsistence agriculture, through commercial single-operator holdings, to large corporate estates and plantations. A situation of this sort also serves to lower the overall impact on productivity likely to be achieved from a given level of investment in national research.

Finally, while much agricultural research is designed to enhance levels of agricultural output, a considerable amount of maintenance research is also required simply to retain gains from previous research. Past output gains are subject to erosion as plant and animal pests and pathogens adapt and overcome research-conferred resistance. In a similar fashion, agricultural policies designed for a certain set of technological and price relationships may no longer be efficient under new institutional and economic relationships (Swallow et al. 1985). Recent evidence (Adusei and Norton 1990) suggests that the US devotes around one-third of its total agricultural production research to maintenance work.

If the research effort required to maintain productivity levels is, among other things, a positive function of the research intensity of a production regime, then the comparatively low research-productivity ratio observed for the more-developed countries (table 7.4) belies the substantial impact of maintenance research, compared to that of output-enhancing research. This is not to argue that the lower research intensities observed for many less-developed countries (table 7.5) necessarily imply that a *correspondingly* lower share of their research is directed toward maintenance. For many of these countries, particularly those situated in the tropics, pest and pathogen adaptation occurs relatively rapidly. This tends to shorten the life of previous research-induced gains, especially when combined with lower levels of on-farm management skills with regard to the identification, monitoring, and timely control of pest and pathogen problems.

To adjust the average product and research intensity ratios in tables 7.4 and 7.5 so that agricultural output and conventional input are paired against appropriately constructed research input variables requires more data than we currently have to hand. The idea would be to use information on the country-specific nature and shape of the process linking a stream of research inputs to output changes, in order to form a weighted aggregate of past and present research expenditures that yields a stock of knowledge or, more appropriately, a measure of the flow of research services.

A crude first approximation has been attempted in table 7.6. In the absence of an alternative set of plausible weights, a simple sum of agricultural research expenditures (and personnel) over a two-decade period (1961-80) has been formed for each region in data

Table 7.6:   Research Spending and Productivity Growth in Agriculture

	Cumulated (1961-80) research inputs		"Corrected" partial research productivity ratio[a]	Change in partial productivity			
	Researchers	Expenditures		Research[b]	Land[c]	Labor[d]	Fertilizer[e]
	(thousands fte years)	(billions 1980 PPP dollars)		($/$)	($/ha)	($/labor)	($/kg)
Sub-Saharan Africa (42)[f]	45.5	5.1	14.0	-3.97	-2.15	-66.26	-27.43
China	242.4	8.7	25.5	1.10	143.98	110.36	-2.08
Asia & Pacific, excl. China (21)	233.8	11.7	29.8	-4.69	159.71	85.38	-7.20
Latin America & Caribbean (33)	97.6	8.7	14.6	-2.48	19.01	342.35	2.39
West Asia & North Africa (18)	80.4	5.0	15.8	-1.27	41.33	324.15	-3.40
Less-Developed Countries (115)	699.6	39.2	21.6	-2.28	53.25	86.11	-3.84
More-Developed Countries (22)	920.8	65.7	4.0	-0.25	22.05	2853.64	0.69
Total (137)	1620.4	104.9	10.6	-0.49	43.35	98.63	0.20

[a] Calculated as the average value of 1981-80 AgGDP relative to cumulated (1961-80) agricultural research expenditures.

[b] Calculated as the difference between the average value of 1981-85 AgGDP relative to cummulative (1961-80) research expenditures and the average value of 1976-80 AgGDP relative to cumulative (1956-75) research expenditures. Research expenditures for 1956-75 were proxied by using the average of expenditures in the 1961-65 period.

[c] Calculated as the average value of 1981-85 AgGDP relative to 1981-85 agricultural land minus the same ratio for the 1976-80 period.

[d] Calculated as the average value of 1981-85 AgGDP relative to 1981-85 agricultural labor minus the same ratio for the 1976-80 period.

[e] Calculated as the average of 1981-85 AgGDP relative to 1981-85 total fertilizer consumption minus the same ratio for the 1976-80 period. The total sample size for this indicator is 130 countries rather than 137 as used elsewhere in this table. The seven countries excluded for lack of data represent less than 1% of total sample AgGDP.

[f] Bracketed figures indicate the number of countries in regional totals.

columns 1 and 2. Taking this aggregate to be proportional to the flow of services derived from the stock of useful knowledge arising from agricultural research, we present a ratio of agricultural output (AgGDP) over cumulated research expenditures in column 3. These "corrected," partial, research-productivity ratios yield a pattern of relativities across regions similar to those presented in table 7.4, with less-developed country ratios over five times higher than those for more-developed. Note, however, that the corrected China and Asia & Pacific ratios are approximately double those of the other less-developed regions — a somewhat larger degree of disparity than was observed when using the contemporaneous 1981-85 research expenditures in table 7.4.

Table 7.6 also presents estimates of the change in the average product of research over the 1976-80 to 1981-85 period as well as the change in output per unit of land, labor, and fertilizer use. Once again sub-Saharan Africa stands alone, with declines across the board in average product ratios for all four inputs and with labor productivity declines being much more pronounced than declines in land productivity. This pattern is in stark contrast with that observed for China. Output per unit of land, labor, and (contrary to the pattern for all other regions of the world) even research services increased, although diminishing returns may well have set in for fertilizer usage over this period — not surprising given the exceptionally high application rates currently prevailing in China (threefold higher than the less-developed country average [table A6.1, chapter 6]).

The fact that research-productivity ratios declined much more rapidly for less-compared with more-developed countries is consistent with the notion that the incremental gains to research occur at a diminishing rate and, as is evident, the corresponding partial productivity *levels* for research are fivefold higher in less- than more-developed countries. Labor-productivity ratios increased at over 100 times the rate of land-productivity ratios for the more-developed countries. The bias toward labor-enhancing versus land-enhancing gains among the less-developed countries is far less dramatic but unevenly distributed across subregions. China and Asia & Pacific showed higher gains in land than labor productivity, while for Latin America & Caribbean and West Asia & North Africa gains in both partial productivity ratios were more equal. The factor bias of these development paths is discussed in more detail in chapter 6.

7.2   AGRICULTURAL RESEARCH IN SUB-SAHARAN AFRICA[8]

It is appropriate to begin consideration of specific regions with commentary on the situation in sub-Saharan Africa. For many reasons, it is at the pinnacle of concern among agricultural research policymakers, both national and international.[9] The tradition of effective agricul-

---

[8]   In writing this section we benefited from the helpful comments of Matt Dagg.

[9]   Contemporary agricultural research policy issues confronting sub-Saharan Africa are discussed variously by Eicher and Baker (1982), Binswanger and Pingali (1988), Lipton (1988), and Lele, Kinsey, and Obeya (1989).

tural research in this region is long, largely dating from successes during colonial times. But, recent achievements have been frustratingly minimal and not sufficient to countervail the effects of rapid population growth throughout the region. This is in spite of the considerable commitment of resources, domestically and externally, as documented in section 7.2.2.

The explanations are several, and doubtless complex, and probably still intrinsically unsatisfactory. Eicher (1990), for instance, places much of the blame with expensive attempts to "go it alone" in nations that are too small to do the needful yet that often seem surprisingly similar to their near neighbors in agricultural research opportunities. Anderson (1991), in an attempt to synthesize the findings of several recent investigators, including Binswanger and Pingali (1988) and Carr (1989), argues in general that research has produced innovations that most African farmers find unprofitable, too risky, or impossible to implement within the agricultural calendar. Such difficulties have, it is argued, arisen in part from less than ideal locational choices for research activities as well as from patchy staff and budget support for both research and extension activities.

Herein lies the major challenge confronting international and national agricultural research policy, at least for the next several decades, as the demands of rapidly growing populations expand rampantly and the failures of the recent past, in organization and implementation of agricultural research, haunt the domestic public-sector investors of tomorrow.

7.2.1 Institutional History

Public agricultural research in sub-Saharan Africa dates back to the late 19th century when Europeans largely colonized the continent. Their agricultural interests in the region arose, in part, from the desire to secure adequate and low-cost supplies of raw materials. These were derived from either indigenous or introduced tropical crops, such as oil palm, cacao, coffee, tea, and cotton, for which no ready substitutes were available in the colonizing country. By the turn of the century, quite a number of botanical gardens, experiment stations, and model farms had been established, particularly by the British and to a lesser extent by the French, in order to screen and propagate tropical export crops. True and Crosby (1902) report that some 24 botanical gardens and experiment stations had been established across 18 contemporary states in sub-Saharan Africa alone. The French had just begun to network a series of botanical gardens and experiment stations through the Colonial Gardens of Vincennes (Headrick 1988), while the Royal Gardens at Kew were a focal point for research activities throughout anglophone Africa from the beginning of Africa's colonization. Indeed, Kew Gardens, which became a public institution in 1841, played a particularly active role in the storage and international transfer of plant genetic material throughout the British Empire.[10]

---

[10] To quote Joseph Chamberlain in the House of Commons on August 1890, "Thousands of letters pass every

These networks of botanical gardens and experiment stations laid the foundation for the emergence of a fledgling agricultural research infrastructure in sub-Saharan Africa, so that by 1920, at least one research station or site had been established in virtually every country in the region (McKelvey 1965; Spencer 1986).

The successful introduction of a wide range of cash crops in sub-Saharan Africa between 1850 and 1940, such as coffee, tea, oilpalm, cacao, groundnuts, rubber, sisal, cotton, and rice, is undoubtedly due to the search-and-screening work by these early stations. At the same time, the successful introduction of these crops contributed to a growing demand for a further expansion of agricultural research in the colonies.

Inter-War Years

The inter-war years saw a consolidation of colonial agricultural research systems throughout Africa. Administrative responsibility for research over this period rested largely with local colonial administrations, with only limited intervention by metropolitan governments such as those in London, Paris, Lisbon, and Brussels. Funding for agricultural research was generally secured from local sources.[11] A good bit of privately funded and executed agricultural research was also undertaken by plantation industries and marketing boards for crops such as oil palm in Zaire and coffee in Kenya.

A major institutional development, which took place in the inter-war years, was the introduction of specialized professional training for the scientific personnel sent to the colonies. Beginning in the early 1920s, the British Colonial Agricultural Service sponsored a two-year course in tropical agriculture for its new recruits — one year at Cambridge, England, and the other in Trinidad at the Imperial College of Tropical Agriculture, which was established in 1921 (Masefield 1972). During the early 1920s, an "Ecole Superieure d'Agriculture Tropicale" was also established in France (Eisemon, Davis, and Rathgeber 1985).

It was not until after World War II, in response to widespread changes in the world's political and economic situation, that metropolitan governments began to take a more active role in the conduct and administration of agricultural research in their colonies. This led to a major reorganization and expansion of the agricultural research capacity in Africa between 1945 and 1960, particularly in the British and French colonies.

---

year between the authorities at Kew and the Colonies, and they are able to place at the service of those Colonies not only the best advice and experience, but seeds and samples of economic plants capable of cultivation in the Colonies" (Masefield 1972, p. 24). See also Brockway (1979).

[11] Jeffries (1964) estimates that total expenditures by the British government on agricultural research in its African colonies between 1920 and 1940 was only £500,000. Eisemon, Davis, and Rathgeber (1985, p. 194) note that "French official support of colonial research was insubstantial before the Second World War" while Headrick (1988, p. 229) states that "Only after World War II did France begin to invest in something other than infrastructures in its colonies."

Post-Second World War Period

Greater involvement in colonial research by the central governments of both France and Great Britain had been called for during the 1930s, but it was not until after World War II that major institutional changes relating to agricultural research in Africa took place. In France, the "Association Colonies-Sciences," established in 1925 to promote colonial research, proposed the establishment of an "Office des Recherches Scientifiques Relatives à la France d'Outre Mer" during its first congress in 1931 and reiterated this proposal at its second congress in 1937 (Eisemon, Davis, and Rathgeber 1985).

In Great Britain, Lord Hailey strongly advocated that the British government provide substantial funds for research in Africa and for more centrally coordinated research activities throughout its African colonies (Jeffries 1964). Acting upon these recommendations, the British government created a Colonial Research Fund with an initial annual budget of £0.5 million. The necessary legislation was enacted in July 1940, and during the war years, several organizational structures, such as the Colonial Research Committee and the Committee for Colonial Agricultural, Animal Health and Forestry Research, were established to initiate and coordinate colonial research.

Implementation of many of the plans made at this time was deferred until after the war. In 1946, the annual budget of the Colonial Research Fund was raised to £1.0 million. The move to a more centralized approach of funding and administering colonial research had important consequences for subsequent agricultural research activities throughout the British colonies in Africa.

Immediately following the war, the British government sought to complement and extend the ongoing local research efforts within its African colonies by reorganizing and expanding its agricultural research endeavors along regional lines. In 1947, the East African Agricultural and Fisheries Research Council was created and headquartered in Nairobi with a mandate to monitor all of the agricultural research carried out by the territorial and interterritorial institutes of Kenya, Tanzania, and Uganda. In the years to follow, the East African Agricultural and Forestry Research Organization (EAAFRO) and the East African Veterinary Research Organization (EAVRO) were established near Nairobi, Kenya. Within the same East African Community, a fisheries research organization was established in Uganda, a marine fisheries station at Zanzibar, and a pesticides research institute in Tanganyika. Further regional research organizations, such as the East African Bureau of Research in Hygiene and Tropical Medicine and the East African Institute of Social Research, were also established around 1950.

The shift towards regional coordination and rationalization of agricultural research in British West Africa was not as pronounced as that experienced in British East Africa. In British West Africa, it was not so much a case of creating new research organizations but of transforming existing agencies into regional institutes. As early as 1944, the Cocoa Research Station of the Gold Coast had been transformed into the West African Cocoa Research Institute, followed in 1949 by the transformation of the rice experiment station in

Sierra Leone into the West African Rice Research Station and the Nigerian oil palm research station at Benin into the West African Institute for Oil Palm Research in 1951 (Masefield 1972).

What evolved throughout the British African colonies was a dualistic organizational structure consisting of ministry-based institutes, largely managed and financed at the local level, along with a series of regionally mandated institutes that were primarily directed and, particularly in East Africa, funded by the metropolitan government. The regional institutes in East Africa were discipline-based operations largely providing specialist research services to all three countries. Thus the soil physicists, virologists, and the like working at EAAFRO complemented the more site-specific work of the breeders, agronomists, entomologists, and so on in the local institutes, while the scientists at EAVRO tended to work in-depth on a selected number of diseases or research problems rather than on issues of a more general veterinary nature. By contrast, the regional agencies formed throughout British West Africa had commodity, not research-specialty, orientations and their work tended to have a higher site-specific component than the corresponding agencies in East Africa. In the absence of any formal regional agricultural research institute, British Central Africa was served by the "unofficial" Agricultural Research Council of Central Africa. This council was operational through the 1950s, during the existence of the Federation of Rhodesia and Nyasaland (Masefield 1972).

Following the recommendations for institutional change advocated prior to World War II, in 1943 the Vichy government created the "Office de la Recherche Scientifique Coloniale" (ORSC), in addition to some specialized agricultural and veterinary research institutes, thereby laying the institutional framework for agricultural research in the French colonies. The organizational model implemented in the years directly after the war entailed centralized metropolitan-based control (Eisemon, Davis, and Rathgeber 1985) — a marked departure from the British structure. By 1960 there were eight specialized institutes that were headquartered in France with satellite research stations in various French colonies.[12] In general they focused on more applied, commodity-oriented research. In 1953 ORSC was renamed as the "Organization de la Recherche Scientifique and Technique d'Outre-Mer" (ORSTOM). It performed some agricultural research but concentrated on more basic research in areas such as geology, oceanography, climatology, and epidemiology.

Perhaps of surprise to some, it was the Belgians, rather than the British or French, who administered the largest tropical agricultural research institute in sub-Saharan Africa prior

---

[12] These eight institutes covered the following commodities/activities: food crops (IRAT); livestock and veterinary medicine (IEMVT); fruit (IFAC); coffee and cacao (IFCC); rubber (IRCA); cotton (IRCT); oil crops (IRHO); and forestry (CTFT). In 1970 these institutes were integrated into "Groupe d'Etudes et des Recherches pour le Developpement de l'Agronomie Tropical" (GERDAT), which in 1985 was restructured and renamed "Centre de Cooperation Internationale en Recherche Agronomique pour le Developpement" (CIRAD). In addition to these eight institutes, GERDAT initiated research programs on practical acridiology and operational ecology in 1975, while in 1984 and 1985, respectively, institutes conducting research on agricultural machinery (CEEMAT) and farming systems (DSA) were added to the group (CIRAD 1987).

to 1960. Building on a long tradition of agricultural research in the Belgian Congo (presently Zaire, Burundi, and Rwanda) dating back to the early days of colonization, the Belgians established the "Institute National pour l'Etude Agronomique du Congo Belge" (INEAC) in 1933.[13] Although headquartered in Belgium, an extensive network of 36 research stations was established throughout the Congo with the central station at Yangambi. Research was undertaken on export crops such as oil palm, rubber, cotton, coffee, and cacao. Contrary to the pattern in other parts of Africa at the time, considerable attention was also given to food crops such as rice, maize, cassava, and groundnuts.

Immediately prior to Zaire's independence in 1960, INEAC employed about 200 researchers and annually spent some 37 million 1980 PPP dollars (Tollens 1987), an impressive research commitment considering that the 1961-65 totals for sub-Saharan Africa were only 1323 researchers with an annual expenditure averaging 150 million 1980 PPP dollars (table 7.1).

Post-Independence Period

Between 1957 and the mid-1960s, 34 countries in sub-Saharan Africa gained political independence. At that time the organizational and administrative agricultural research structures inherited from colonial times were deemed inappropriate for the political realities confronting these newly independent nation states.

Throughout British Africa responsibility for the local as well as the regional research institutes was transferred to the newly formed governments at independence. The regional research centers in former British West Africa were either dissolved or absorbed into national structures. In former British East Africa, regional research structures such as the East African Agricultural and Forestry Research Organization continued to function well beyond independence,[14] but not without problems. Likely asymmetries in the incidence of research benefits and costs may account in large measure for the different fates of these regional organizations. The West African organizations relied more heavily on local funding while pursuing commodity-focused research. The site-specific characteristics of their research programs made it more probable that one or other country in the region captured a disproportionate share of the benefits. By contrast, the regional institutes in East Africa were not as dependent on local funding[15] and undertook research that was generally

---

[13] For a description of research activities in the Belgian Congo prior to the establishment of INEAC see Drachoussof (1989).

[14] Data taken from Pardey and Roseboom (1989a) and Jamieson (1981) show that at the time of independence, the regional organizations within East Africa accounted for about 20% of the total number of agricultural researchers then working in Kenya, Tanzania, and Uganda.

[15] For example, approximately two-thirds of the total (i.e., agricultural plus nonagricultural) research expenditures by the meteropolitan government in British Africa during the period 1940-60 went to British East Africa (Jeffries 1964). By contrast, the regional organizations in West Africa were financed to an apparently large degree by funds derived from cesses on export crops.

less commodity and/or site oriented. A qualitative assessment of this benefit-cost calculus suggests that there was consequently more incentive to maintain these regional organizations after independence in British East rather than West Africa (chapter 1 and section 8.5.1).

Nevertheless, aspirations for national control over the regional public agencies in general, rather than over these agricultural research organizations in particular, created frictions with regard to their funding, administration, and operation. Such frictions resulted in the eventual collapse of the East Africa Community in 1978 and, in most instances, the integration of the remaining interterritorial research services into national structures. The disintegration of these pan-territorial research organizations eventually led to a decline in the number of British agricultural researchers and a substantial reduction in British financial support for agricultural research in these countries.

This pattern of post-independence institutional development in former British Africa contrasts markedly with that in former French Africa. Rather than transferring responsibility for the colonial research institutes to the newly formed national governments, as occurred throughout British Africa, the French retained administrative responsibility, operating out of France for many years following independence. The maintenance of French property rights and/or long-term rights of access to the local research infrastructure throughout Africa formed part of the cooperative agreements that France signed with all the French African territories that gained independence during the late 1950s and early 1960s. It was not until the 1970s that the local research structures affiliated with GERDAT (transformed into CIRAD in 1985) were gradually taken over by the client governments and integrated into their newly established national agricultural research organizations.

CIRAD, headquartered in France, now negotiates and implements its African research activities on a bilateral basis. In 1985, CIRAD's total budget was around $84 million with a total staff of 1718 employees, 909 of whom were scientists and technicians. Approximately 56% (513) of CIRAD's professional staff were posted overseas, and 70% (362) of these were located in sub-Saharan Africa. ORSTOM's 1985 budget was around $75 million, of which 36% was spent on Africa. Around 25% (311) of its 1264 scientists and senior technicians were out-posted in Africa, though not all worked on issues of direct relevance to agriculture. Clearly, the administration and conduct of French research activities in post-independence Africa has retained a largely France-based institutional character, in contrast to the project-based support mechanisms increasingly favored by others assisting in the region.

The extensive research infrastructure in the Belgian Congo that was in place by the late 1950s (INEAC) deteriorated rapidly following Zairian independence in 1960. Political upheavals and civil strife led to the complete withdrawal of all Belgian agricultural researchers soon after independence and ended all contributions by the Belgian government to INEAC. In the mid-1960s, a joint effort was mounted by Italy, West Germany, France, and Belgium to revive INEAC. Five of INEAC's seven major research stations were staffed by a total of 37 expatriate researchers, but after several years the program lost ground and was discontinued (Webster n.d.).

Other aspects of institutional history are taken up in chapter 9, especially concerning bilateral efforts directed towards agricultural research in Africa. From these historical notes we turn to the quantitative evidence to examine aspects of the development of sub-Saharan Africa's agricultural research capacity over the quarter century following independence.

7.2.2   Contemporary Developments in Agricultural Research

Research Personnel and Expenditures

The agricultural research capacity within sub-Saharan Africa that was in place and survived the transition to independence was thin indeed. During the 1961-65 period, no country in sub-Saharan Africa employed more than 180 researchers, while over one-half of the 43 countries in the region had fewer than 15 researchers. The number of agricultural researchers in sub-Saharan Africa subsequently grew 2.7-fold over the following two decades, which was faster than Asia & Pacific and Latin America & Caribbean, but slower than China and West Asia & North Africa.[16] However, the continental pattern of development was markedly uneven (table 7.7). By 1981-85, only six national systems, Nigeria, Côte d'Ivoire, Kenya, Mali, Sudan, and Tanzania employed more than 200 agricultural researchers each, and together they accounted for around 50% of the regional total. This contrasts with only about 10% of the region's agricultural researchers who were collectively employed by its 10 smallest systems.

The Nigerian system dominates the regional totals. Although experiencing substantial difficulties of late, it grew at an annual rate of 9.2%, from 172 researchers in 1961-65 to about 1000 in 1981-85, thereby increasing its share of the region's cadre of researchers from 13% to 20% (table 7.7). Most of Nigeria's growth in personnel occurred during the 1970s and early 1980s, with a net total of 655 researchers added to the system over the 1971-75 to 1981-85 period. Unfortunately, this growth in personnel was not matched by a corresponding growth in real expenditures. With a public sector buoyed by an expanding and profitable oil sector, real research expenditures in Nigeria grew during the 1960s at roughly double the rate of research personnel, peaked in the 1976-80 period, and declined in absolute terms by 25% over the last decade of our sample. As a consequence, real expenditures per researcher peaked in the early 1970s at $180,000 then declined precipitously to $80,000 by 1981-85 — in real terms, some 33% lower than the level that had prevailed two decades earlier (table 7.8).[17]

The regional totals presented in table 7.7 show that, as for Nigeria, Eastern and Western Africa also exhibited substantial rates of increase in research personnel over the

---

[16] Judd, Boyce, and Evenson (1986, table 1) report a 4.2-fold increase for Africa over the 1959-1980 period. The reasons for the substantially lower rate of growth reported here are discussed in section 5.4.

[17] For a more detailed account of the development of the Nigerian agricultural research system see Idachaba (1980, 1987).

Table 7.7: *Agricultural Research Personnel and Expenditures in sub-Saharan Africa*

(Sub)region	1961-65	1966-70	1971-75	1976-80	1981-85	Growth rate[a]
	Agricultural research personnel (full-time equivalents)					%
Nigeria	172	306	348	903	1,003	9.2
Western Africa[b] (17)[c]	356	487	667	1,023	1,636	7.9
Central Africa (7)	108	115	169	177	255	4.4
Southern Africa (10)	312	391	478	518	732	4.3
Eastern Africa (8)	375	543	755	906	1,316	6.5
Sub-Saharan Africa (43)	1,323	1,841	2,416	3,526	4,941	6.8
Less-Developed Countries (130)	19,753	28,829	37,004	55,143	77,737	7.1
	Agricultural research expenditures (millions 1980 PPP dollars per year)					%
Nigeria	21	37	63	104	80	7.0
Western Africa (17)	43	59	71	97	125	5.5
Central Africa (7)	14	14	17	16	18	1.4
Southern Africa (10)	34	50	53	58	65	3.3
Eastern Africa (8)	38	67	73	84	85	4.1
Sub-Saharan Africa (43)	149	227	227	359	372	4.7
Less-Developed Countries (130)	1,094	1,604	2,201	2,997	3,630	6.2

Note: Data may not add up exactly because of rounding.

[a] Compound annual average between 1961-65 and 1981-85.
[b] Excludes Nigeria.
[c] Bracketed figures indicate the number of countries in regional totals.

1961-65 to 1981-85 period. This compares with annual growth rates of 4.4% in Central Africa and 4.3% in Southern Africa, which were significantly below the corresponding regional and less-developed country averages. The continental pattern of growth in real expenditures paralleled but failed to match the growth in research personnel. A notable exception was Central Africa, which, as a region, virtually stagnated in terms of its absolute levels of research expenditures over the entire post-1960 period. This was largely in response to a sustained contraction in Zaire's spending from $6.8 million in 1961-65 to $4.0 million in 1981-85.

As shown in table 7.8, these persistent asymmetries between growth in personnel and expenditures have resulted in an accelerating decline in levels of spending per scientist since the early to mid-1970s. In fact, 65% of the countries in the region experienced a decline over the 1976-80 to 1981-85 period, with 42% experiencing reductions in excess of 30% (sections 7.1.2 and 8.3).

Transferring the effective control and execution of agricultural research in sub-Saharan Africa from foreign to local national administrations has been a lengthy process

Table 7.8: *Real Expenditures per Researcher in sub-Saharan Africa*

(Sub)region	Expenditures per researcher per year				
	1961-65	1966-70	1971-75	1976-80	1981-85
	(1980 PPP dollars)				
Nigeria	120,000	119,900	180,500	115,800	79,900
Western Africa[a] (17)[b]	120,700	121,300	107,200	94,600	76,200
Central Africa (7)	125,200	121,400	98,300	91,200	70,500
Southern Africa (10)	109,300	128,900	111,300	112,200	88,500
Eastern Africa (8)	102,100	123,700	96,500	92,200	64,500
Sub-Saharan Africa (43)	113,000	123,400	114,600	101,800	75,300
Less-Developed Countries (130)	55,400	55,600	59,500	54,400	46,700

Note: Data rounded to nearest hundred dollars.

[a] Excludes Nigeria.

[b] Bracketed figures indicate the number of countries in regional totals.

that has yet to run its course. The protracted nature of this process can be traced to several factors. While 1960 marks the year of political independence for 17 of the region's 43 countries, roughly one-third of the countries in the region did not gain independence until 1965 or later (Europa Publications 1990). As mentioned earlier, throughout former French Africa in particular, administrative control and legal ownership of the colonial research infrastructure was ceded to national governments only gradually — in many cases well after independence. The industrial-led development strategies that many of the newly independent African countries pursued during the 1960s and 1970s (chapter 1; Bates 1983) accelerated the transfer of resources out of their agricultural sectors through an implicit taxation of the sector. These strategies also served to bias public-sector investments in favor of urban constituencies, thereby further slowing the rate of the development of public-sector research systems throughout the region.

Human Capital Development

In table 7.9, the proportion of expatriate researchers working within the region over the 1981-85 period is shown to be around 30%, significantly more than the 12% average for less-developed countries as a whole (section 8.3). It is substantially lower, however, than the expatriate ratio of 90% or so witnessed during the early 1960s. Despite this decline in relative terms, the overall number of expatriates increased from around 1,200 immediately following independence in the 1961-65 period to about 1,400 by the mid-1980s.[18]

But staffing patterns have varied markedly across NARSs. In the larger systems, the

---

[18] The number of expatriates is estimated by applying the expatriate ratio for each period to the corresponding total number of researchers.

Table 7.9: *Nationality and Qualification Levels of sub-Saharan African Researchers, 1981-85 Average*

Country	Expatriates[b]	Qualification ratio[a]	
		Nationals only	Expatriates and nationals
	%	%	%
Angola	46	0	46
Benin	7	71	73
Botswana	55	36	71
Burkina Faso	43	na	na
Burundi	44	73	85
Cameroon	36	100[c]	100[c]
Cape Verde	21	45	57
Chad	29	na	na
Comoros	50	0	50
Congo	46	na	na
Côte d'Ivoire	74	na	na
Ethiopia	6	41	45
Gabon	58	30	71
Gambia	27	na	na
Ghana	6	69	71
Guinea-Bissau	13	71	75
Kenya	16	na	na
Lesotho	50	33	67
Liberia	28	57	69

Country	Expatriates[b]	Qualification ratio[a]	
		Nationals only	Expatriates and nationals
	%	%	%
Madagascar	10	34	40
Malawi	7	25	29
Mali	11	20	29
Mauritania	na	na	92
Mauritius	na	na	28
Mozambique	83	0	83
Niger	56	na	na
Rwanda	29	5	37
Senegal	30	na	na
Seychelles	39	0	38
Somalia	12	na	9
Sudan	na	na	81
Swaziland	33	17	44
Tanzania	22	na	na
Zaire	na	na	23
Zambia	51	22	62
Zimbabwe	na	na	44
Weighted average[d]	29	45	57
Simple average	33	36	56

Note: Excludes Central African Republic, Guinea, Nigeria, Sierra Leone, Uganda, Sao Tome & Principe, and Togo because of lack of data.

[a] Measures the proportion of national or total (expatriate & national) researchers holding a PhD or MSc. All expatriates were presumed to hold at least an MSc.
[b] Proportion of expatriates working with "line responsibilities" in the NARSs, not those working on short-term development projects. [c] These data represent the Cameroonian, not necessarily the chapter 5, definition of a researcher. [d] Weighted by the share of each country's researchers in the regional total.

number of expatriates has declined substantially, both in absolute and, to an even greater extent, in relative terms. By contrast, although the relative number of expatriates in the smaller and often more rapidly growing systems has declined, in absolute terms the number has increased substantially. The reliance on expatriate researchers in Africa is still strongest in the former Portuguese countries (41% on average), followed by the former French/Belgian countries (35% on average). The proportion of expatriate staff is the lowest in the former British countries (averaging 26% — Nigeria not included), which, however, is still rather high when compared with most other less-developed regions (see section 8.3.2).

Researcher counts represent a far from satisfactory yardstick of the development of scientific capacity within a NARS. More adequate measures would, at a minimum, adjust for differences in the quality of researchers. But systematically adjusting for differences in skill levels among researchers goes well beyond the available data and the current understanding of the factors that distinguish more- from less-productive researchers. Although qualification levels and levels of research experience are clearly dimensions of human capital that are relevant here, the data, even for these indicators, are still rather sketchy. The qualification index presented in table 7.9 shows that approximately 57% of all the agricultural researchers working in sub-Saharan Africa during 1981-85 held a postgraduate degree. But, only 45% of the nationals working within these systems were similarly qualified. Whether or not this represents an appropriate qualification profile for the NARSs of the region is a moot point and the subject of further comment in section 8.3.

The limited data available with regard to the experience levels of researchers within sub-Saharan African NARSs have been assembled in table 7.10. Taking the data for these seven NARSs as representative for the region as a whole during the early 1980s,[19] they suggest that, on average, 59% of the region's researchers had less than six years of research experience, while only 14% had more than 10 years. Unfortunately, analogous data for other regions of the world are virtually nonexistent, so it is difficult to judge the significance of these figures on an internationally comparative basis, except intuitively.

The rapid rate of growth in the region's research personnel (table 7.1) would, to a large degree, shape these experience profiles, and it suggests that they are probably low in comparison with more-developed regions. Certainly the large number of researchers with exceptionally low levels of research experience would substantially downgrade the region's researcher count if attempts were made to adjust for "quality" when measuring the scientific capacity of the region.

It bears emphasizing that researcher counts measure net rather than absolute changes in the number of researchers. Factoring in rates of staff turnover would result in changes in the absolute number of researchers that are even higher than the net changes reported here. Casual observation suggests that turnover rates in excess of 5% per annum were widespread throughout the region during the early 1980s (Bennell 1986). Turnover rates may be even higher for those well-qualified and skilled researchers who face high opportunity earnings,

---

[19] They collectively account for 1,991 (40%) of the region's researchers in 1981-85.

Table 7.10:   *Experience Profiles of Researchers in sub-Saharan Africa*

Country	Survey year	Years of research experience		
		0 to 5	6 to 10	11 plus
		%	%	%
Kenya	1982/4	76	15	10
Madagascar	1982/4	44	50	6
Nigeria	1982/4	46	31	23
Senegal	1982/4	69	21	10
Zaire	1982/4	52	35	14
Zimbabwe[a]	1982/4	67	11	22
Average	*1982/4*	*59*	*27*	*14*
Gambia	1990	56	26	17
Kenya[b]	1988	43	36	21
Senegal[c]	1989	20	47	33

Source: Bennell (1986) for all 1982/4 data, Quirino (1989) for Kenya 1988, Zuidema (1990) for Gambia 1990, ISRA/ISNAR (1989) for Senegal 1989.

Note: Percentages may not sum to 100 because of rounding.

[a] Data for research units within the Ministry of Agriculture.
[b] Data only for KARI's National Agricultural Research Center Mugaga.
[c] Data for Institut Senegalais pour la Recherche Agricole, ISRA.

either overseas or in local, nonresearch positions, relative to their prospective civil-service remuneration. It is relative, rather than absolute, remuneration levels that drive attrition rates. The recent difficulties experienced by many economies throughout sub-Saharan Africa mean that attrition rates need not necessarily be high, even if research salary levels and promotion prospects are perceived as dismal. Both Kenya and Senegal, which report somewhat comparable data for both the early and late 1980s, point to increasing levels of experience in their cadre of researchers, thereby reflecting either slower rates of growth in research personnel and/or lower attrition rates.

7.3   AGRICULTURAL RESEARCH IN CHINA[20]

We began our regional survey with the most problematic, and we now switch to a region that, in several respects, is the largest and thus most important for the concerns of agricultural research policy analysis. Output and productivity developments with regard to China's agricultural, and indeed nonagricultural, sectors over the coming decade will not only have profound effects on the country's own development prospects but also significant

---

[20] This section is written with the generous help of Shenggen Fan and draws heavily on material presented by Fan (1991b). For additional insights into the nature and impact of the Chinese agricultural research system see Stavis (1978), Lu (1985), Fan (1991a), and Lin (forthcoming).

global ramifications as well. As Kym Anderson (1990) reminds us, this is because even relatively small changes in China's degree of agricultural self-sufficiency will have substantial effects on international markets, given the country's actual and potential importance in world agriculture. China currently accounts for about one-sixth of the world's production and consumption of grain and livestock products, one-quarter of the global cotton market, and one-eighth of the world's usage of wool.

Contemporary agricultural production growth in China has been rapid despite several periods of relative stagnation. The efforts of the national agricultural research system have contributed to this growth. It has pioneered several important biological innovations, particularly with regard to rice. China released the first semi-dwarf improved rice variety in 1959, some seven years before IRRI released its similar IR8 variety, and in the 1970s China was the first nation to develop and widely adopt hybrid rice varieties. China has also developed techniques for rapidly stabilizing varietal characteristics in new plant material, technologies that have been widely adopted in other countries. Advanced work has also been done on wheat with regard to dwarfing, cold tolerance, rust resistance, and early maturity, while hybrids are sown on over 70% of the area planted to maize (World Bank 1981b).

After more than 30 years of development, particularly in recent years, the agricultural research system of China has emerged as one of the largest and more complex public research systems in the world. However, the system is still characterized by large numbers of relatively small units and poorly trained personnel when compared with the more-developed countries. The current research system could well benefit from further reform and strengthening to promote continued growth in agricultural output.

7.3.1 Characteristics of Chinese Agriculture

A large population, scarcity of land, and an unequal regional distribution and varying quality of natural resources have helped to shape the pattern of development of Chinese agriculture. Land-saving and labor-using technologies are commonly used throughout much of the country, yet farming systems vary markedly among regions, even among those working with similar levels of technology.

In terms of land area, China is the third largest country with 9.6 million km^2. Nevertheless, its agricultural, and especially arable land, is limited. About 66% of its land consists of mountains and only 11% of the total area is arable. The rough topography in mountainous areas means that soil erosion is a constant threat to the environment. It also makes it difficult to install irrigation systems and in many areas makes the use of agricultural machines virtually impossible.

Overall, rainfall is abundant but distributed unequally among regions. The southern and eastern parts of China receive monsoonal summer rains with around 32% of the country classified as wet (on the basis of standard evapotranspiration indices). Semi-dry areas comprise 19% of the total land area, where crop yields are low and unstable in the absence of irrigation. Dry regions make up over 31% of the total land area and are classified as semi-desert, where cultivated agriculture is possible only with irrigation.

Soil types also vary greatly throughout China. The dryer northern and western parts of the country are dominated by two soil types: the very thin, poor mountain soils found on the Tibetan plateau and the prairie, steppe, and desert soils in the regions of Xinjiang, Nei Monggol, the Loess plateau, and parts of Northeastern China. Eastern China is humid; its soils are generally podzols and are distinguished by the moisture and temperature regimes in which they developed. These podzols support the more productive agricultural pursuits of the country.

The varying natural conditions dictate the farming systems that are employed throughout the country. Beginning in the north and moving southward, cultivation is distinguished by four systems: (a) single cropping, spring-grown temperate cereals in the northeast, (b) a winter wheat/summer cropping cycle (three crops in two years) in the North China Plain, (c) double cropping with a summer rice crop in the Changjiang Basin of Central China, and (d) double (occasionally triple) cropping in the tropical southern coastal area (World Bank 1981b).

7.3.2   Patterns of Agricultural Production and Productivity Growth

Over the 1965 to 1985 period, total agricultural production grew at an annual rate of 5.0% (table 7.11). Until 1980, however, the rate averaged only 4.0% but thereafter increased to 7.5% (Fan 1990). This growth rate is not only the most rapid for all socialist countries but is also more rapid than those for most other countries in the world (Hayami and Ruttan 1985; Wong and Ruttan 1988).

China's labor productivity level is low compared with that of other less-developed or socialist countries, in part because of its rather low land-to-labor ratio.[21] Nevertheless, despite a decrease in land area per unit labor over the 1965 to 1986 period, the growth rate of China's labor productivity was quite high. It averaged about 3.9% per annum — 2.7% prior to 1980 and 7.1% thereafter. Comparing labor and land productivity growth indices (see also figure 6.4, chapter 6), it is observed that land productivity grew more rapidly than labor productivity, suggesting that land-saving and labor-using technologies were adopted in the main.

7.3.3   Evolution of the Chinese Agricultural Research System

The origins of the present Chinese agricultural research system can be traced to the beginning of this century. Despite China's reluctance to adopt Western technological and institutional innovations at that time, the first agricultural experiment station was estab-

---

[21] Substantial differences across countries in data coverage, data quality, and variable definition can confound international comparisons of productivity. See chapter 5 for more details.

Table 7.11: *Indices of Agricultural Production and Land and Labor Productivity for China, 1965 = 100*

Year	Total production[a]	Labor productivity[b]	Land productivity[c]
1965	100	100	100
1970	118	98	116
1975	153	116	144
1976	147	112	142
1977	148	113	142
1978	160	139	155
1979	173	153	167
1980	180	150	176
1981	192	159	189
1982	213	170	211
1983	230	181	229
1984	257	201	255
1985	267	218	267
1986	276	225	275

Source: Fan (1990).

[a] Agricultural production is based on total value of agricultural production measured in constant 1980 prices. Rural industry is excluded from agricultural production.

[b] Labor productivity is measured as total value of agricultural production (in constant 1980 prices) divided by total labor input measured in stock terms.

[c] Land productivity is measured in terms of total value of agricultural production (in constant 1980 prices) per unit of land. Land is measured here in terms of sown area. Grassland equals one sown-area equivalent.

lished in 1902 in Hebei province, and a second (as a central government agency) in Beijing in 1906 (CAAS 1987). Soon after the 1911 Revolution,[22] which paved the way for modernizing Chinese society, an additional national institute and some local experiment stations were established. Still more agricultural experiment stations were developed during the 1920s in the northern and northeastern parts of the country, followed by several agricultural improvement research institutes in Jiangsu and Sichuan provinces (Lu 1985). In 1932 a Central Experiment Institute was founded near Nanjing, then the capital of China.

The first agricultural research initiatives within the universities date back to 1914 when Jinling University began experiments related to the selection of wheat varieties. Nantong, Dongnan, and Guangdong Universities, among others, were also engaged in research on cotton, pest control, and the selection of improved seeds before 1949.

These significant but rather limited agricultural research initiatives during the first half of this century formed the genesis of a national agricultural research effort, but political instability and war impeded the development of a truly functioning research *system*.

---

[22] The Qing dynasty was overthrown by the so called "bourgeois democratic revolution" led by Dr. Sun Yat-Sen.

The "First" Period (1949 to 1966)

Immediately following the establishment of the People's Republic of China in 1949, the new government began to devote significant attention to the development of the nation's agricultural research capacity. The Chinese Academy of Sciences (CAS), which largely undertakes basic research in the general and agricultural sciences, was established in 1949. In 1952, the existing agricultural research institutions were reorganized into seven institutes — one each in the north, northeast, east, central, south, southwest, and northwest — and two specialized research institutes (veterinary and sericulture). By 1952, the number of technical personnel engaged in agricultural research throughout the country was around 1000, with multidisciplinary agricultural experimental stations located in 18 provinces, municipalities, and autonomous regions (Lu 1985).

It was not until 1957, when the Chinese Academy of Agricultural Sciences (CAAS) was established, along with academies of agricultural sciences in each province, municipality, and autonomous region, that agricultural research in China could be said to have taken on the characteristics of a functioning agricultural research system. CAAS served not only as a national agricultural research institution with multiple disciplines, but also as an academic center for agricultural research in China.

The period during the Great Leap Forward (1958) and the Anti-Rightist Campaign (1959) was an unsettling time within China. There was a tendency for cadres at various levels to strive for overambitious research objectives and to inject a degree of arbitrariness in the work schedules set for agricultural researchers (SSTC 1986). This resulted in a major dislocation in the performance of institutionalized agricultural research in China, which, in combination with a series of natural calamities (widespread drought in some areas as well as severe flooding in others), prompted readjustments in the national economic policy, including the science and technology policy. These adjustments, which were implemented in 1961, set in motion a more realistic and stable pattern of development in the NARS, at least until the Cultural Revolution in 1966.

The "Second" Period (1967 to 1978)

During this second period, which began with the Cultural Revolution, agricultural research administrators sought to pursue a practical, problem-solving approach. Most national research institutes were disbanded and many resources and personnel were transferred to rural areas. Although this new approach was successful in some respects, the system suffered considerable damage and the consequences continue to affect it even today. This so-called "open-door" research policy was put into effect at all administrative levels, including the institutes of the academies of agricultural sciences, along with thousands of field testing groups in provinces, prefectures, counties, communes, brigades, and production teams throughout the country (IRRI 1978). Concomitant with these changes, a four-level agricultural science and technology network below the provincial level was formed in the early 1970s to extend new techniques to the field. The network had corresponding agricultural research and extension organizations at each of the four levels, involving an agricultural research institute, an extension station, a modern cultivar seed station, a plant protection station, and a soil and fertilizer station at the county level; an agricultural scientific station and a veterinary station at the commune level; an agricultural scientific team at the brigade level; and a scientific research group at the production team level.

The open-door policy required that researchers move off-station and undertake a significant part of their research in the field with the direct participation of peasants and farmers (Stavis 1978). Although the policy sought to foster a problem-solving approach and enhance the links between the research and extension systems, little systematic, controlled, and replicated experimentation was conducted during this period. Basic research in the agricultural sciences was virtually ignored.

The Contemporary Period (1979 to Present)

China entered a new era in 1979. The government increasingly took the view that a key to promoting further growth in agricultural production lay in an effective research system. It was therefore deemed necessary to redress some of the developments during the second period and pay more attention to formal experimentation and basic research. After 1979, many research institutes were reconstituted and relocated back to the cities. In addition, the anti-intellectual climate that prevailed during the Cultural Revolution was largely removed and more attention was given to fostering improvements in the formal scientific capacities of the research staff.

A NARS can be classified according to its mode of organization and structure. In contrast to the integrated research, extension, and education model of the US, the autonomous or semi-autonomous publicly and privately supported research model of the UK (Beck 1987), and the agricultural research council model that is typical of some Asian countries such as India and Bangladesh (Jain 1989), the Chinese system is best thought of as a multi-ministry model. Agricultural research at the national level is conducted mainly by academies and institutes within the Ministry of Agriculture, complemented by the research efforts of various institutes under the administrative control of other ministries. The Chinese Academy of Agricultural Sciences is administered by the Ministry of Agriculture and constitutes the principal national-level research agency. Its research is conducted at over 30 national commodity, resource, and disciplinary research institutes located throughout the different agroclimatic regions. Its research program is mostly focused on issues of national significance, such as the development of hybrid rice.

The Chinese Academy of Forestry (CAF) was founded in 1958, integrated with the Chinese Academy of Agricultural Sciences in 1970, and reestablished as a separate entity in 1978 when the Ministry of Forestry was separated from the Ministry of Agriculture. The academy undertakes all basic and applied research and development relating to forestry. The Chinese Academy of Fishery Sciences (CAFS), founded in 1978, is administered by the

General Bureau of Fisheries of the Ministry of Agriculture. It currently supports 18 research units of its own, which undertake basic and applied research along with fishery technology development. Other research institutions at the national level include several research institutes of agricultural modernization under the Chinese Academy of Sciences; the Chinese Academy of Agricultural Engineering and Design, the South China Academy of Tropical Crops, the Agricultural Environment Protection Institute, and the Biogas Institute, all under the Ministry of Agriculture; the Institute of Agricultural Mechanization Sciences under the Ministry of Machine Building; and the Institute of Water Construction under the Ministry of Water Conservation and Power.[23]

The provincial agricultural research academies conduct research that is targeted more specifically to their local conditions. Upon establishment, these academies functioned as branches of CAAS, but since the Cultural Revolution they have all been placed under the jurisdiction of the provincial ministries of agriculture. They are currently linked to the national institutes through a series of collaborative programs, with leadership in areas such as rice breeding located at the national centers. Provinces are so large that provincial academies cannot readily provide new varieties or other technologies targeted for the specific agricultural conditions in each agroclimatic zone. Hence, the prefectural institutes have responsibility for adaptive research. Their research endeavors have been fruitful and many new varieties have been released at this level. County-level technical staff conduct relatively little research, other than demonstration trials, and are primarily involved in technology transfer activities.

7.3.4 Quantitative Development of Chinese Agricultural Research

During the Great Leap Forward of the late 1950s to early 1960s, the central government raised its investment levels throughout the Chinese economy to unrealistic and unsustainably high levels. Investment in agricultural research was more than doubled or tripled in a matter of several years. The ensuing policy readjustments, which were instigated in 1961, reduced agricultural research expenditures by more than 50% between 1960 and 1962. From this lower level, agricultural research capacity in China steadily increased until the Cultural Revolution in 1966, which again disrupted the development of the NARS. As depicted in figure 7.5, the agricultural research capacity, especially as measured by research expenditures, contracted sharply during the first few years of this period. It was not until 1972 that the system returned to more stable and balanced growth. Particularly since 1979, the central government has made a fairly sustained effort to strengthen the nation's agricultural research capacity. Both agricultural research expenditures and personnel have increased substantially since that time, with a more rapid increase in research personnel.[24]

---

[23] A reasonably comprehensive listing of the contemporary agricultural research institutes (including forestry and fisheries) operating in China is given in Fan (1991b).

[24] Chinese universities resumed their undergraduate and graduate training programs in 1977. As a

Figure 7.5: *Development of agricultural research expenditures and personnel in China, 1961-85*

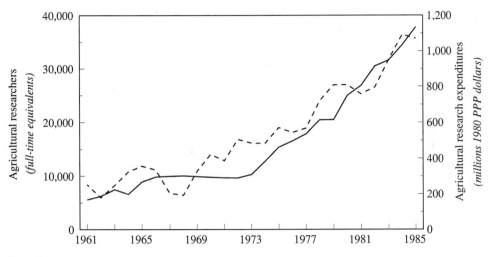

Legend: ——— number of researchers; - - - research expenditures.

Source: Constructed from data in Fan (1991b)

One of the salient characteristics of the Chinese NARS is that research resources in the agricultural universities, particularly levels of funding, are quite limited. In the West, scientific research has often been closely identified with universities. In the Chinese system, however, well-trained university professors often lack the resources to engage in research. In 1987, expenditures per researcher at the universities were only about 23% of the levels prevailing within the research institutes.[25] This underutilized human capital points, hopefully, to an important opportunity for increasing the efficiency and capacity of agricultural research in China.

China's investment in agricultural research in relation to AgGDP was at a relatively high level in the early 1960s in comparison with other less-developed countries (table 7.12). Even during the Cultural Revolution in the second half of the 1960s, China maintained a respectable official level of investment in agricultural research. Between the late 1970s and early 1980s, however, the Chinese agricultural research-intensity ratio dropped considerably and, for the first time in this period of observation, fell below the less-developed country average. This decline was not the result of a contracting or stagnating level of agricultural research expenditure; it had its principal origins in the high rate of growth in AgGDP that occurred in the early 1980s, largely in response to institutional reforms within the agricultural sector (Fan 1991a).

---

consequence, a considerable number of domestically trained graduates entered the agricultural science profession beginning in 1981.

[25] Calculated from data taken from SSTC (1988).

Table 7.12: *Quantitative Aspects of Agricultural Research in China*

	1961-65	1966-70	1971-75	1976-80	1980-85	Growth rate[a]
Expenditures	*(millions 1980 PPP dollars)*					%
Research Institutes	271.4	296.2	485.4	689.3	933.7	6.4
Universities	3.8	3.3	4.2	11.7	27.5	10.4
Total [b]	*275.2*	*299.5*	*489.6*	*701.0*	*961.2*	6.5
Research Personnel	*(full-time equivalents)*					
Research Institutes	6,966	9,900	11,563	20,048	32,224	8.0
Universities	504	425	521	1,541	4,110	11.1
Total	*7,470*	*10,325*	*12,084*	*21,589*	*36,335*	8.2
Expenditures per Researcher[c]	*(1980 PPP dollars)*					
Research Institutes	19,000	29,900	42,000	34,400	29,000	
Universities	7,600	7,800	8,100	7,600	6,700	
Total	*36,800*	*29,000*	*40,500*	*34,300*	*26,500*	
Expenditures as a percentage of AgGDP	*(percentages)*					
China	0.41	0.31	0.39	0.47	0.39	
India	0.11	0.13	0.18	0.27	0.27	
Less-developed countries	0.24	0.29	0.34	0.41	0.41	

Source: Fan (1991b).

[a] Compound annual growth rate between 1961-65 and 1981-85.

[b] Data may not add up exactly because of rounding.

7.4   AGRICULTURAL RESEARCH IN ASIA & PACIFIC[26]

Although growth and productivity gains for agriculture in Asia have been substantial over the past few decades, the nature of these gains has undergone significant change. While there have been fairly sustained increases in agricultural output per hectare and per worker, the trend, at least until recently, was for greater gains in land productivity compared with that of labor (figure 6.4). As a result, contemporary levels of land productivity for agriculture throughout the region are high, while levels of labor productivity are low, especially by international standards. Moreover, land-per-worker ratios are generally low and have declined, in quality unadjusted terms, by 17% since 1961.

The likely sources of future growth in output in the region are of a different character from what they were two decades ago. Further additions to the agricultural land base are likely to be minimal if not negative, with little new land available to agriculture and increasing pressure to divert existing agricultural land to nonagricultural uses. Over the past

---

[26] The Asia & Pacific region, as defined here, excludes China and Japan.

two decades, there has been a continuing and substantial investment in terracing, irrigation, and drainage facilities that have enhanced the quality of agricultural land. Still further land-improving investments are possible (particularly expanding or rehabilitating irrigation facilities). But these are unlikely to result in output gains that match those from previous investments, given that the less-difficult and less-expensive sites have already been developed (Byerlee 1990). Fertilizer use is widespread, with application rates at substantial but not, in general, abnormally high levels. While diminishing returns to fertilizer use are apparent, particularly in some areas such as the Indian and Pakistan Punjabs (Byerlee 1990), higher application rates — with closer attention being paid to fertilizer composition, placement, timing, etc. — may become evident and contribute to further but less dramatic gains. Ruttan (chapter 12; 1989) and others have also noted that raising the yield potentials of cereal crops appears to be increasingly difficult, while the "economically recoverable gap" between on-station versus on-farm yields is now quite small for some crops, such as maize and irrigated wheat.

In summary, it appears that the additional gains to be had from intensifying the use of conventional inputs, such as fertilizer or irrigation, are declining throughout Asia, but other sources of growth are likely to emerge. Certainly the improvement in management skills that comes with improved education levels will enable farmers to make better use of existing inputs, including more effective use of disembodied sources of technical change, such as information on the optimal timing of input use, fertilizer and seed placement, crop selection and management (particularly in the intensive cropping systems practiced throughout the region), and so on.

If the nonagricultural sectors of the region continue to expand at a relatively rapid rate and increasingly shift resources (particularly labor) off-farm, there may eventually be less political pressure to pursue self-sufficiency policies in some countries, and the policy stance may shift to one of food self-reliance, where, possibly, increased levels of imports of some commodities are seen to be consistent with a country pursuing its overall comparative advantage. Increasing per capita incomes are also likely to shift the structure of demand away from basic staples, such as rice, to horticultural and livestock products (with concomitant increases in feedgrains such as maize and soybeans).

Generalizations about complex phenomena are inevitably false in some respects and this is also the case here. The above remarks apply well to large tracts of South and Southeast Asia but not to much of the Pacific. Even within the Pacific there is great diversity, with American Samoa being effectively part of the US system and French Polynesia and New Caledonia still being served by the French system in much the same way as noted for the ex-French African nations. The special difficulties in mounting effective agricultural research in the many very small Pacific Island States are necessarily given little attention in this broad review (although see section 7.4.2) but, fortunately, are well elaborated by Hardaker and Fleming (1989).

7.4.1 Institutional History

For thousands of years, humans have selected plant and animal species for agricultural production, and have improved and adapted agricultural practices. Formal agricultural research, as we now know it, is only a recent phenomenon. It began in Europe less than two centuries ago and pushed its agricultural technology well ahead of that in most other regions of the world. If we look back a further century, however, and compare 18th century agriculture in Europe with that in Asia, there was no clear European technological superiority. On the contrary, several areas in Asia, such as China, Japan, and India, supported higher population densities than those of Europe and were using agricultural practices that were considerably more complex (Boserup 1965, 1981).

Colonial Inheritance

Botanical gardens played an important role in the transfer and screening of tropical crops in Asia throughout the 19th century.[27] Although their activities and mode of operation would perhaps not be considered "research" by present-day standards, they can well be considered the forerunners of agricultural research in the region. Building directly on these institutional precedents, formal agricultural research practices began to take root throughout the European (particularly British and Dutch) colonies, as well as those countries under Japanese influence, by the turn of the century.

The first British initiatives date from 1868 when a model farm was established at Saidapet, near Madras, in India. This model farm eventually evolved into an agricultural college and research institute and was joined by several other research institutes formed around the turn of the century, such as the Indian Veterinary Research Institute and the Indian Agricultural Research Institute, established in 1889 and 1905, respectively (Menon 1971).

Constitutional changes in 1919 moved the primary responsibility for agriculture within the Indian public sector from the central to provincial governments. In the case of agricultural research, however, a dual system was maintained with agricultural research institutes operating at both levels of government. To oversee the research activities at these different levels, the Imperial Council of Agricultural Research (subsequently the Indian Council of Agricultural Research, ICAR) was established in 1929. For more than three decades, however, ICAR had only a coordinating role. It was only after the Council's reorganization in 1964, that the central research institutes were brought under its direct jurisdiction. Until this reorganization, however, the so-called "central commodity committees," although formally under the umbrella of ICAR, played a dominant role in defining the nation's agricultural research agenda. Each of these committees was established to promote

---

[27] Botanical gardens in the Asian colonies were established as early as 1768 in India, 1796 in Malaysia, 1810 in Sri Lanka, 1817 in Indonesia, 1822 in Singapore, and 1864 in Vietnam (Headrick 1988).

research, development, extension, and marketing of specific — in all cases commercial — crops such as cotton, sugarcane, and tobacco. The first of these central commodity committees was established as early as 1921 for cotton. They financed a good deal of the agricultural research activities undertaken in India with funds derived from legislatively sanctioned cesses on agricultural products along with special grants-in-aid provided by the government. Menon (1971) has argued that this crop-specific approach led to a considerable fragmentation of the nation's agricultural research effort and diverted attention from research on noncommodity problems.

The commodity orientation of these early research initiatives carried over to the other British colonies in the region. In Sri Lanka, for example, research institutes for rubber, tea, and coconut were established in 1911, 1925, and 1929, respectively (CARIS 1978), while in Malaysia research institutes for rubber and forestry were established in 1925 and 1929 (Abdullah 1990).

In Indonesia (called the Dutch East Indies at the time), the first agricultural research initiatives got underway in 1876 when the State Botanical Garden established an Economic Garden or "Cultuurtuin" at Bogor (Mangundojo 1971). Its research program, which gained an international reputation, involved the acquisition, screening, and field testing of new plant material that was then released to private estate corporations. Eventually these estate companies invested in and undertook their own, often crop-specific, research, and were complemented by a variety of public research agencies established by the colonial administration during the early 1900s. This basic institutional structure, involving a mixture of public- and private-sector research initiatives, remained virtually unaltered for several decades to follow.

A more formal institutional approach to agricultural research throughout the French colonies in Indochina did not get underway until the 1920s and developments thereafter generally lagged behind these in other countries in the region. What technologies were used, particularly with regard to estate crops, spilled in from neighbouring Asian colonies (FAO 1986a; Headrick 1988).

In parallel with developments in sub-Saharan Africa, the majority of research institutes established in the European colonies throughout Asia focused on export crops and were commodity-based operations. The emphasis on export crops continued throughout the colonial period with research on food crops being given greater priority only after national independence in the late 1940s and early 1950s. In general there were only limited numbers of national agricultural researchers at the time of independence. Agricultural colleges, training nationals through to graduate level in the agricultural sciences, had only been established in the former British colonies.[28] By contrast, the first agricultural college in Indonesia was not established until 1941 in Bogor, only to close in 1943 with the Japanese

---

[28] By the beginning of the 20th century, agricultural colleges had been established in Kanpur, Poona, Coimbatore, Nagpur, and Lyallpur (now Faisalabad in Pakistan). By 1947, India had 17 agricultural colleges with an annual enrollment of about 1500 students (Yadav 1985). In Sri Lanka, formal agricultural education began around 1916 with the establishment of the Peradeniya Farm School (Gunasena 1985).

occupation. A substantial commitment and investment in training Indonesian research personnel did not materialize until after national independence in 1949 (Mangundojo 1971).

Not all Asian countries were colonized by the Europeans. China and Japan, for example, fended off European colonization by largely closing their borders to the outside world. With the coming to power of the Meiji government in 1868, however, Japan opened its frontiers to European technology and ideas.[29] It wholeheartedly embraced the concepts of formal agricultural research and was one of the first countries in the world to establish a public agricultural research system.[30] In Korea and Taiwan, which between 1895 and 1945 were both under Japanese occupation, public agricultural research institutes were established around the turn of the century.[31]

Agricultural research in the Philippines was initiated in 1909, some 11 years after Spain handed the country over to the US, with the establishment of the University of the Philippines College of Agriculture at Los Baños. Almost from the beginning, Cornell University played an important role and for years the college was run by an American faculty. Responsibility was taken over only gradually by the Philippines (Boyce 1980).

Public agricultural research, where it exists at all, in the Pacific island states is a fairly recent development. Population sizes are often too small to support agricultural research. Only in Fiji and Papua New Guinea does agricultural research date from before World War II.[32] Private research by plantation companies seems to have been relatively important in this region.

Post-Colonial Era

The period during and the decades immediately following World War II were unsettling for many Asian countries. Some were ravaged by the war and/or were undergoing a transition to independence from colonial rule. There were also added pressures from a burgeoning population, widespread poverty, food and nutritional deprivation, and attendant economic, social, and political upheavals. These events eventually helped stimulate a restructuring and reorientation of national agricultural research efforts away from essentially isolated, ad hoc structures toward system-based efforts that were national in scope and organizational

---

[29] Hayami and Yamada (1975), Boserup (1981), and Hayami and Ruttan (1985). See section 7.3 for an overview of historical developments concerning the Chinese agricultural research system and section 7.7 for more details on the Japanese system.

[30] In most European countries agricultural research began under private sponsorship and only later did governments assume greater responsibility in this area. This contrasts with Japan, where, from the onset of formal research activity, government took the initiative in establishing agricultural research institutes (Hayami and Yamada 1975).

[31] These were established in Taiwan in 1895 (Wan 1971) and in Korea in 1905 (Kim 1971). Institutional developments within Japan are presented briefly in section 7.7.1.

[32] The first research station in Fiji was established in 1933 (FAO 1986a) and in Papua New Guinea in 1928 (CARIS 1978).

structure (Senanayake 1990).

The post-war decades through to about 1960, however, saw no fundamental structural and organizational changes in most of the region's NARSs other than an increase in the resources committed to agricultural research. Many post-colonial governments retained the earlier organizational structures and orientation of their research agencies with a continuing predominance of export and cash-crop research. Consequently, the ad hoc character of the expansion of agricultural research in these years acted to fragment further national agricultural research efforts. As noted, demands by organized producer groups in India, for example, led to an increase in the number of commodity research institutes financed by taxes on producers. Moreover, Pray and Ruttan (1985) argued that the establishment of provincial agricultural universities based on the US land-grant university model exacerbated the degree of fragmentation of the nation's research effort.

This increasing fragmentation of the public agricultural research effort, particularly in the larger Asian countries like India and Indonesia, resulted in inefficiencies, duplication of effort, lack of coordination, and the like, while at the same time domestic food shortages continued to grow in many countries. The potential gains from economies of size and scope in these national research systems became manifest and, in conjunction with the more general social forces already noted, led to increased pressure for institutional change. Eventually this contributed to a major restructuring of almost all the Asian NARSs after 1960. A common element in all these restructurings was an attempt to gain more central coordination in order to achieve a more holistic, national, system-based approach.

In India, Pakistan, Bangladesh, the Philippines, and, more recently, in Taiwan and Sri Lanka, agricultural research councils were established and/or empowered with more central control. In Malaysia, Indonesia, and South Korea, central autonomous institutes with national mandates were established. According to Pray and Ruttan (1985), this institutional change did not occur in some of the larger Asian countries until political power had become more centralized. The strengthening of ICAR, for example, coincided with the centralization of political power in India. In Indonesia, the first steps towards the creation of AARD were taken after General Suharto gained power and put an end to a long period of political instability. And in Pakistan and the Philippines, the creation of PARC and PCARRD, respectively, took place under martial law.

The US, complemented by the efforts of Australia, New Zealand, and, increasingly, Japan, has played an important role in the institutional development of agricultural research capacity throughout much of Asia & Pacific. While not denying the critical role of national scientists and administrators in the many significant post-colonial agricultural research initiatives, the several American institutions involved must be singled out as having contributed signally to both institutional and programmatic innovations and achievements.

On the presumption that appropriate technologies to increase agricultural production in the region were readily available, initial USAID-sponsored efforts in the immediate post-war period focused on extension and rural-development programs. The limitations of these initiatives became increasingly apparent by the late 1950s, and the emphasis shifted to establishing agricultural universities throughout Asia modeled on the US land-grant university system. This often entailed US universities working in a variety of sisterhood relationships (including the provision of advanced training overseas) to build up the capacity of Asian universities. This approach also fell short of expectations, largely because the research component of many of the newly constituted universities remained weak (Pray and Ruttan 1985). Since the late 1960s and early 1970s, USAID-sponsored efforts focused increasingly on more broadly based assistance to NARSs — including both their university and nonuniversity components — providing assistance and advice on organizational matters, substantial funding for overseas training programs, and investments in research facilities. The role of these extensive development programs in human capital in particular would be difficult to overestimate in the modern history of Asian agricultural development.

Other external institutions that were important were the charitable foundations, most notably the Rockefeller and Ford Foundations, through their agricultural program staffs assigned to Asian nations. The history of these activities merges into that of the international initiatives that led to the establishment of IRRI and CIMMYT, among others (chapter 9). Suffice to note for the present that foundation personnel, working alongside those of the respective NARSs, played vital roles in getting the green revolution going as early as it did (Baum 1986; Anderson, Herdt, and Scobie 1988).

7.4.2 Contemporary Developments in Agricultural Research

Research Personnel and Expenditures

Public investment in agricultural research has increased considerably over the period 1961-85 throughout the Asia & Pacific region (table 7.13). With about 29% of all less-developed country agricultural researchers and 32% of the less-developed country agricultural research expenditures, the Asia & Pacific NARSs together represent a large part of the global agricultural research capacity.

On average research expenditures in the region have increased faster throughout the 1961-85 period than in the rest of the less-developed world (6.7% versus 6.2%), while the number of agricultural researchers has increased more slowly (6.3% versus 7.1%). The growth of agricultural research personnel as well as of expenditures, however, slowed considerably in the most recent decade.

Expansion patterns of agricultural research in the Asian subregions have differed in that the South Asian rate of increase in the number of researchers has been considerably slower over the past two decades than the rate of increase in expenditures (5.8% against 7.0%), while the reverse was the case in Southeast Asia (7.2% against 6.3%).

In contrast to the populous nations of Asia, the Pacific is an extremely small subregion, consisting mainly of small island states. Most of these states are too small to support anything but a modest research effort and tend to depend heavily on their (former)

Table 7.13: *Agricultural Research Personnel and Expenditures in Asia & Pacific*

(Sub) region	1961-65	1966-70	1971-75	1976-80	1981-85	Growth rate[a]
	Agricultural research personnel					
	(full-time equivalents)					%
South Asia (8)[b]	4,337	6,342	8,329	11,738	13,502	5.8
Southeast Asia (9)	2,205	3,013	3,932	6,645	8,824	7.2
Pacific (11)	98	125	177	177	251	4.8
Asia & Pacific, excl. China (28)	6,641	9,480	12,439	18,559	22,576	6.3
Less-Developed Countries (130)	19,753	28,829	37,004	55,143	77,737	7.1
	Agricultural research expenditures					
	(millions 1980 PPP dollars per year)					%
South Asia (8)	165	223	330	531	642	7.0
Southeast Asia (9)	144	242	307	371	487	6.3
Pacific (11)	8	10	15	26	30	7.2
Asia & Pacific, excl. China (28)	317	475	651	928	1,160	6.7
Less-Developed Countries (130)	1,094	1,604	2,201	2,997	3,630	6.2

Note: Data may not add up exactly because of rounding.

[a] Compound annual average between 1961-65 and 1981-85.

[b] Bracketed figures indicate the number of countries in regional totals.

"mother" countries or other donors for support and execution of agricultural research.[33]

Expenditures per Agricultural Researcher

Asia & Pacific as a whole has experienced a stable expenditure-per-researcher ratio throughout the 1961-85 period (table 7.14). For each of the subregions, however, the picture is quite different. South Asia began with low expenditures per researcher in the 1960s but then experienced a steady increase in this indicator. This experience contrasts with that of Southeast Asia where expenditures per researcher were higher than the less-developed country average until 1971-75. But these were then reduced as the number of researchers grew faster than expenditures. The high ratio in the Pacific region reflects in part the high proportion of expatriate researchers working in this region, as well as heavy donor involvement in general.

Compared with other regions in the world, agricultural research expenditures per researcher in South Asia are among the lowest. Translating nominal research expenditures using annual average official exchange rates rather than PPP exchange rates would act to

---

[33] The seven Pacific NARSs for which data were available received, on average, 45% of their research budget from external donors.

Table 7.14:   *Real Expenditures per Researcher in Asia & Pacific*

(Sub)region	Expenditures per researcher per year				
	1961-65	1966-70	1971-75	1976-80	1981-85
	(1980 PPP dollars)				
South Asia (8)[a]	38,100	35,200	39,600	45,200	47,600
Southeast Asia (9)	65,200	80,300	78,000	55,900	55,200
Pacific (11)	77,400	83,400	84,300	147,700	121,500
Asia & Pacific, excl. China (28)	47,700	50,100	52,400	50,000	51,400
Less-Developed Countries (130)	55,400	55,600	59,500	54,400	46,700

Note: Data rounded to the nearest hundred dollars.

[a] Bracketed figures indicate the number of countries in regional totals.

lower even further the measured expenditures per researcher in South Asia. This reflects the relatively low salaries in the region as well as low operating budgets (see section 8.2).

Size and Organizational Structure

Compared with the rest of the less-developed world, most of the Asian NARSs are quite large in terms of researchers as well as expenditures. Of the 17 Asian NARSs in our sample, nine had more than 500 researchers in 1981-85 and seven more than 1000. By contrast, the Pacific comprises 11 micro NARSs and counts for less than 260 researchers (table 7.15).

The size of the NARS and its organizational structure seem closely correlated. In all micro-, small-, and medium-sized NARSs in Asia & Pacific, research is predominantly carried out by one or more ministries — except for Bangladesh, which has organized its

Table 7.15:   *Asia & Pacific NARSs Classified According to the Number of Researchers Employed*

Micro (<25)	Small (25-99)	Medium (100-999)	Large (≥1000)
Brunei	Afghanistan	Bangladesh	India
Cook Islands	Fiji	Malaysia	Indonesia
French Polynesia	Laos	Myanmar	Korea, Republic of
Guam		Nepal	Pakistan
Hong Kong		Papua New Guinea	Philippines
New Caledonia		Sri Lanka	Taiwan
Singapore			Thailand
Solomon Islands			
Tonga			
Tuvalu			
Vanuatu			
Western Samoa			

Note: Classified on the basis of the 1981-85 average number of full-time researchers.

agricultural research according to a council model, and Sri Lanka where semi-autonomous institutes and universities play an important role in agricultural research in addition to a range of ministries. In the large Asian NARSs, however, the ministry model has been replaced by an agricultural research council model (Bangladesh, India, Pakistan, Philippines, and Taiwan) or a (semi-)autonomous institute model (Indonesia, South Korea, and Malaysia). In all cases, those systems evolved from situations where a large number of entities were conducting agricultural research and where the need for coordination was apparent (see also section 7.4.1).

7.5   AGRICULTURAL RESEARCH IN LATIN AMERICA & THE CARIBBEAN[34]

Despite the recent pattern of stagnation throughout the economies of Latin America & Caribbean (figure 6.2), in many respects this area remains the most advanced of the less-developed regions. In 1981-85, average per capita income throughout the region was slightly higher than that in West Asia & North Africa, and approximately three-, four- and fivefold higher than in Asia & Pacific, China, and sub-Saharan Africa, respectively. Agricultural output accounts for just over 10% of the region's GPD and, although the agricultural sector still employs some 31% of the economically active population, there has been only a modest increase in the agricultural labor force over the past two decades. This contrasts with Asia & Pacific, China, and sub-Saharan Africa where agriculture still accounts for 20% to 40% of GDP and employs between 50% to 80% of the economically active population, and where the agricultural labor force has increased by 37% (237 million) over the past two decades.

AgGDP per hectare has increased at a slower rate in Latin America & Caribbean than in all other less-developed regions except sub-Saharan Africa over the past 25 years. But the modest increase in the agricultural work force has been matched by a limited increase in agricultural land so that land-labor ratios within agriculture have, on average, remained quite stable.

There is nevertheless a marked degree of heterogeneity across and within the region's agricultural sectors. Regional land-labor ratios are quite varied. They range from an average of two hectares of agricultural land per worker in the Caribbean to 10 in Central America and 25 in South America and, in general, the variation around the average is quite large too. There is also a substantial degree of variation in the region's agroecology as well as a diversity of cultural practices between more traditional and modern farmers. This sectoral diversity in particular, and the episodic but somewhat turbulent social and economic history in general, continue to challenge the NARSs of the region.

---

[34] The Latin America & Caribbean region, as discussed here, excludes Cuba because of data limitations. The authors thank Carlos Valverde for his comments on this section.

7.5.1 Institutional History

Latin America

Contrary to the experience in most other regions of the world, very little public agricultural research activity was underway in Latin America by the turn of this century (Marcano 1982).[35] Most Latin American countries emerged from colonial rule well before agricultural experiment stations were established through much of Europe (but not, notably for the present discussion, in the Iberian Peninsula) during the latter half of the 19th century. As a result, the agricultural research structures and expertise that were subsequently transferred to many less-developed countries under colonial rule essentially by-passed Latin America.

According to Trigo and Piñeiro (1984, p. 76), the early agricultural research initiatives in Latin America that did get underway were often privately sponsored, strongly discipline-oriented, and "unfolded in institutional structures which were generally unstable." Often the stimulus for the establishment of these early experimental farms was the need to address a particular problem in a specific crop (Scobie 1987). A major exception, however, was coffee research in Brazil. Being the major agricultural commodity of the country, formal research on coffee was initiated as early as 1887, with the foundation of the Agricultural Research Institute of Campinas, São Paulo. The strong problem-orientation of the research secured, within a decade, substantial financial support and much prestige for the institute (Yeganiantz 1984). In most Latin American countries, however, formal agricultural research did not take hold until after 1930.[36]

Three reasons for the delayed development of agricultural research in Latin America stand out: (a) as noted, colonial rule in the region had ceased prior to the colonially sponsored agricultural research that was initiated elsewhere in the world, (b) agricultural land and labor in Latin America were relatively abundant, so the need for these resources did not place as much pressure on society to develop improved agricultural technologies as it did in Europe and other parts of the world, and (c) the ineffective government and state structures that characterized many Latin American countries at the time impeded the provision of public goods and services in general, of which agricultural research was but one.

By 1930, however, the need for agricultural research had become increasingly apparent. The rapid commercialization of agriculture, in particular the increasing importance of agricultural exports, induced the need for improved agricultural technologies that could sustain and improve the region's competitive position in international markets. The Great Depression of the 1930s exacerbated these concerns over competitive advantage. At

---

[35] True and Crosby (1902) identified only a limited number of experiment stations in Argentina and Brazil (more particularly in the states of São Paulo and Minas Gerais) that were established during the 1880s and 1890s. In Colombia, an agricultural experiment station was established as early as 1879 in the famous Botanical Garden of Bogota (Arnon 1989).

[36] In addition to Argentina, Brazil, and Colombia, agricultural experiment stations were established prior to 1930 in Uruguay (1914), Peru (1920), and Chile (1925) (Valverde 1990).

the same time, limits began to appear concerning the ongoing expansion of agricultural land, while an expanding population led to a substantial increase in the domestic demand for food. Consequently, there were many efforts by governments in the region to establish units within the central government bureaucracies that either promoted or executed agricultural research. For instance, in 1930, Chile established the Department of Genetics and Plant Improvement, while Brazil established the Directorate of Scientific Research, both within their respective ministries of agriculture. In 1932-33 Mexico reestablished the Department of Experimental Fields within the General Bureau of Agriculture.[37] This process of institutional development continued during the 1940s and 1950s with efforts to expand the commodity, disciplinary, and geographic coverage of state-sponsored and managed agricultural research throughout the region (Samper 1980).

These national efforts to broaden public agricultural research endeavors were reinforced by technical and financial support from foreign, particularly US, agencies and foundations. Under the auspices of a cooperative agreement between the Government of Mexico and the Rockefeller Foundation, an "Office of Special Studies" was created in 1944 within the Mexican Ministry of Agriculture with the objective of increasing yields of basic crops (including, among others, maize, wheat, and potatoes) and training Mexican agricultural scientists (Venezian and Gamble 1969). This initiative laid the institutional and human capital foundations for the International Center for Improvement of Maize and Wheat (CIMMYT) that was eventually constituted in 1966 (chapter 9). The Rockefeller Foundation formed similar cooperative agreements with the governments of Chile, Colombia, and Ecuador. A complementary program under the financial sponsorship of USAID, and with the technical and training assistance of several US universities, established the "Cooperative Agricultural Research Services" in Guatemala, Honduras, Panama, and Peru in an attempt to strengthen their national agricultural research capabilities (Valverde 1990). Notwithstanding the immediate research and, particularly, training accomplishments of these cooperative programs, it has been argued that the centralized ministerial governance that was characteristic of public agricultural research throughout the region at the time fostered an overly bureaucratized approach to the management and execution of agricultural research. In general, the scientific cadre remained poorly trained, subject to limited and often unstable financial support, and by and large continued past research practices that were structured along strict disciplinary lines, thereby often failing to address the contemporary production problems facing farmers (Samper 1980).

In an attempt to overcome these constraints, efforts to rehabilitate national research agencies were initiated in the late 1950s and led to the wide adoption of a "semiautonomous national agricultural research institute" model.[38] In most cases, these new institutional

---

[37] The first agricultural research station in Mexico was established in 1906 (INIA, n.d.). After only a few years, however, its activities were interrupted by the Mexican Revolution of 1910-11 and were apparently not resumed until 1932-33 (Venezian and Gamble 1969).

[38] Establishment dates for these semiautonomous institutes are as follows: INTA, Argentina in 1957; INIAP,

structures featured centralized decision making with respect to setting strategic priorities and allocating resources but sought to decentralize the execution of research through a network of experiment stations and commodity-based programs. The basic objectives, not always successfully attained, were to confront the problems created by the bureaucratic environment of the ministries and to develop mechanisms that might improve the funding situation and the conditions of service for research personnel. At the same time, they tried to keep research in the public domain and ostensibly responsive to agricultural development policies (Trigo 1986). Substantial technical and financial assistance was provided by international (particularly US) aid and development organizations to help to facilitate the establishment of these national research institutes.

Beginning in the mid-1980s, several countries in the region sought to reorganize their NARSs or parts thereof along quasi-private or foundation lines. These developments may be viewed as an attempt to maintain or, in some instances, revive the institutional changes initiated during the 1950s and 1960s. The sources of support, mode of operation, management, and governance of these new organizations, as well as the stimulus for change vary considerably. Many of the initiatives that got underway in the mid- to late-1980s (e.g., in Honduras, Jamaica, Peru, Dominican Republic, and Ecuador) were in large measure a response to donor, often USAID, incentives to provide financial backing to these foundations.[39] In another, even more recent, instance (Argentina) the stimulus for change came largely from domestic interests. In either case the stated objectives were to increase the level and diversity of financial support for agricultural research, increase the client orientation of the research program, and remove the management and operational impediments arising from compliance with public-service regulations — many of the same motivations noted earlier that prompted moves toward the national agricultural institute model.[40] It is to be hoped that this current round of initiatives will be successful in mobilizing domestic constituencies in support of agricultural research as well as increasing the efficiency by which agricultural technologies are made available to domestic markets.[41] To the extent these objectives are achieved, such changes may well provide the institutional basis for the productivity growth that has of late eluded many of the region's agricultural sectors.

Ecuador in 1959; FONAIAP, Venezuela in 1961; INIA, Mexico in 1961; ICA, Colombia in 1962; INIA, Chile in 1964; EMBRAPA, Brazil in 1973; ICTA, Guatemala in 1973; IDIAP, Panama in 1975; INIA, Peru in 1978; and INIA, Uruguay in 1989 (Valverde 1990).

[39] See Sarles (1990) for more details.

[40] Indeed EMBRAPA (Brazil), which was not formed until 1973, appears to have addressed many of these concerns in ways that more closely conform to the current round of quasi-private initiatives.

[41] In essence many of these recent changes are an attempt to increase the degree of *contestability* in domestic agricultural R&D markets. See section 8.5 for more details.

The Caribbean[42]

In marked contrast to Latin America, the early institutional development of agricultural research in most Caribbean countries was largely shaped by their colonial experiences. Only Cuba, Haiti, and the Dominican Republic had gained political independence prior to the local inception of formal agricultural research, and as a result, they have institutional histories more akin to those of Latin America. The remaining 20 or so Caribbean islands inherited colonial agricultural research structures of one form or another.

The first botanical garden in the region was established in 1764 on St. Vincent, and by the late 19th century a dense network of botanical gardens was in place throughout the British Caribbean. These gardens were not research stations in the style of today, but here as elsewhere in the British Empire, they laid the institutional foundations for the agricultural research activities to follow. Formal agricultural research in the British Caribbean colonies dates to 1898 when the Imperial Department of Agriculture for the West Indies (IDA) was established in Barbados. In 1921-22 IDA was transferred to St. Augustine, Trinidad, and formed the nucleus of the Imperial College of Tropical Agriculture (ICTA), whose staff engaged in an integrated program of research and teaching. Its major research thrusts involved breeding programs for the principal plantation crops (sugar, cacao, bananas, and citrus) as well as a strong soil science program. Over the 1921-51 period, the college was apparently the only center in the British Commonwealth offering postgraduate training in tropical agriculture.

Beginning around 1940, ICTA witnessed a functional separation of its research and teaching operations, and in 1955 a Regional Research Center (RRC) was founded that absorbed the research functions of ICTA. The college continued as a teaching institution until 1960 when it was closed and its assets transferred to the newly constituted University College of the West Indies (UCWI). UCWI opened on the St. Augustine campus in 1961 and became the University of the West Indies (UWI) with its own charter in 1962. RRC continued as an autonomous research institute except for the period between 1965 and 1975 when it was integrated into UWI's Faculty of Agriculture. Subsequently, it was institutionally, but not physically, separated from the university to form the Caribbean Agricultural Research and Development Institute (CARDI). CARDI is an autonomous institute, with a regional research and development (including extension) mandate whose budget is funded, in part, by the 12 member states of the Caribbean Community (CARICOM).[43]

These institutional changes reflect a transfer of control and, correspondingly, financial responsibility for the conduct of (publicly sponsored) regional research initiatives to the respective governments. ICTA's (1924-60) operations were entirely in the hands of the British Colonial Office and the private commodity associations and boards that provided

---

[42] This section draws heavily on Wilson (1985) and Parasram (1990).

[43] The member countries are Antigua and Barbuda, Barbados, Belize, Dominica, Grenada, Guyana, Jamaica, Montserrat, St. Kitts/Nevis, St. Lucia, St. Vincent and the Grenadines, and Trinidad and Tobago.

the institute with substantial resources. The regional governments gained increasing control with the formation of RRC and then assumed full control of the research agenda with the formation of CARDI in 1975. CARDI's governing body is the council of ministers responsible for agriculture, while its board of directors is drawn from member governments and regional agencies.

Although CARDI and its predecessors have played and, indeed, continue to play an important regional role in agricultural research throughout the Caribbean, other agencies have also contributed to this effort. Jamaica and Trinidad & Tobago, the two larger countries in the Caribbean Commonwealth, established their first research units within their ministries of agriculture in 1933 and 1945, respectively (Davis 1975). Complementing the efforts of the public sector is a substantial amount of research on (plantation-based) export commodities by quasi-public commodity boards or private producer associations. Davis (1975) reports a total of 13 agricultural research establishments of this nature, seven of which were established before 1950. The total grew to some 27 by the late 1980s (Parasram 1990).[44] The oldest is the West Indies Sugar Cane Breeding Station at Barbados, which dates back to 1888 and is still operating.

Guadeloupe, Martinique, and French Guyana, have the status of French Overseas Departments and are served by a regional research center of INRA, the Paris-based National Institute for Agronomic Research of France. This regional research center was founded in 1949, with its main station located in Guadeloupe and smaller units in French Guyana and Martinique (Anais 1985). Puerto Rico and the Virgin Islands were incorporated into the US agricultural experiment station system in the 1930s (Carrasquillo 1984), while the Netherlands Antilles, a country with only a small agricultural sector for much of this century, has seen only sporadic agricultural research initiatives.[45]

7.5.2   Contemporary Developments in Agricultural Research

Research Personnel and Expenditures

As in other regions, public agricultural research has expanded considerably over the past 25 years in Latin America & Caribbean (table 7.16). Annual average growth rates of both the number of agricultural researchers as well as agricultural research expenditures have, however, been somewhat lower than the less-developed country average.[46]

The growth of the public agricultural research capacity in terms of expenditures and

---

[44] For reasons of data compatibility, this figure excludes five Cuban-based agencies identified by Parasram (1990).

[45] By 1930 less than 10% of the Netherlands Antilles population was employed in the agricultural sector (Anon 1983).

[46] The annual average growth in the number of agricultural researchers throughout the 1961-65 to 1981-85 period was 6.3% in Latin America & Caribbean, compared with 7.1% in all less-developed countries. For expenditures, these percentages were 5.8% and 6.2%, respectively.

Table 7.16: *Agricultural Research Personnel and Expenditures in Latin America &*
Caribbean

(Sub)region	1961-65	1966-70	1971-75	1976-80	1981-85	Growth rate[a]
	Agricultural research personnel					
	(full time equivalents)					%
Caribbean (18)[b]	282	331	363	415	502	2.9
Central America (8)	370	449	763	1,262	1,723	8.0
South America (12)	2,014	3,342	4,713	5,314	6,774	6.3
Latin America & Caribbean (38)	2,666	4,122	5,840	6,991	9,000	6.3
Less-Developed Countries (130)	19,753	28,829	37,004	55,143	77,737	7.1
	Agricultural research expenditures					
	(millions 1980 PPP dollars per year)					%
Caribbean (18)	17	24	29	31	30	2.8
Central America (8)	24	26	56	106	159	9.9
South America (12)	188	305	402	542	520	5.2
Latin America & Caribbean (38)	229	355	487	679	709	5.8
Less-Developed Countries (130)	1,094	1,604	2,201	2,997	3,630	6.2

Note: Data may not add up exactly because of rounding.

[a] Compound annual average between 1961-65 and 1981-85.

[b] Bracketed figures indicate the number of countries in regional totals.

number of researchers, however, has been quite different in the three different subregions.
The Caribbean has experienced relatively slow growth in both indicators, 2.8% and 2.9%
annually, respectively, while agricultural research in Central America has grown at 9.9%
annually in terms of expenditures and 8.0% annually in terms of researchers.[47] In the past
decade especially, Central America (in particular Mexico) has made substantial progress,
while the growth of research expenditures in the Caribbean has stagnated. Although public
agricultural research in South America did not grow particularly slowly or rapidly before
the late 1970s, in the early 1980s agricultural research expenditures in real terms declined,
while the number of researchers increased sharply.

Nearly two-thirds of the region's agricultural researchers are employed by its three
largest NARSs, namely Argentina, Brazil, and Mexico (see table 7.17). The majority (12 of
21) of the Latin American NARSs falls in the medium-sized range, and only three in the
small range.

The relatively small Caribbean subregion consists of a large number of small island
states, and this has fragmented agricultural research into a large number of small research
efforts. Many of these economies are too small to support, or indeed justify, anything but a

---

[47] The Central American figures are dominated by Mexico, which represents nearly 80% of the research
capacity in the region.

Table 7.17:   *Latin America & Caribbean NARSs Classified According to the Number of Researchers Employed*

Micro (<25)	Small (25-99)	Medium (100-999)	Large (≥1000)
Antigua	Barbados	Bolivia	Argentina
Bahamas	Guyana	Chile	Brazil
Belize	Haiti	Colombia	Cuba[a]
Bermuda	Honduras	Costa Rica	Mexico
Dominica	Jamaica	Dominican Republic	
Grenada	Nicaragua	Ecuador	
Guadeloupe	Paraguay	El Salvador	
Martinique	Puerto Rico	Guatemala	
Montserrat	Trinidad & Tobago	Panama	
Netherlands Antilles[a]	Uruguay	Peru	
St. Kitts-Nevis		Venezuela	
St. Lucia			
St. Vincent			
Suriname			
Virgin Islands (US)			

Note: Classified on the basis of the 1981-85 average number of full-time researchers.

[a] Not included in the statistics presented elsewhere in this chapter because of incomplete data.

modest and highly focused program of agricultural research. They collaborate regionally, as is the case of the Commonwealth countries in CARDI, or depend heavily upon the mother country, as do the French and the Dutch Caribbean islands. Only five, or six if we include Cuba, of the 20 or so island states in the Caribbean have been able to establish a NARS large enough to be able to stand on its own.

Expenditures per Agricultural Researcher

Over the period 1961-85, expenditures per agricultural researcher in the three Latin America & Caribbean subregions have ranged from $60,000 to $100,000, (table 7.18). This level is in the higher ranges for both the less- and more-developed countries. The use of the PPP index as the exchange rate may have overvalued the Latin America & Caribbean exchange rates somewhat. The hyperinflation experienced by many Latin American countries over this period also compounded the difficulties of converting expenditure items into constant currency units. But even when the Atlas exchange rate is substituted for a PPP convertor, thereby lowering the real expenditure aggregrates for the region by roughly 40%, expenditures per researcher in Latin America & Caribbean are still in the middle range compared with other regions. And despite the difficulties in generating internationally comparable aggregates in this instance, the temporal pattern that has emerged is anything but similar to the steady and continuous increase of the more-developed countries.

The severe economic and financial crises most South American countries have

Table 7.18: *Real Expenditures per Researcher in Latin America & Caribbean*

(Sub)region	Expenditures per researcher per year				
	1961-65	1966-70	1971-75	1976-80	1981-85
	(1980 PPP dollars)				
Caribbean (18)[a]	61,300	72,100	79,100	74,300	59,500
Central America (8)	65,000	58,100	73,700	84,100	92,300
South America (12)	93,200	91,300	85,200	102,000	76,700
Latin America & Caribbean (38)	85,900	86,200	83,300	97,200	78,800
Less-Developed Countries (130)	55,400	55,600	59,500	54,400	46,700

Note: Data rounded to nearest hundred dollars.

[a] Bracketed figures indicate the number of countries in regional totals.

experienced during the last period seem to have had a profound effect on the growth of agricultural research budgets. Trigo and Piñeiro (1984) noticed that, after 1970, the upward trend of the 1950s and 1960s faltered throughout Latin America and that, in certain countries, agricultural research funding even suffered a cutback. This contractionary trend seems to have continued during the early 1980s, especially in South America. A similar trend has also been observed by Elias (1985) with regard to total government expenditures on agriculture in a study of nine Latin American countries. In short, expenditures per agricultural researcher at the aggregate level have thus been rather erratic in the Latin America & Caribbean region, especially in the South American subregion (table 7.18). Such severe fluctuations in real expenditures are indicative of a highly unstable research environment — one that can hardly have been conducive to productive scientific achievement in this region.

7.6   AGRICULTURAL RESEARCH IN WEST ASIA & NORTH AFRICA

Of the four less-developed regions defined for this study, West Asia & North Africa (WANA) is exceptional in its own way. Because of oil revenues, some of the countries in the region have per capita incomes that are among the highest in the world, while others, which have only limited or no oil revenues, fall in the range of the lower- and middle-income less-developed countries.

In no other region of the world has the share of the agricultural population in the total population declined as fast as in WANA, reflecting the rapid expansion of the region's oil industry. Many jobs were created over the past 20 years in both the oil sector and the service sector of the high-income oil-exporting countries. This effect has not been limited to these countries only. Labor-surplus countries have profited by exporting their labor to the oil-exporting countries (Khaldi 1984). However, the Gulf Crisis of 1990-91 has revealed just how fragile such employment can be.

The conditions under which agricultural production in the WANA region takes place can be characterized as rather difficult. Rainfall is the most limiting factor, and it is so constraining that only about 10% of the total area in the region is in use as agricultural land. Before the oil boom in 1973, the WANA region was already one of the major food-importing regions of the world, and since then, food self-sufficiency has declined even further.

7.6.1 Institutional History

Agricultural research in West Asia dates from rather recent times. For 12 of the 15 countries in West Asia, public agricultural research was not initiated until after World War II. Only in Israel, Iran, and Turkey do the first agricultural research initiatives date back as far as the 1920s and 1930s.[48]

Although the influence of the European colonial powers in West Asia was substantial during the second half of the 19th and the first half of the 20th centuries, most of the countries in the region were never colonized but rather were granted (or had imposed on them) protectorate status. This guaranteed only a minimum of interference by the metropolitan authorities in areas such as defense, foreign relations, and law and order. The agricultural potential of most countries in West Asia was considered of little importance to the colonial powers whose presence in the region was based more on geopolitical than economic concerns. This helps explain why the Europeans did not establish agricultural research institutes in the region as they did in nearly all their colonies in Asia and Africa.

Compared with West Asia, agricultural research in North Africa has a much longer history. Research on cotton in Egypt was initiated in 1898 by the British, who at that time occupied the country (Watson 1964). The establishment of several other agricultural research institutes followed during the first 20 years of this century while Egypt was still under British rule (Watson 1964; Hariri 1988).[49] Since then, and particularly after 1960, Egypt has built one of the largest agricultural research systems in the less-developed world.

Agricultural research in Algeria, Morocco, and Tunisia was initiated by the French during the first two decades of the 20th century (Casas 1988; Arnon 1989) mainly to support French and Italian settlers. Although Italian farmers also settled in Libya after Italy invaded in 1911, it appears that agricultural research was not initiated there until after complete independence in 1951.

7.6.2 Contemporary Developments in Agricultural Research

Research Personnel and Expenditures

In terms of agricultural researchers, the WANA total is dominated by Egypt, which had nearly one-half of the region's agricultural researchers in 1981-85. In terms of expenditures,

---

[48] For more details see CARIS (1978), Sardar (1982), Arnon (1989), and FAO (1990d).

[49] Egypt was under direct British rule between 1882-1922.

however, Egypt's share was far smaller, only some 10%, which represents an extreme situation. In light of this disparity, we present and discuss Egypt separately from the rest of North Africa.[50]

As can be seen in table 7.19, the growth in agricultural researchers over the past two decades for countries in the WANA region other than Egypt has been relatively slow compared with most other less-developed regions (5.6% annually against 7.1% for the whole less-developed country group). Egypt, however, experienced a rapid expansion of researchers over the period 1961-85 — more than 10% annually. West Asia also experienced a rapid increase in researchers between 1976-80 and 1981-85, mainly in Turkey.

Table 7.19: *Agricultural Research Personnel and Expenditures in West Asia & North Africa*

(Sub)region	1961-65	1966-70	1971-75	1976-80	1981-85	Growth rate[a]
	Agricultural research personnel (full time equivalents)					%
Egypt	569	1,431	2,070	2,748	4,246	10.6
North Africa[b] (4)[c]	301	371	444	616	770	4.8
West Asia (15)	1,287	1,683	2,232	2,655	3,980	5.8
West Asia & North Africa (20)	2,157	3,485	4,746	6,019	8,995	7.4
Less-Developed Countries (130)	19,753	28,829	37,004	55,143	77,737	7.1
	Agricultural research expenditures (millions 1980 PPP dollars per year)					%
Egypt	17	28	23	32	45	5.0
North Africa[b] (4)	30	39	58	62	81	5.1
West Asia (15)	80	183	219	248	329	7.3
West Asia & North Africa (20)	127	250	301	341	455	6.6
Less-Developed Countries (130)	1,094	1,604	2,201	2,997	3,630	6.2

Note: Data may not add up exactly because of rounding.

[a] Compound annual average between 1961-65 and 1981-85.
[b] Excluding Egypt.
[c] Bracketed figures indicate the number of countries in regional totals.

Most of the NARSs in the WANA region are of medium size (table 7.20). Only Egypt and Turkey employ more than 1000 agricultural researchers in their public sectors.

---

[50] The available statistics on investment in agricultural research (as well as more general economic indicators) for the WANA region are quite weak. Therefore, the quantitative data presented here must be viewed with even more reservation than for other regions.

Table 7.20:   *West Asia & North African NARSs Classified According to the Number of Researchers Employed*

Micro (<25)	Small (25-99)	Medium (100-999)	Large (≥1000)
Kuwait	Cyprus	Algeria	Egypt
Qatar	Jordan	Iran	Turkey
United Arab Emirates	Lebanon	Iraq	
	Oman	Israel	
	Yemen, A.R.	Libya	
	Yemen, P.D.R.	Morocco	
		Syria	
		Tunisia	

Note: Classified on the basis of the 1981-85 average number of full-time researchers.

Expenditures per Agricultural Researcher

Even considering all the possible distortions that may have been introduced by less than ideal currency convertors, deflators, and so on, expenditures per researcher in Egypt are among the lowest in the world and have steadily declined over time. This has been caused particularly by the rapid expansion in the number of researchers in Egypt. The apparent over-employment in relation to research budgets presumably must have had devastating effects on the productivity of the Egyptian NARS. By contrast, average expenditures per researcher in North Africa (excluding Egypt) and West Asia are among the highest in the world (table 7.21). This doubtless reflects the relatively high costs of research in these countries, where human resources are scarce and where, for some, ample state revenues from oil sales are available.

Table 7.21:   *Real Expenditures per Researcher in West Asia & North Africa*

(Sub)region	Expenditures per researcher per year				
	1961-65	1966-70	1971-75	1976-80	1981-85
	(1980 PPP dollars)				
Egypt	29,600	19,200	11,300	11,600	10,500
North Africa[a] (4)[b]	99,800	105,600	130,600	100,200	105,600
West Asia (15)	62,200	108,800	98,300	93,300	82,800
West Asia & North Africa (20)	58,800	71,700	63,400	56,700	50,600
Less-Developed Countries (130)	55,400	55,600	59,500	54,400	46,700

Note: Data rounded to nearest hundred dollars.

[a] Excluding Egypt.

[b] Bracketed figures indicate the number of countries in regional totals.

7.7   AGRICULTURAL RESEARCH IN THE MORE-DEVELOPED COUNTRIES

With the exception of the China "region," the less-developed regions considered above represent somewhat arbitrary aggregations of political units in some geographic proximity. As our discussion has indicated, much of the interest and commentary has been on intraregional differences rather than on outstanding unifying features. This is also the case with our final "region," that of the more-developed world.

The increasing, if not accelerating, degree of international integration of domestic factor, product, and technology markets means that the agricultural research policies of more-developed countries will have an increasing impact on agricultural research policies in less-developed countries and vice versa (chapter 2). This is not only because the comparative advantage and international competitiveness of less-developed country agriculture are inextricably linked to that in the more-developed countries, but it is also the case that progress in the agricultural sciences of the more-developed countries helps shape the research potentials of many of the less-developed countries. An additional rationale for including this particular "regional" aggregation is that not too many decades ago, all the nations involved were less-developed in the same sense as this qualifier has been applied above. Thus, a review of how things are now done in this part of the world may give guidance as to where the presently less-developed nations may be heading.

In this regard, the major focus of attention in reviewing the "more-developed" experience is the institutional history, as in some respects it may be more important to comprehend how and why these countries proceeded to their present arrangements rather than to see just where they have arrived.

7.7.1   Institutional History

The origins of contemporary agricultural research practices can be traced to the early 19th century, but on-farm experimentation, tinkering, and learning by doing have been salient characteristics of farming practices since the dawn of agriculture. Although this process of informal research continues to the present, the series of institutional innovations initiated during the early 19th century meant that, by the end of that century, virtually all the more-developed countries supported institutions whose primary task was to conduct agricultural research. These developments not only heralded a shift toward a more systematic and scientific approach to agricultural experimentation but also resulted in a division of labor between those primarily engaged in agricultural production and those primarily engaged in agricultural research.

These institutional innovations were derived from changes in both the supply of and demand for new knowledge and agricultural technologies. On the supply side, Russell (1966), for example, stressed the role played by the rapid accumulation of agricultural knowledge during the 17th and 18th century in Great Britain — by 1800 around 440 books dealing with agricultural production problems and practices had been published in Great Britain alone. Unfortunately, these books often conveyed information and advice of untested and uncertain quality (Salmon and Hanson 1964), but these reservations notwithstanding, at the beginning of the 19th century, Great Britain was generally "regarded by those interested in agricultural improvement as the school for agriculture" (Ruttan 1982, p. 71).

A further factor, which has been given some emphasis by both Ruttan (1982) and Grantham (1984), concerns the educational innovations that first took hold in Germany at the beginning of the 19th century. The traditional European universities focused on the classical disciplines of theology, philosophy, medicine, and law, rather than on scientific research. In 1809 what can be considered the first modern research university was established in Berlin and, in the decades to follow, many more such universities were created throughout Germany. The integration of teaching and research within a university turned out to be extremely fruitful and within half a century placed Germany well ahead of Great Britain in nearly all the sciences, particularly physics and chemistry (Ruttan 1982).[51] By the mid-19th century, Germany had built up a sizable stock of formally trained researchers. However, few received any training directed specifically to agriculture and most who went on to play important roles during the early days of the agricultural sciences were formally trained as chemists.[52]

The demand for establishing institutions that formalized the process of agricultural research was given additional impetus with the creation of agricultural societies, the earliest of which seems to have been set up in Dublin in 1731 by a group of local landowners. Many more societies were established throughout Great Britain and the Continent in the decades to follow, particularly in the years around 1790 (Russell 1966). Some of them were little more than local social clubs that, according to Grantham (1984, p. 201), "busied themselves with holding fairs and awarding prizes and medals." Other societies had regional rather than local mandates and played an active role in seeking to solve farmers' production problems and protect them against fraudulent practices.

In fact, the growth in the use of purchased inputs in agriculture played a key role in the demand for agricultural societies and, in turn, agricultural experiment stations. Dubious practices by fertilizer and seed merchants were widespread and farmers were quite vulnerable (Arnon 1989). Consequently, the increased use of nontraditional (commercial) fertilizers triggered an enormous increase in demand for fertilizer and soil analysis services,

---

[51] It was not until the end of the 19th century that this new concept of university training was adopted in Great Britain. The strong laissez faire sentiment in British society throughout the 19th century gave rise to little public investment in education and research.

[52] During the first half of the 19th century, agricultural sciences were not included in the German university system, although instruction in the agricultural sciences at a relatively high level was provided through separate agricultural academies. The establishment of these academies was initiated by Albrecht Thaer in the early 19th century, but in most instances, they lacked a strong research orientation. It was not until 1863 that agricultural faculties were established within German universities (Ruttan 1982).

while increased use of purchased seed led to a demand for seed testing and certification services. Thus, the agricultural societies either functioned as pressure groups to lobby governments and others to fund and execute agricultural research or established laboratories themselves. In this respect, the societies represent an early attempt to institutionalize a process of collective action in pursuit of the interests of those directly involved in production agriculture (chapter 1).[53]

Early Agricultural Experiment Stations

History records several claims to the establishment of the first agricultural experiment station. We will mention only four of the more widely cited cases here. The earliest traces back to 1834 when J. B. Boussingault constructed a chemical laboratory on his estate at Bechelbronn in the Alsace, France. Boussingault, generally considered one of the founders of modern agricultural science, instigated a series of carefully planned and controlled field experiments in conjunction with soil and plant analyses conducted in his laboratory. Although less well known than his contemporary, Justus von Liebig, the scholarly tenor of his writings and the quality of his findings are judged by some to have had a more significant and lasting impact in agricultural science (Russell 1962; Salmon and Hanson 1964). But von Liebig had a far greater influence on his contemporaries. In particular, his book, *Organic Chemistry in Its Application to Agriculture and Physiology*, published in 1840 in both Germany and Great Britain, triggered widespread demand for the application of science to agriculture.[54] Many regard von Liebig's book as the critical dividing line in the evolution of modern agricultural research (Ruttan 1982).

The second case involves an early example of collective demand for the application of scientific experimentation to agricultural production. This was brought about in 1841 by a group of landowners in Scotland, who appealed to the Scottish Highland and Agricultural Society to establish an agricultural laboratory. The society turned down their request on the grounds of expense. The promoters thereupon founded the Agricultural Chemistry Association of Scotland in 1842, established an agricultural laboratory in Edinburgh, and hired a chemist. The laboratory, however, was dissolved in 1848 because of its inability to respond to the association members' demands for immediate practical results (Russell 1966; Ruttan 1982).

In 1843, an agricultural experiment station was established by John Bennet Lawes at

---

[53] Nearly a century went by before agricultural interests in Europe were represented at a ministerial level within government. Europe's first ministry of agriculture was established in Austria in 1868, while many other European countries did not follow suit until the period between 1880 and 1900. Germany and Great Britain did not establish a separate ministry of agriculture until 1919 (Ruttan 1982, p. 74) and 1921 (Russell 1966, p. 280), respectively.

[54] Prior to 1848, von Liebig's book had gone through 17 different editions, translations, and revisions, mostly in Germany, England, France, and the US, but also in Denmark, Italy, the Netherlands, Poland, and Russia (Russell 1962).

Rothamsted, near London, and it lays claim to being the oldest agricultural experiment station still in operation today. Lawes took over the running of his ancestral estate at Rothamsted in 1834 and, in the late 1830s, began manufacturing phosphatic fertilizers from bones. In 1842, he took out what ended up being a very profitable patent for the manufacture of superphosphate and a year later initiated its production. In 1843 he also hired Henry Gilbert, who had studied in Germany under von Liebig, and placed him in charge of experimental work. Financial support for the Rothamsted agricultural experiment station was covered by the profits of the Lawes phosphate enterprise until 1889. In that year, Lawes endowed the station through the Lawes Agricultural Trust, although today government support underwrites most of the station's expenditures.

The publication of von Liebig's book also had a large and immediate impact within Germany. Until 1840 "[most] agricultural chemists believed in the humus theory of plant nutrition, in vitalism, and in the transmutation of water or earth into plant tissue" (Salmon and Hanson 1964, p. 21). Although von Liebig introduced numerous false assertions of his own concerning plant nutrition, one of the major contributions of his proposed mineral theory was to demystify plant nutrition.

The potential to improve farmers' economic returns by using nontraditional fertilizers in accordance with von Liebig's principles of plant nutrition did not go unnoticed. In a manner reminiscent of the collective demands of the Scottish landowners several years earlier, Friedrich Riessner and his associates petitioned the Saxon government in 1845 to appoint and fund several agricultural chemists to undertake research and perform soil and fertilizer analyses. At this time the request was denied, but seven years later a publicly supported agricultural experiment station was opened at Mockern, Saxony.

This event brought together many of the elements that had played a role in the establishment of earlier agricultural experiment stations. As in the Scottish case, it was established in close cooperation with a group of local agricultural societies and also secured the help of a wealthy landowner who conveyed a portion of his estate to the station, as had Lawes at Rothamsted. However, the new element in this particular undertaking was that government financial support (in the form of an annual appropriation) was secured and the station's charter was legalized by public statute. The Mockern station can thus lay claim to being the first *publicly supported* agricultural experiment station in the world, and it functioned as a model for the establishment of agricultural experiment stations throughout the more-developed countries during the second half of the 19th century.

The Agricultural Experiment Station Boom

Throughout the 19th century, but especially during its latter half, Germany was at the forefront of new ideas and institutions related to agricultural teaching and research. Students from around the world came to study at German universities and agricultural institutes and, once back home, often became founding fathers of the early agricultural experiment stations in their own countries. Moreover, German scientists were frequently appointed as university professors and as directors of newly established agricultural experiment stations in other European countries. Thus, German concepts and ideas on agricultural research spread rapidly throughout the Western world as did the institutional innovation of publicly supported agricultural research stations staffed and directed by professional scientists.[55] By 1875, over 90 agricultural research experiment stations had been established throughout Europe, and by 1900, the number of stations had grown to more than 500, employing about 1500 professional scientists with an agricultural research budget roughly totaling $2 million (Grantham 1984). In the same tradition, the first agricultural experiment station was established as early as 1874 in Canada, 1876 in the US, and 1872 in Japan.[56]

In 1887 the passage of the Hatch Act provided a major impetus to the establishment of agricultural experiment stations in the US. This act provided federal funding for the support of experiment stations located within the land-grant colleges and greatly facilitated the rapid buildup of a unique network of state-level agricultural experiment stations throughout the country.[57] By 1900, every state in the US had its own agricultural experiment station and many supported a network of substations throughout their state.

The only non-Western country to establish a network of publicly funded agricultural experiment stations before the end of the 19th century was Japan. With the Meiji Restoration in 1868, Japan was transformed from an essentially closed, inward-looking country to one that was open and outward looking. The increasingly stronger central government took the lead in encouraging the introduction and adoption of Western technologies. Being primarily an agricultural nation at that time, major efforts were made to import agricultural technology from Europe and the US (Hayami and Yamada 1975). However, it soon became apparent to the Japanese government that many of the latest Western agricultural technologies were not well suited to Japanese agricultural conditions. Local adaptation and development of agricultural technologies was deemed necessary.

As early as 1872, the Naito Shinjuku Agricultural Experimental Station began operating, followed by a Tree Experiment Station in 1878. In 1886, an Experimental Farm of Staple Cereals and Vegetables was established near Tokyo, which, in 1893, was redesignated the National Agricultural Experiment Station along with six regional branch stations located throughout the country. The nation's research capacity, particularly at the prefectural level, increased quite rapidly following the enactment, in 1899, of the "Law Providing State Subsidies for Prefectural Agricultural Experiment Stations." Hayami and

---

[55] On the basis of a listing of agricultural experiment stations in True and Crosby (1902), establishment dates for the first agricultural experiment stations throughout Europe are as follows: Germany (1852), Austria (1859), Sweden (1861), France (1864), Czechoslovakia (1865), Hungary (1869), Italy (1870), Denmark (1871), Belgium (1872), Switzerland (1875), the Netherlands (1877), Spain (1890), Yugoslavia (1894), and Ireland (1898).

[56] Establishment dates come from True and Crosby (1902) for Canada, Knoblauch et al. (1962) for the US, and Hayami and Yamada (1975) for Japan.

[57] For a comprehensive description of the origins of the US agricultural research system, see True (1937) and Knoblauch et al. (1962).

Yamada (1975) report that total agricultural research expenditures tripled in real terms between 1897 and 1902.

Into the 20th Century

The first 50 years of public agricultural experimentation was built on the progress made by von Liebig and others in the understanding of plant nutrition. Agricultural chemistry and soil science were the dominant disciplines of the agricultural sciences during these early years. A common complaint of many working in the agricultural experiment stations at the time was that they were overloaded with doing analyses for farmers and had little time left for experimental work; however, their incomes were largely dependent on contributions by farmer organizations and on fees for analyses. In Europe in particular, the "public" agricultural experiment stations received only a small proportion of their budget from the government. True and Crosby (1902) report that in Germany, for example, only one-third of station expenditures were covered by the state in 1900. In this respect, the US and Japanese governments seem to have been more important stakeholders in the establishment and financing of agricultural experiment stations than those in Europe.

By the end of the 19th century, major breakthroughs in agricultural and related sciences began to reshape the work at the agricultural experiment stations. Darwin's theory of evolution, the pure-line theory of Johannson, the mutation theory of de Vries, and the rediscovery of Mendel's Laws all contributed to the rise of plant breeding. Pasteur's germ theory of disease and the development of vaccines opened up lines of research in the veterinary sciences. The same is true for Smith and Kilbourne's discovery of insect transmission of diseases. These are but a few of the many examples that have underpinned the technical progress of agriculture in the earlier part of this century.[58]

The evolution of the science of genetics gathered pace around the middle of this century with discoveries by Hersey and Chase, Watson and Crick, and others uncovering the role and structure of DNA. These findings led directly to the modern biotechnologies based on recombinant DNA technology, monoclonal antibodies, and new cell and tissue culture techniques that are just beginning, and no doubt will continue, to reshape fundamentally the science of agriculture well into the 21st century.

7.7.2   Contemporary Developments in Agricultural Research

Research Personnel and Expenditures

The disparities in public research investments between less- and more-developed countries have narrowed considerably over the past two decades, at least in quality-adjusted terms. While the more-developed countries still account for well over one-half of the global

---

[58] See Salmon and Hanson (1964) for an extensive description of these discoveries.

expenditures on public agricultural research — and certainly a much higher proportion of total (public and private) expenditures — they now employ slightly fewer than one-half of the world's public-sector agricultural researchers.

Among the more-developed regions there is considerable diversity in the rate of growth in research personnel (table 7.22). Over the past 25 years, both the Japanese and North American NARSs increased their staffing levels at moderate rates (about one percent annually).[59] Western and Northern Europe expanded at about twice this rate, while Southern Europe plus Australia & New Zealand grew roughly four times faster. Indeed, Australia & New Zealand, until at least the mid-1980s, grew to the point that the number of agricultural researchers working in these two systems is particularly high relative to their total and agricultural populations as well as AgGDP, although the Australian ratio of

Table 7.22: *Agricultural Research Personnel and Expenditures in More-Developed Countries*

(Sub)region	1961-65	1966-70	1971-75	1976-80	1981-85	Growth rate[a]
	Agricultural research personnel					
	(full-time equivalents)					%
Japan	12,535	13,123	13,798	13,747	14,779	0.8
Australia & New Zealand	2,627	3,278	4,294	5,392	5,902	4.1
Northern Europe (5)[b]	1,519	1,753	1,996	2,317	2,711	2.9
Western Europe (8)	7,639	8,733	9,887	10,384	11,396	2.0
Southern Europe (4)	2,135	2,132	2,584	3,542	4,485	3.8
North America (2)	13,940	15,020	15,565	16,220	17,103	1.0
More-Developed Countries (22)	40,395	44,039	48,123	51,602	56,376	1.7
	Agricultural research expenditures					
	(millions 1980 PPP dollars per year)					%
Japan	404	573	781	891	1,022	4.7
Australia & New Zealand	161	209	290	259	313	3.4
Northern Europe (5)	90	122	135	156	182	3.6
Western Europe (8)	454	714	980	1,059	1,135	4.7
Southern Europe (4)	88	97	142	190	317	6.6
North America (2)	994	1,342	1,399	1,617	1,845	3.1
More-Developed Countries (22)	2,191	3,057	3,726	4,171	4,813	4.0

Note: Data may not add up because of rounding.

[a] Compound annual average between 1961-65 and 1981-85.

---

[59] Most of the growth in these systems occurred prior to our sample period. For instance, the state agricultural experiment station system in the US grew rapidly, in terms of full-time equivalents, in the immediate post-World War II period (averaging 3.9% per annum throughout the 1950s), but it slowed to 0.1% per annum over the 1980-85 period (Pardey, Eveleens, and Hallaway 1991).

researcher to (quality-unadjusted) agricultural land remains extremely low by more-developed country standards.

The level of resources commited to and spent by public agricultural research systems is determined by a complex set of demand- and supply-side influences. On the demand side there are the political economy forces that seem to bias governments in favor of subsidizing rather than taxing their agricultural sectors as the development process unfolds and the size, but not necessarily political influence, of agricultural constituencies shrinks relative to the nonagricultural sectors. However, the evidence in chapter 1 suggests that these forces do not induce governments in more-developed countries to make disproportionately large investments in agricultural research vis-à-vis other forms of public expenditures on agriculture. Rather, agricultural research systems in more-developed countries spend about the same share of direct agricultural expenditures on agricultural research as do less-developed countries. But, the size of overall and research-specific public expenditures on agriculture, when indexed over total and, especially, agricultural populations, increases dramatically when moving from low- to higher-income countries — a trend that is consistent with the fairly rapid increase in agricultural research expenditures experienced by Japan as it underwent a transformation from a middle-income to high-income country during the post-1960 period.

On the supply side, there are many factors that affect the cost structures of NARSs. These relate principally to the size and diversity of the agricultural research system as well as the agricultural sector it serves. As diversity in its various dimensions (including the commodity, agroecological, technological, and problem orientation of a system's research program) increases, the average cost structure of a NARS would also be likely to increase. This gives rise to diseconomies of scope that interact with more familiar notions of economies of size to determine the costs involved in providing a given bundle of research services. The diversity issues make it difficult, if not impossible, to form impressionistic judgments about the relative cost structures of more-developed NARSs that in aggregate expenditure or personnel terms are all medium- to large-sized systems (section 8.4).

Expenditures per Researcher

In addition to the supply and demand forces that shape the nature and overall level of NARS expenditures, there are other, not unrelated, influences that determine the factor proportions and, hence, levels of spending per scientist in agricultural research. These additional influences include the available quantities, qualities, and in particular relative prices of the factors of production.

As evident from the long-run US data presented in section 8.2 of the following chapter, stage-of-development considerations are also likely to play a significant role in determining differences in spending per scientist, particularly between more- and less-developed countries. Having relatively mature systems, all the more-developed subregions except Australia & New Zealand show a fairly sustained increase in expenditures per

Table 7.23: *Real Expenditures per Researcher in More-Developed Countries*

(Sub)region	Expenditures per researcher per year				
	1961-65	1966-70	1971-75	1976-80	1981-85
	(1980 PPP dollars)				
Japan	32,300	43,700	56,600	64,800	69,100
Australia & New Zealand	61,300	63,700	67,600	48,000	53,000
Northern Europe (5)[a]	59,000	69,600	67,600	67,300	67,000
Western Europe (8)	59,400	81,700	99,100	102,000	99,600
Southern Europe (4)	41,400	45,500	54,900	53,600	70,600
North America (2)	71,300	89,400	89,900	99,700	107,900
More-Developed Countries (22)	54,200	69,400	77,400	80,800	85,400

Note: Data rounded to nearest hundred dollars.

[a] Bracketed figures indicate the number of of countries in regional totals.

researcher (table 7.23). In Australia & New Zealand the number of researchers increased more rapidly than did expenditures and, contrary to the US experience at least, they retained a fairly large component of researchers trained only to the BSc level.[60]

7.8 CONCLUDING COMMENTS

In reviewing these highly contrasting regional experiences, it is not surprising that there is as yet little agreed wisdom on just what constitutes the best approach in a particular situation. In short, the existence of such diversity is surely a strong *prima facie* case for a continuing study of alternative agricultural research policy options. We hope that a small step in this direction has been taken through this compilation of contrasting situations and the varied investment patterns described above.

Not only is this a richly diverse experience, but it is also highly evolutionary. And even 1990, as an example of but one specific year, has seen many events that will have profound effects on the future patterns of agricultural research work around the world. These developments include the yet-uncertain consequences of the Gulf War, the major expansion of the CGIAR, the major economic and political changes in Central and Eastern Europe, and the continuing emphasis on privatization of agricultural research and extension activities in many parts of the more-developed world. Surely such phenomena will continue to evolve and change in sometimes highly unpredictable ways. Research administrators need to know what is happening in several respects, including what others are doing, as they seek guidance for their own decision making. Information such as we have assembled on at least some quantitative aspects of research systems should be informative for these research investment decisions. What is still sorely needed is further formal analysis of such data, and

---

[60] See table 8.10 (chapter 8) for details on degree status.

this will be an important item on the research agenda for those concerned with agricultural research policy through the 1990s and beyond. A few suggestive themes are picked up in the following chapter.

Chapter 8

Topical Perspectives on National Agricultural Research

Philip G. Pardey, Johannes Roseboom, and Jock R. Anderson

A host of agricultural research policy issues receives repeated attention in both the literature and the on-going policy dialogue surrounding public agricultural research. These include concerns about the commodity orientation, factor mix, human capital composition, size, scope, spillovers, and sources of support for agricultural research systems in less-developed countries. A detailed breakdown of personnel and expenditure data is needed to provide more than qualitative impressions about these policy concerns. The sections that follow represent an initial attempt at systematically disaggregating and scrutinizing the agricultural research personnel and expenditure aggregates (presented in chapter 7) for those countries for which tolerably comparable data could be assembled. To place the variability observed in this contemporary, but unavoidably incomplete international data set in a broader perspective, the data have, where appropriate, been juxtaposed against a newly developed longitudinal data series for the US agricultural experiment station system.

8.1   THE COMMODITY FOCUS OF AGRICULTURAL RESEARCH

One of the more important policy and management dimensions of a NARS is its overall commodity orientation. Research has the ability to generate scientific advances that translate into productivity gains for particular commodities or agricultural production systems. It is these gains that are one of the principal benchmarks against which publicly sponsored research systems are ultimately judged. Unfortunately, dissecting aggregates of research personnel or expenditures along individual commodity lines is far from simple. For example, universities tend to provide information by discipline, while research institutes tend to categorize information by problem area or commodity aggregates. A further difficulty arises when seeking to identify the commodity focus of a research system: some agricultural research is either targeted to multiple crop or commodity environments or has secondary impacts that go beyond the targeted commodity. In any case, some standardization of the data is a prerequisite to any international comparison of activity-specific input data. Very little such standardization has yet taken place, either nationally or internationally, perhaps because the problems to be solved are complex and in some instances insuperable.

8.1.1   Measurement Options

We begin this section with a brief review of a well known attempt to measure the commodity focus of NARSs[1] and then describe our own measurement efforts on this score. Instead of directly decomposing research expenditure aggregates into their commodity components, Judd, Boyce, and Evenson (1986) employed a rather more indirect approach. Specifically, they first calculated ratios of commodity-level spending per publication using detailed Brazilian data.[2] They then applied this ratio to corresponding publication data for a group of 26 countries. The resulting expenditure estimates were used, inter alia, to develop regional estimates of research expenditures expressed as a share of the value of agricultural product for a total of 21 commodities.[3]

The assumptions underlying this measurement procedure seem rather extreme. It is questionable whether the Brazilian data on spending per publication can be considered representative of the other 25 countries in the sample. Research operations in general, and commodity research programs in particular, differ markedly between countries as well as within countries over time. This is likely to result in quite different expenditure-research output relations from those found in Brazil. Indeed, econometric work using state-level data in the US to relate published agricultural research output to current and lagged research expenditures found a good deal of residual variation in published output that was accounted for by site-specific factors other than research expenditures (Pardey 1989). These site-specific factors reflect cross-state differences in the propensity to publish as well as efficiency differences in each state's agricultural research program — differences that are most likely magnified in an international context. Although surely a creative step, the shortcut method used by Judd, Boyce, and Evenson for measuring the commodity orientation of agricultural research expenditures cannot be considered to represent a reasonable solution to this problem, and perhaps it is not even a very reliable one.

Rather than attempt to fine-tune the Judd, Boyce, and Evenson idea and expand the data coverage to a larger group of countries, we opted for a more direct measurement approach. The nearly 1000 sources of information used for the construction of the Indicator

---

[1] The data reported in this section and throughout this chapter refer only to the public-sector component of NARSs.

[2] The discussion in Evenson (1984, p. 248) suggests that the detailed commodity-level data used subsequently as a global scaling factor by Judd, Boyce, and Evenson (1986) in fact relate only to 1960-76 data from the state of São Paulo, and not to Brazil in general.

[3] These same commodity-level estimates have also been used by Evenson, Pray, and Scobie (1985), Evenson (1987), and Bengston and Gregersen (1988).

Series contain a great deal of detailed data at the research entity level, from which total expenditures and total numbers of researchers have been aggregated.[4] Data on agricultural research input at the commodity level, however, are less comprehensive and often fail to account for all agricultural researchers or expenditures in the NARSs. This lack of detailed data constrained us to disaggregating research inputs into a few broad commodity categories, namely, crop, livestock, forestry, and fisheries research, plus a residual nonallocable category.[5]

Research input indicators based on expenditures are more widely encompassing than those based on personnel, but unfortunately there are substantial difficulties in prorating total research expenditures to specific commodity groups. These difficulties relate both to variable and especially fixed (i.e., overhead) costs. For practical reasons we opted to use the more restrictive research personnel indicator in the belief that the available data for this indicator yield a more accurate representation of commodity relativities. Even so, variations across research entities in their organizational structure and/or level of reporting made it more difficult in some cases than others to allocate researchers to specific commodity programs. This was especially the case for those entities in which research is not the only or primary task, such as universities and in organizations where research and extension, or research and production, are combined. We have, therefore, introduced three categories of research entities: namely primary research organizations, and two types of secondary research organizations, specifically, universities and nonuniversities.

We compiled (and below report) data covering the period 1981-85 for as many less-developed countries as possible, with the provisos that (a) the total number of researchers at the system level was deemed complete and (b) when more than 20% of the researchers in the primary research organizations were not allocable to a specific organization, the country in question was excluded from the sample. In the end, 83 of the 130 less-developed countries in our sample are included here, representing 58% of the estimated total number of agricultural researchers in less-developed countries (excluding China) in 1981-85. The sample is somewhat biased towards the smaller NARSs because for these there were fewer allocation problems. The total numbers of researchers in each category as well as the relative shares in the total are reported in table 8.1.

8.1.2 Commodity Orientations

We took these 83-country sample averages to be representative of the 130 less-developed countries included throughout this volume. Table 8.2 presents the totals of less-developed country researchers and expenditures (excluding the nonallocable category) decomposed into commodity groups. Of course, this decomposition procedure assumes that nonallocable staff have the same broad proportional commodity orientation as those for whom data

---

[4] A research entity refers to the institutional unit by which the data were reported.

[5] In this scheme, pasture and fodder crop research falls under livestock research.

Table 8.1:   *Agricultural Researchers Classified by Commodity Orientation and Type of Organization, 1981-85 Sample Average*

Research orientation	Primary research agency		Secondary research agencies				Total	
			Nonuniversity		University			
	Number	Share	Number	Share	Number	Share	Number	Share
		%		%		%		%
Crops	12,936	65.1	261	25.6	1,621	29.1	14,818	56.0
Livestock	2,942	14.8	101	9.9	1,011	18.2	4,054	15.3
Forestry	1,303	6.6	13	1.3	271	4.9	1,587	6.0
Fisheries	1,006	5.1	25	2.5	96	3.5	1,227	4.6
Nonallocable	1,675	8.4	620	60.8	2,463	44.3	4,758	18.0
Total	19,862	100.0	1,020	100.0	5,562	100.0	26,444	100.0

Note: These data pertain to 83 of our total sample of 130 less-developed countries for which this information was available. Data may not add up exactly because of rounding.

were available and that expenditures per researcher are more or less the same for each research activity category. To the extent that there are systematic differences across broad commodity classes in the ratios of spending per scientist, it is possible that using researcher-based proportionalities to prorate research expenditures biases the expenditure estimates.[6]

Agricultural research is predominantly oriented towards crops. In this sample, just

Table 8.2:   *Estimated Total Research Expenditures and Researchers in Less-Developed Countries by Research Orientation, 1981-85 Average*

Research orientation	Share	Expenditures	Researchers[a]
	%	*(million 1980 PPP $)*	*(full-time equivalents)*
Crops	68.3	2,480	53,100
Livestock	18.7	679	14,500
Forestry	7.3	266	5,700
Fisheries	5.7	205	4,400
Total	100.0	3,630	77,700

Note: The commodity shares in data column one were derived after excluding the nonallocable residual from the last column of table 8.1. These rescaled shares, representing 83 countries, were then applied to the expenditure and researcher totals for 130 less-developed countries in table 7.1 to give the corresponding commodity totals in data columns two and three.

[a] Data rounded to the nearest hundred.

---

[6] For instance, ratios of spending per scientist are likely to be higher on average for livestock research than for crop programs. Higher ratios are also more likely for those crop programs researching perennial crops versus seasonal crops or those more heavily biased toward breeding versus crop management and protection.

over two-thirds of the agricultural researchers are classified as engaged in research related to crops or crop production. Livestock research represents a far smaller share of agricultural research, with nearly four crop-oriented researchers for every livestock-oriented researcher. Regional differences in the breakdown of agricultural research in different categories are presented in table 8.3. Of the four less-developed regions, Asia & Pacific and Latin America & Caribbean represent the extremes in the distribution of agricultural research over product orientations. Research on forestry and fisheries production has a larger share in agricultural research in Asia & Pacific than in Latin America & Caribbean, with the share of fisheries research differing the most between these two regions. Crop-oriented research is more dominant in West Asia & North Africa than in the other regions.

Table 8.3: *Agricultural Researchers in Less-Developed Regions by Research Orientation, 1981-85 Average*

Region	Crops	Livestock	Forestry	Fisheries
	%	%	%	%
Sub-Saharan Africa (29)[a]	67.3	20.0	7.3	5.4
Asia & Pacific, excl. China (18)	63.7	17.4	9.4	9.6
Latin America & Caribbean (22)	68.7	24.1	5.4	1.8
West Asia & North Africa (14)	75.4	16.2	5.7	2.7
Less-Developed Countries (83)	68.3	18.7	7.3	5.7

Note: Regional breakdown based on the allocable component of the 83 country sample used to construct the estimates in table 8.1. Data may not add up exactly because of rounding.

[a] Bracketed figures represent number of countries included in the regional samples.

8.1.3   Congruence Comparisons

A more relevant comparison may be a congruence test, for example, between the share of crop research in agricultural research and crop production's share of value-added in agriculture (AgGDP). Data on a breakdown of AgGDP in all four production categories, however, are not presently available. UN National Account statistics decompose it into three categories (crops & livestock, forestry, and fisheries) for a limited, but still reasonable, number of countries. The degree of congruence between production and research for these three can be assessed from the data in table 8.4.

Assuming the samples are representative enough to justify the comparison, it can be concluded that, in the less-developed world, the share of crop & livestock research is smaller than might be expected on the basis of its share in production. Conversely, in all four regions, forestry research accounts for a larger than congruent share of agricultural research; for fisheries research, this is the case in only two of the four regions.

Table 8.4: *Congruence between AgGDP and Agricultural Research Personnel*

Region	Crops & Livestock		Forestry		Fisheries	
	AgGDP	Research	AgGDP	Research	AgGDP	Research
	%	%	%	%	%	%
Sub-Saharan Africa (22)[a]	88.6	87.3	4.7	7.3	6.6	5.4
Asia & Pacific, excl. China (10)	89.7	81.1	5.2	9.4	5.0	9.6
Latin America & Caribbean (20)	94.2	92.8	2.9	5.4	2.8	1.8
West Asia & North Africa (7)	95.9	91.6	2.4	5.7	1.7	2.7
Less-Developed Countries (59)	90.7	87.0	4.6	7.3	4.6	5.7

Source: Sectoral AgGDP data adapted from UN (1986).

Note: Data may not add up exactly because of rounding.

[a] Bracketed figures represent number of countries included in the regional samples on which the AgGDP breakdown is based. As shown in table 8.3, the research breakdown is based on regional samples which include a somewhat larger number of countries.

Forestry Research

The conclusion that forestry research absorbs a relatively larger share of agricultural research capacity than do crops & livestock contradicts the conclusion of Mergen et al. (1986, 1988), a study based on a global inventory of forestry research.[7] As depicted in table 8.5, this contradiction does not appear to stem from differences in forestry research expenditures per se.

An explanation for the observed differences in research intensity ratios must therefore be sought in the way output is measured. The value-added measure of output used throughout this volume measures the value of gross agricultural output, net of inside and outside inputs (chapter 5). Moreover these value aggregates are based on quantities of agricultural commodities and producer prices that, in principle, are both measured in farm-gate terms. By contrast, it appears that the study by Mergen et al. juxtaposed forestry research expenditures against a gross, not value-added, output measure that was formed using quantity and price data often measured well beyond the "farm-gate level." Certainly there is nothing intrinsically wrong, and for some purposes it may well be preferable, to use a gross-output rather than a value-added measure. But adding a processing cost component to a farm-gate price can substantially inflate the price used to form a value-of-output aggregate, especially in the case of wood products where the off-farm cost component is particularly large. This results in a value measure that is several times larger than ours. Thus, when a final-goods aggregate is used to normalize forestry research expenditures on the one hand and a farm-gate aggregate is used to normalize agricultural research expendi-

---

[7] For additional quantitative and institutional information on forestry research in less-developed countries see FAO (1984a, b, c) and Lundgren, Hamilton, and Vergara (1986).

Table 8.5: *Forestry Research Expenditures by Region*

Region	Mergen et al.[a]		Pardey & Roseboom[b]	
	Expenditures	Share	Expenditures	Share
	(million 1980 $)[c]	%	*(million 1980 $)*[c]	%
Sub-Saharan Africa (43)[d]	23.3	23.8	26.5	23.5
Asia & Pacific, excl. China (28)	34.4	35.2	43.8	38.7
Latin America & Caribbean (38)	30.8	31.4	25.7	22.7
West Asia & North Africa (20)	9.4	9.6	17.1	15.1
Less-Developed Countries (129)	97.8	100.0	113.1	100.0

Source: Mergen et al. data adapted from Mergen et al. (1986, 1988).

Note: Data may not add up exactly because of rounding.

[a] Data for the period 1980-81.

[b] Data represent the average of the 1976-80 and 1981-85 observations.

[c] Currency conversions using annual average exchange rates.

[d] Bracketed figures represent the number of countries included in the Pardey & Roseboom sample. Efforts were made to ensure that the Mergen et al. data, as reported here, corresponds as closely as possible to the Pardey & Roseboom sample.

tures (excluding forestry and fisheries) on the other, then systematic biases are introduced into the research intensity ratios so formed. This appears to be the case with the gross output measures used by Mergen et al.[8]

Comparisons made on the basis of our sample data, however, are also less than ideal. Although we attempted to exclude postharvest research to ensure consistency with our value-added measure of agricultural output, it was not possible to exclude all such research. This may be particularly the case for forestry research. The forestry research capacity in many countries is often, in absolute terms, too small to support separate institutes for both preharvest and postharvest research. In most instances these research specialties are combined in a single institute, which precluded the possibility of identifying the postharvest component in our data. Mergen et al. (1986), however, provide information on this point. Based on their sample data, they calculated that the postharvest component within forestry research amounts to 39% in Africa, 20% in Asia, and 2.2% in Latin America. Even if it is assumed (conservatively) that our estimate of forestry research capacity includes 40% postharvest research, it merely scales the share of forestry research in agricultural research down to about the same level as the share of forestry production in AgGDP. There is still no basis to conclude that, as a percentage of value-added at the "farm level," forestry research is getting less congruent effort than other types of agricultural research.

---

[8] The estimate made by Mergen et al. of the value of gross forestry output is apparently based on data reported in the *FAO Yearbook of Forest Products* (1980). This source reports quantities and prices of a wide variety of forestry products in various forms ranging from "wood in the rough" through to paper and wood-based panel products.

One of the major issues brought forward by Mergen et al. is that there is little forestry research capacity in the less-developed countries. This is, of course, true: of the 130 less-developed countries in our sample, 95 supported NARSs that in 1981-85 had fewer than 200 researchers. On average, 7.3% of these researchers were engaged in forestry research. In other words, roughly 75% of the public systems in less-developed countries employed fewer than 15 forestry researchers, and 50% employed even fewer than seven. It does not appear to be the case, however, that forestry research is getting less congruent attention than crop, livestock, or fisheries research;[9] rather, the problem is more that the agricultural research capacity in general (including forestry) is still rather limited in the majority of less-developed countries.

Another thing that should be taken into consideration is that there seems to be a distinct division of labor between public and private research in the sense that public agricultural research concentrates largely on preharvest production, while private research focuses much of its effort on the postharvest stage. If this is also true for forestry research, then only considering public forestry research while using a production concept that includes the postharvest stage constitutes a mismatch of concepts.

What do these results hold for agricultural research policy? It seems that the invisible hand of resource allocation has produced a set of broad commodity-allocative decisions that (a) are remarkably congruent with values of production and (b) are remarkably consistent between the less-developed regions. Whether such allocations are wise and good depends, of course, on many other factors, notably including the scope for benefits (such as those captured in proportional cost reduction opportunities) and the chances of achieving these benefits through the research programs that are funded. It seems unlikely that these factors would be constant across commodity groups or less-developed regions; hence, the extent of congruence noted here should be taken as a challenge for research decision makers to seek new allocations that are socially more profitable. Investment in more disaggregated analytical methods for priority setting may be the most urgently needed initiative for research planners. As well as looking more closely into the nature of the research process itself, detailed and disaggregated data on particular commodities and their associated research efforts will clearly be required for such work.

8.2   FACTOR SHARES IN AGRICULTURAL RESEARCH

Efficient use of available resources is also one of the most critical management issues in agricultural research. The question at the heart of the matter is how to combine human resources, capital, and other inputs in such a way as to get the most for each dollar spent on

---

[9] In this regard it is worth noting that neither the value added nor the gross value of output measures discussed here take the nontimber services of forests into account. Although these services can be valued (Graham-Tomasi 1990a, b), in the present discussion it is only relevant to do so to the extent that forestry research is addressing these aspects.

research. Such an optimal factor mix is not static but differs across countries and over time, depending on relative prices, the available quantities and qualities of the factors of production, the types of research being performed, the problems being addressed, and so on. Although theoretically a useful concept, in day-to-day life a precise calculation of this optimal factor mix is not realistic. There are certainly no standard rules that can safely be invoked.

Nevertheless, information about actual factor mixes used in different NARSs may provide some guidance to the ranges in which an optimal factor mix might be found. To this end, agricultural research expenditures have been split into three categories (salaries, operating costs, and capital) for as many less-developed country NARSs as possible for the period 1981-85. This was done for 43 of 130 less-developed country NARSs for one or more years during this period. Together these NARSs represent about one-third of total agricultural research expenditures by less-developed country NARSs (excluding China) over this period. In table 8.6 the weighted averages of the factor shares for each of the four less-developed regions are given, as well as corresponding data for the US. Features of the comparisons shown in table 8.6 are elaborated in the following three subsections.

Table 8.6:   *Agricultural Research Expenditures by Factor Share, 1981-85 Average*

Region	Recurrent expenses		Capital	Total
	Salaries	Operating		
	%	%	%	%
Sub-Saharan Africa (17)[a]	60	25	15	100
Asia & Pacific, excl. China (9)	51	27	23	100
Latin America & Caribbean (11)	56	25	19	100
West Asia & North Africa (6)	68	15	17	100
Less-Developed Countries (43)	57	25	19	100
United States	69	23	8	100

Note: Data may not add up exactly because of rounding.

[a] Figures in brackets represent number of countries in sample.

8.2.1   Physical Capital

Capital Expenditures

In addition to the 43 less-developed countries, a smaller sample of 16 countries, representing about 16% of total agricultural research expenditures by less-developed countries (excluding China), was also examined. Although for this smaller sample it was only possible to distinguish between recurrent and capital expenditures, it corroborates the 19% capital share for less-developed systems witnessed in the 43-country sample. Compared with the average capital share for the US over the period 1981-85, the capital expenditure component within less-developed systems appears inordinately high. There are several reasons why the share of capital in agricultural research expenditures in less-developed countries may well be higher than in the contemporary US.

First, the historical data compiled by Pardey, Eveleens, and Hallaway (1991) for the US show that the share of capital in agricultural research expenditures has been constant at around 8% for the past three decades. But this is in stark contrast with the early establishment phase of the US experiment station system at the beginning of this century, during which time the capital share grew steadily, peaking at 29% in 1912 — a level that has not been matched since. In the consolidation phase of a NARS, capital expenditures typically cover the replacement of the capital stock used in the course of a year, while in the build-up or expansion phase capital expenditures must cover replacement plus the extra capital needed for expansion. All other things being equal, the relatively high share of capital in agricultural research expenditures in less-developed countries suggests that they are still in an expansionary phase, an interpretation in line with the rapid growth of the NARSs in less-developed countries that was noted in chapter 7.

Second, capital items are, in general, relatively more expensive in less-developed than in more-developed countries. Even if the factor mix in both less- and more-developed countries were the same when measured in terms of real aggregates or comparable volumes (chapter 5), the share of capital in total *expenditures* would be higher in less-developed countries. A counterbalancing effect can be expected from factor-substitution induced by relative factor-price differences. Much contemporary agricultural research technology has, of course, been developed by more-developed countries and strongly reflects a more-developed country factor bias.

Other influences may also operate to modify the contrast between less- and more-developed countries. One that may tend to reduce capital expenditures in less-developed NARSs is the composition of capital. Several influences (not the least of which is scarcity of foreign exchange) can lead less-developed NARSs to spend relatively more on bricks and mortar and less on state-of-the-art electronic equipment. Notwithstanding the observed vacant space in many NARS buildings in less-developed countries, there are probably some natural curbs to constructing new research facilities, and those already constructed tend to be quite durable.[10] On the other hand, the obsolescence rates of many laboratory instruments are notoriously high, and research facilities in more-developed countries that are kept well up to date will naturally spend more on modernizing their capital stock than those that are not able to do so.

Working against this tendency is the repair and maintenance capacity, which varies systematically with the stage of economic development. Many less-developed countries

---

[10] For instance, during the five years centered on 1912, the US experiment station system spent around 14% of its annual research budgets on land and buildings and 9% on equipment, while over the 1981-85 period, the land and buildings share had dropped to only 2.7% relative to an equipment share of 5.1% (Pardey, Eveleens, and Hallaway 1991).

find it difficult indeed to keep delicate laboratory equipment fully and effectively functional. The less-developed world's research laboratories are littered with research equipment that is unusable because either spares are unavailable or the skilled technical expertise required to do the maintenance cannot be tapped. All this has the effect of leading to relatively high rates of complete replacement of such capital items.

Capital Service Flows

Whereas salaries and operating costs in agricultural research expenditures represent flows, capital expenditures represent additions to a stock. A service-flow measure of capital would improve data comparability over time and across countries. But aggregating a stream of capital expenditures or investments into a real measure of capital services requires estimating the (perhaps time-dependent) flow of services to be had from these capital investments. This in turn requires detailed information on capital prices, utilization rates, economic depreciation rates, and the life span of different capital types. Assuming, for purposes of simplicity, that service-flow profiles can be proxied by a "One-Hoss Shay"[11] assumption with a zero salvage value and a serviceable life for land and building investments and equipment expenditures of 40 and 10 years, respectively, it is possible to generate a real measure of capital services from research capital expenditures by the US land-grant system over the past century (Pardey, Eveleens, and Hallaway 1991). The resulting differences over time between the capital expenditure and service-flow measures are dramatic. During the system's first 20 to 25 years, beginning in 1889, the expenditure measure consistently overestimated the more preferred real service-flow measure. In fact, for 28 of the first 30 years, the difference was greater than 20%, and in eight years, it was greater than 100%. For many of the years following World War II, capital expenditures were consistently lower than the corresponding service-flow measure, except for several years immediately following the war and a three-year period during the late 1960s that witnessed an acceleration in US experiment station capital investments, which paralleled a "baby boom"-induced expansion of the US land-grant university system.

The relatively young age and rapid growth of many less-developed NARSs suggests that their capital expenditures may similarly be an inflated measure of the contemporary flow of capital services within these systems. Because of the lack of detailed long-term data on agricultural research capital items in less-developed countries, it is, unfortunately, impossible to estimate this flow.

8.2.2 Noncapital Inputs

Turning our attention to recurrent expenditures and assuming for the sake of discussion that the factor mix across noncapital items in more- and less-developed countries is identical,

---

[11] One-Hoss Shay profiles represent a constant flow of services for the life of a capital item.

then the share of operating costs in recurrent expenditures would generally be higher in less- than in more-developed countries. This is because salaries tend to be significantly lower in the former while operating costs tend to be higher. Factor substitution may again counterbalance this effect somewhat, but it is not likely that it will outweigh it. The higher average share of operating costs we found for less-developed countries in comparison with the US (table 8.7) is consistent with this reasoning. The West Asia & North Africa figure of 18% is an exception, caused by the dominance of Egypt in the weighted average with its remarkably low figure (14%) for the share of operating costs in total recurrent expenditures. Excluding Egypt gives a weighted average of 25%, a figure that is more in line with the shares observed for the other less-developed regions.

Table 8.7: *Salaries and Operating Costs and Their Shares in Recurrent Agricultural Research Expenditures, 1981-85 Average*

Region	Expenditure per researcher	Recurrent expenses		
		Salaries	Operating	Total
		(1980 PPP dollars)		
Sub-Saharan Africa (17)[a]	75,300	45,400	18,800	64,200
Asia & Pacific, excl. China (9)	51,400	26,000	13,800	39,800
Latin America & Caribbean (11)	78,800	44,500	19,600	64,100
West Asia & North Africa (6)	50,600	34,600	7,400	42,000
Less-Developed Countries (43)	59,200	33,600	14,500	48,100
United States	99,100	68,400	22,800	91,200
		(percentages)		
Sub-Saharan Africa (17)		71	29	100
Asia & Pacific, excl. China (9)		65	35	100
Latin America & Caribbean (11)		69	31	100
West Asia & North Africa (6)		82	18	100
Less-Developed Countries (43)		70	30	100
United States		75	25	100

Source: United States data taken from Pardey, Eveleens, and Hallaway (1991).

Note: Data may not add up exactly because of rounding.

[a] Figures in brackets represent number of countries included in the (regional) samples.

Once again, drawing on historical evidence from the US land-grant system adds a temporal dimension to the necessarily snapshot perspective given in table 8.7. Figure 8.1 plots various cost components of the US system on a per researcher or, more accurately, full-time-equivalent (fte) basis. All figures are expressed in units of 1980 US dollars. During the system's earlier years, the rate of increase in research personnel outpaced that of research expenditures, so total expenditures per fte fell steadily over the first two decades. These figures also show operating expenditures per fte (but not capital expendi-

Figure 8.1: *Real expenditure per researcher in US agricultural experiment stations, 1890-1985*

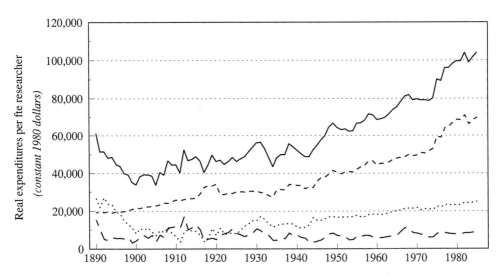

Legend: - - - salaries; ····· operating costs; – — capital; ——— total expenditures.

Source: Pardey, Eveleens, and Hallaway (1991).

Note: Here as elsewhere in this chapter the total service flow from the stock of capital in period t' is given by
$$k \sum_{t=t'}^{t'-(T-1)} C_{0t}, \quad \text{where } C_{0t} \text{ is the gross (undepreciated) market value of a capital asset purchased in period}$$
t with a service life of T, $k = (r / (1+r)) [1/(1-[1/(1+r)]^T)]$, and r is the discount rate.

tures other than for the first several years) falling over the corresponding period from around $25,000 to $8,500 some 30 years later. It took until the mid-1970s for real operating expenses per fte to reach the levels experienced by the system during its first few years. This temporal pattern of spending per fte experienced by the US system during its earlier years is not at all dissimilar to that presently observed in those less-developed systems for which comparable data are available.

Factor Bias within Agricultural Research

Putting operating expenses to one side for the moment, figure 8.2 shows the long-run trend in labor-capital ratios for the US system. Using expenditure ratios gives a pattern of change that exhibits three phases: an initial establishment phase where the labor-capital ratio trends downward at about 2.6% per annum; an inter-war growth phase in which the ratio trends upward at about the same rate; and finally a post-World War II consolidation phase in which the ratio continues to trend upward but at the slower compound rate of 0.7% annually. When capital expenditures are converted to service-flows (and are thereby more

Figure 8.2: *Research labor-capital ratios for the US agricultural experiment station system, 1890-1985*

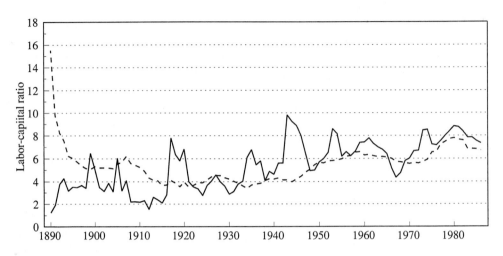

Legend: —— expenditure ratio; - - - service-flow ratio.

Source: Pardey, Eveleens, and Hallaway (1991).

comparable with corresponding measures of labor expenditures) a somewhat different picture emerges.

For one thing, the ratio is far less volatile when labor and capital are measured in comparable service-flow rather than mismatched expenditure terms. During the first 30 or so years of the series, the expenditure ratio appears to significantly underestimate the labor-capital relativities when compared with the equivalent service-flow ratio.[12] Moreover, the service-flow series preserves the decreasing trend in the labor-capital ratio over the initial establishment phase (and, in fact, extends this phase for a decade or so beyond that of the expenditure ratio), but this is followed by a rather steady, long-run increase in the labor-capital ratio rather than the two-phased increase that was apparent in the expenditure series. The US research system has experienced a sustained increase in its use of human capital relative to physical capital and, although this is continuing, it is doing so at a slower rate. Today the system performs with around 14 cents of physical capital for each dollar of human capital compared with about 27 cents 50 years ago.

These long-run data have important lessons for our contemporary snapshot across a sample of less-developed country systems. In particular, the expenditure data, which by necessity we are forced to work with in these instances, may not be a very representative

---

[12] The service-flow measure also removes the large spike in the series observed in the expenditure ratio for the very first couple of years when these new experiment stations began constructing their facilities but presumably deferred recruiting researchers until they had buildings.

measure of the relative flows of labor and capital services actually used to perform research. In fact, as many of these systems are in their early establishment and growth phases we are likely to underestimate the real contribution of labor vis-à-vis capital.

Finally, the pattern of contemporary capital expenditures for research in the US clearly reflects a mature system in which expenditures on maintenance and replacement dominate new additions to the physical stock of capital, with the result that labor-capital expenditure ratios more nearly match those of service flows. But these long-run data serve to emphasize that contemporary US relativities may well be an inappropriate benchmark against which to judge the situation in less-developed countries. Such comparisons, at a minimum, need to be conditioned by the stage of development in which the system operates, as well as the economic environment in which it finds itself.

An outstanding issue that can seriously compromise cross-country comparisons of relative factor *shares* concerns the difficulty of finding appropriate convertors that properly reflect cross-country differences in absolute price levels for each of the specific expenditure components. Because cost shares represent preaggregated combinations of different prices and quantities, comparisons of cost shares can only be made on the basis of inevitably inadequate assumptions about these quantities and prices. In the comparison above, for example, fixed quantities were assumed. The next comparisons deal with aspects of the prices.

8.2.3 Cross-Country Price Differentials

One approach is based on the notion that there exist systematic differences in purchasing power between specific sectors across countries. By converting research expenditures denominated in local currency units using purchasing power parities (PPPs) over GDP,[13] we implicitly assume that the basket of goods and services used in agricultural research is similar to that which relates to the whole economy (Pardey, Roseboom, and Craig *forthcoming*). PPPs over GDP are aggregates in which the sectoral differences in purchasing power have been averaged out. Summers and Heston (1988), however, have calculated two sets of sectoral PPPs that may be of use here, namely PPPs for government goods and services (PPP^G) and PPPs for investment (PPP^I). Drawing on the Summers and Heston data, weighted averages of the PPP^G / PPP^{GDP} and PPP^I / PPP^{GDP} relativities were calculated for each of the four less-developed regions (Pardey and Roseboom 1989a).

In all regions, investments were, on average, considerably more expensive than the general price level, while government goods and services were considerably cheaper. Assuming that PPP^G is more or less representative for the salary component of agricultural research expenditures and PPP^I is similarly representative for the operating cost component,

---

[13] PPPs represent a synthetic exchange rate that seeks to compare the relative cost in local currencies of a specific basket of (traded *and* nontraded) goods and services. See chapter 5 for more details and additional references.

we attempted to adjust these expenditure categories for sectoral differences in purchasing power and to recalculate the regional average recurrent costs per researcher and the salaries and operating cost shares (table 8.8).

Table 8.8: *Salaries and Operating Costs and Their Shares in Recurrent Agricultural Research Expenditures, Adjusted for Differences in Sectoral Purchasing Power, 1981-85 Average*

Region	Recurrent expenses		
	Salaries	Operating	Total
	(1980 PPP dollars)		
Sub-Saharan Africa (17)[a]	79,700	11,500	91,100
Asia & Pacific, excl. China (9)	52,000	9,800	61,800
Latin America & Caribbean (11)	57,000	16,800	73,800
West Asia & North Africa (6)	52,500	5,000	57,500
Less-Developed Countries (43)	56,900	10,400	67,300
United States	68,400	22,800	91,200
	(percentages)		
Sub-Saharan Africa (17)	87	13	100
Asia & Pacific, excl. China (9)	84	16	100
Latin America & Caribbean (11)	77	23	100
West Asia & North Africa (6)	91	9	100
Less-Developed Countries (43)	85	15	100
United States	75	25	100

Source: United States data taken from Pardey, Eveleens, and Hallaway (1991).

Note: Data may not add up exactly because of rounding.

[a] Figures in brackets represent number of countries included in the (regional) samples.

Because, in general, salaries take up the larger part of the recurrent expenditures, the purchasing power adjustment results in considerably higher recurrent (real) expenditures per researcher in less-developed country regions. The application of specific PPPs for each of the cost components induces the recalculated cost shares to represent more comparable quantity or volume aggregates. When the recalibrated cost shares in table 8.8 are compared with those in table 8.7, all less-developed regions are seen to be more labor-intensive relative to the contemporary situation in the US, although this adjustment is less pronounced in Latin America & Caribbean than in sub-Saharan Africa and Asia & Pacific. As mentioned, the West Asia & North Africa observation is somewhat exceptional.

In all four less-developed regions, operating expenditures per researcher are much smaller than the US counterpart. Agricultural researchers in sub-Saharan Africa, Asia & Pacific, Latin America & Caribbean, and West Asia & North Africa work with only 50%, 43%, 74%, and 22%, respectively, of the operating resources provided to a US researcher.

The situation is, however, somewhat different in the relativities concerning the labor-service component of recurrent costs. Agricultural researchers in sub-Saharan Africa, Asia & Pacific, Latin America & Caribbean, and West Asia & North Africa are provided, on average, with 117%, 76%, 83%, and 77%, respectively, of real labor-services (including both scientific and support staff) that a researcher in the US enjoys. The generally high labor-service situation relative to the US may cause one to question the validity of our implicit assumption that salary levels in public agricultural research are more or less in line with government salaries. But the relatively high number of expatriate researchers working in African NARSs may account in large measure for the particularly high labor-service expenditures in this region. Expatriate researchers are generally paid according to their home-country standards plus an extra bonus for living abroad. The presence of a relatively large group of expatriates could also have an impact on the salaries of (scarce) national researchers. It may place them in a position to secure salaries that are higher than the average government salary. Indeed a relationship between the presence of expatriate researchers and the level of expenditures was observed in the discussion of temporal trends in chapter 7. Expenditures per researcher in sub-Saharan Africa declined after the late 1960s while at the same time the share of expatriate researchers dropped significantly. A further rationale for the relatively high labor-service component observed in many less-developed countries is the preponderance of support staff within these systems. Many agricultural research agencies must adhere to government employment policies that frequently lead to overemployment in the public sector.

Another approach to dealing with cross-country differentials in average price levels is based on the notion that operational inputs consist mainly of tradable goods such as cars, fuel, fertilizers, etc. From this perspective, international price differences between these items would be less pronounced and would be considerably smaller than for salaries, in particular. When annual average exchange rates are used, as opposed to PPPs, the average regional operating costs are $18,900 in sub-Saharan Africa, $6,400 in Asia & Pacific, $13,300 in Latin America & Caribbean, and $5,700 in West Asia & North Africa, which, as already noted, are all well below the contemporary US level of $22,800.

8.3   THE HUMAN CAPITAL COMPONENT OF AGRICULTURAL RESEARCH

Original research is a skilful and creative endeavor and, while there is certainly an element of routine to the applied and adaptive nature of a sizable portion of agricultural research, the whole exercise is clearly, in economists' jargon, a human-capital intensive undertaking. As a consequence, one of the fundamental strengths (and unfortunately, in all too many instances, weaknesses) of national agricultural research systems lies in the quality, composition, and deployment of their research staff. Developing meaningful indicators of this human capital component is challenging from both a conceptual and practical point of view.

Conceptually, it is not at all clear what the relevant level and combination of factors that distinguish more- from less-productive researchers are. While qualifications and levels of research experience are informative on this score, they alone are of only limited value and need to be interpreted with care when the research capacity of NARSs is assessed over time or across countries. Observed differences in the composition of the research labor force within a NARS reflect, inter alia, the research mission of a system and/or its components. It would be unlikely for small systems focusing on search, screening, and highly adaptive research to have, or indeed seek, a researcher profile that would mirror the profile of larger systems that often confront an altogether different scale and set of research problems.

Demand and supply forces that influence local (and international) markets for scientific expertise also shape the researcher profile observed for any NARS at a particular point in time. The substantial opportunity costs for skilled researchers, particularly those with training in specialties that are in high demand, gives them a good deal of international mobility or may lead them to forsake the public sector in favor of more lucrative jobs in the private (not necessarily research) sector. It is also not uncommon to find systems where the few well-qualified personnel with strong research potential are quickly promoted to administrative and management roles that do not directly call upon their research expertise. Thus, while *average* experience levels may be on the increase, attrition rates among more skilled researchers may be at levels that compromise the research leadership and depth of a particular research program or indeed of an entire system.

The practical difficulties in constructing comparable human capital indicators are no less formidable than the conceptual difficulties. Qualification levels were primarily used to differentiate research from nonresearch personnel in the series reported throughout this volume, such that NARS scientific personnel with at least a BSc degree were classified as researchers (chapter 5). The obvious difficulty with this approach — especially at the BSc level — is that what is a technical or research assistant position in one system may well be a researcher position in another. Without detailed information with regard to job descriptions, some mismeasurement problems are unavoidable.[14]

Moreover, combining BSc-, MSc-, and PhD-holders in one total is like adding apples and oranges. The different degrees do not overtly represent the same amount of human capital or potential research capacity. It is, however, not only the degree level that is relevant in this respect. Degree specialization and subject-matter content, research experience, and perspicacity, for example, are also important. The focus here is on degree level only because it is the most "objective" and tractable method of differentiating researchers for their contribution to the human capital component of agricultural research.

---

[14] In those cases where this approach led to patently misrepresentative estimates, an attempt was made to rectify the problem. In other cases the classification made by each NARS was accepted at face value.

8.3.1 Qualification Profiles

To make maximum use of the disparate observations that have been assembled on the degree status of agricultural researchers, we constructed, on the basis of sample data, weighted averages of the share of postgraduates in the subregional total number of agricultural researchers. These qualification ratios have only been calculated for the less-developed countries during the 1981-85 period. For the pre-1980 years, there were too few observations per region to construct usefully representative indices.

An additional complicating factor is that expatriate researchers were registered separately as often as possible, but with no degree specification. It seems reasonable to assume, however, that all expatriate researchers hold at least an MSc degree. This raises another important issue, namely the extent to which expatriate researchers represent the more qualified component of the research capacity in NARSs. Two qualification ratios are thus presented in table 8.9, one including and one excluding expatriates. Leaving data problems aside, the latter index is, presumably, the better indicator of the local level of agricultural research expertise.

Table 8.9 shows that the regional qualification ratios, including expatriates, range between 40% and 60% in the majority of the less-developed regions. Only three subregions, namely Western Africa, Eastern Africa, and South Asia, score slightly higher than 60%, while Central America and West Asia score lower than 40%. On average, these sample data suggest that about one-half of the agricultural researchers in the less-developed countries hold a postgraduate degree. Although the number of more-developed countries for which a qualification ratio could be constructed is too small to constitute a representative group average, it seems to range between 70% and 90%. Some such countries proclaim a policy that only those with postgraduate qualifications, preferably PhD-holders, can obtain research positions.

It may seem surprising that some of the poorer less-developed regions have relatively high qualification ratios. For most, however, the ratio drops substantially when expatriate researchers are excluded (table 8.9). The dependence upon expatriate researchers is particularly high in sub-Saharan Africa, West Asia, and the island states of the Pacific and Caribbean. Excluding expatriates, none of the less-developed regions has a qualification index above 60%, while five subregions (namely, Southern Africa, Pacific, Caribbean, Central America, and West Asia) scored lower than 40%.

A strong relationship is usually argued to exist between average educational levels and the level of economic development. It is thus natural to expect to find relatively higher national qualification indices in the richer less-developed regions, such as Latin America & Caribbean and West Asia & North Africa, than in the poorer less-developed regions, such as sub-Saharan Africa and Asia. The data presented in table 8.9, however, fail to support this expectation. On the contrary, the lowest national qualification indices were found in rather well-off less-developed regions such as Central America and West Asia, while Western Africa and South-East Asia scored among the highest national qualification ratios.

Table 8.9: *Nationality and Degree Status of Agricultural Researchers, 1981-85 Average*

Region	Expatriates[b]	Qualification ratio[a]	
		Nationals only	Nationals & Expatriates
	%	%	%
Western Africa	31	54	62
Central Africa	44	46	57
Southern Africa	41	25	49
Eastern Africa	17	45	61
Sub-Saharan Africa	*29*	*45*	*57*
South Asia	3	47	63
South-East Asia	11	55	45
Pacific	44	29	53
Asia & Pacific, excl. China	*11*	*53*	*52*
Caribbean	18	32	42
Central America	3	23	28
South America	1	53	49
Latin America & Caribbean	*2*	*51*	*46*
North Africa	6	44	49
West Asia	26	17	39
West Asia & North Africa	*18*	*27*	*47*
Less-Developed Countries	12	48	50

Note: These percentages represent weighted averages. Because of data limitations, the number of countries included in our sample differs across data columns. Reading from left to right, the less-developed totals are based on 56, 42, and 79 countries, respectively, which in turn represent 25%, 20%, and 65% of the total number of researchers in less-developed countries (excluding China). This explains why, for instance, the first two data columns do not sum to the third.

[a] Measures the proportion of national or total (expatriate plus national) researchers holding a PhD or MSc degree. All expatriates were presumed to hold at least an MSc degree.

[b] Proportion of expatriates working with "line responsibilities" in the NARSs, not those working on short-term development projects.

A plausible explanation for this may be that, under the constraints of civil-service regulations and conditions, public-sector NARSs in the richer less-developed countries are often unable to offer salaries attractive enough to retain highly qualified researchers. As other sectors of the economy develop, the competition for qualified staff becomes more vigorous. For example, instances were found where there were a considerable number of PhD-holders in the agricultural sciences within the universities, while agricultural research institutes employed no PhD-holders at all. The salary levels and perhaps other terms and conditions of employment of the agricultural research institutes for well-qualified staff were seemingly not competitive with those of the universities.

Another possible explanation may be that it is largely a structural, not a remunerative, problem in the sense that the research positions available within the NARSs of the upper-middle-income countries are classified at the BSc level in the majority of cases. This seems to be particularly the case in countries such as Mexico and Argentina.

The discussion in chapter 1 highlights the fact that rules of thumb concerning appropriate research investment levels are rather vacuous and generally offer little in the way of meaningful policy guidance. So it is when using contemporary data concerning NARSs in more-developed countries (especially the US) as a yardstick against which to assess the human capital component, and in particular qualification profiles, of NARSs in less-developed countries.

For one thing, it is only recently that the US land-grant system has had a majority of its researchers trained to PhD level (figure 8.3). There were relatively equal proportions of researchers holding PhDs, MScs, and BScs during the system's earliest years. Indeed, for the period up to and including the 1920s, when researcher numbers increased at a particularly rapid rate, the system became substantially *less* PhD-intensive, with the share of researchers holding PhDs declining from 29% in 1890, to 20% in 1910, and reaching only 27% by 1920. The research that contributed significantly to the rapid rates of productivity growth seen in US agriculture during the middle part of this century (Ball 1985) was carried out by a cadre of researchers heavily biased in favor of MSc- and BSc-holders.

Figure 8.3: *Degree status of US experiment station researchers, 1890-1980*

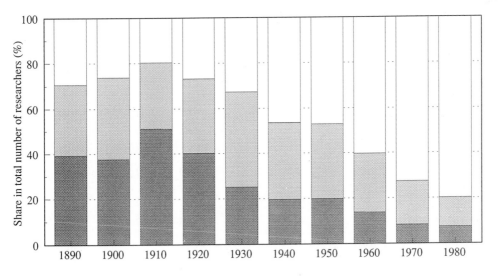

Legend: ▨ BSc; ▨ MSc; ☐ PhD.

Source: Huffman and Evenson (1987, table 4.5).

Parenthetically, the post-war transition to a PhD-intensive system in the US also saw a substantial shift from full- to part-time researchers. During the late 1940s about 45% of experiment station researchers held a full-time research position, but that figure has declined steadily so that by 1985 only 26% of these researchers were engaged full-time in research. The rest held joint research-teaching (59%), research-extension (6%), and re-search-teaching-extension (10%) appointments (Pardey, Eveleens, and Hallaway 1991).

Comparative cross-country data serve to caution further against using contemporary US data as a basis for establishing qualification "norms" with meaningful policy im-plications for less-developed countries. As just noted, while nearly 80% of the researchers in the US land-grant system are presently trained to a PhD level and over 90% hold at least an MSc degree, this is certainly not the case in other more-developed country systems, even over relatively recent times (table 8.10). By 1981, the latest year for which comparable data have been compiled, the PhD intensity of the Canadian system averaged just under 70%, while the New Zealand and Australian systems were around 55% and 41%, respectively. Moreover, only 17% of researchers in the state-level component of the Australian system were qualified to PhD level, and an equivalent percentage were qualified at MSc level. This component accounts for 44% of the country's agricultural researchers and carries out much of the nation's very applied and adaptive (farm-level) research. Given that these Austral-asian systems appear to have generated substantial returns on the public investment made

Table 8.10: *Degree Status of Agricultural Researchers in Selected More-Developed Coun-tries*

Institution	Australia		New Zealand		Canada		US	
	1969	1981	1969	1981	1969	1981	1970	1980
	%	%	%	%	%	%	%	%
PhD								
Federal[a]	48.4	50.6	38.6	52.0	51.4	66.9	na	na
State	5.7	16.9	—	—	36.9	31.8		
University	61.4	74.1	42.2	62.4	62.8	75.2	72.5	79.8
Total	28.7	40.7	39.3	54.7	54.4	68.7	na	na
PhD + MSc								
Federal[a]	66.5	67.9	66.8	76.4	80.9	88.7	na	na
State	20.2	33.8	—	—	63.1	72.3		
University	79.5	88.1	72.1	80.7	94.0	95.3	91.7	92.6
Total	44.9	57.2	67.9	77.5	84.3	90.7	na	na

Source: Australia, New Zealand, and Canada—CAB (1969, 1981); US—Huffman and Evenson (1987, table 4.5).

[a] Consists of Commonwealth Scientific and Industrial Research Organization (Australia), Ministry of Agriculture and Fisheries and Department of Scientific and Industrial Research (New Zealand), and Agriculture Canada (Canada), along with other relevant federal agencies within each country.

in them,[15] a highly PhD-intensive cadre of researchers, in and of itself, is surely not a necessary precondition for a productive research system.

There are many matters that deserve more careful scrutiny in assessing the value of PhD training for NARSs. One of those implicit in the previous paragraph is the research orientation and training within BSc and especially MSc programs in different university systems. Another is the influence of holding a PhD in contributing to the quality of research — for example, as indicated by its publishability in leading international journals, and thus in the international transmissibility of research findings and their eventual spillover benefits. Research policy analysts do not appear to have given such matters the attention they seem to deserve.

8.3.2 Expatriate Researchers

As mentioned in the previous section, expatriate researchers in some regions represent a significant portion of the human capital available for agricultural research. This seems to be particularly so in sub-Saharan Africa, the Pacific, the Caribbean, and West Asia. Based on sample data for the period 1981-85,[16] on average, 12% of the researchers working in the NARSs were expatriates (table 8.9). Knowing that there is a strong bias in this sample towards (a) countries that have expatriate researchers and (b) small countries with a relatively high proportion of expatriate researchers, the actual global percentage is probably somewhat lower. For example, assuming that the numbers of expatriate researchers working within the Chinese and Indian systems are negligible would lower the (weighted average) percentage of expatriate researchers in the less-developed countries to 2.9%, while increasing the sample coverage to 66%.

For the pre-1980 period, insufficient quantitative data are available to make realistic estimates about the number of expatriate agricultural researchers working in less-developed countries. But even without this quantitative evidence, there is no doubt that expatriate researchers have played an important role to date in the agricultural research of the less-developed world. During the first half of this century, agricultural research in the colonies (as in the New World) was largely staffed by expatriate researchers. This, coupled with the marked absence of trained local researchers, had important implications for agricultural research after independence. In cases where there was an abrupt break with the colonizing country, such as between Zaire and Belgium or Indonesia and the Netherlands, whole agricultural research systems collapsed because there was effectively no trained local

---

[15] For Australia, rates of return in the 50% to 80% range were reported for pasture research (Duncan 1972), while research projects undertaken by the Commonwealth Scientific and Industrial Research Organization's Division of Entomology over the 1960-75 period yielded benefit-cost ratios in the 0.9 to 4.4 range (Marsden et al. 1980). Scobie and Eveleens (1987) report a rate of return around 30% for 1926/7-1983/4 investments in the New Zealand system.

[16] Representing 60 of 130 less-developed countries and 25% of the total number of agricultural researchers in the less-developed countries.

staff to take over the work. But there have also been serious problems in cases where the transition to independence took place rather smoothly.

In sub-Saharan Africa, for example, France and Great Britain had implemented agricultural research structures that were coordinated at a transnational level. The British colonies hosted several regional agricultural research institutes, while in the French colonies, all agricultural research was coordinated by institutes headquartered in France (section 7.2.1). Many of these organizational setups, however, became obsolete after independence. Nationalization and a complete reorganization and restructuring of the existing research capacity took place. An additional complicating factor for French colonies was that France considered all research structures in former colonies as French property and insisted on continuing research under French direction. The process of nationalizing agricultural research in most cases was only gradually initiated some time after independence, and in some instances, it has yet to be completed. Although the percentage of expatriate researchers in sub-Saharan Africa declined from roughly 90% in 1961-65 to 29% in 1981-85, the absolute number has not been reduced. On the contrary, it has increased from an estimated 1,200 in 1961-65 to 1,400 in 1981-85. It seems, however, that, while the larger African countries have almost completely replaced expatriate agricultural researchers with nationals, the smaller countries, where agricultural research was nearly nonexistent in the early 1960s, have considerably increased the numbers of expatriates attached to their NARSs.

In regions consisting of a large number of relatively small countries, such as the Caribbean and the Pacific, high percentages of expatriate researchers are still working in the NARSs. Heavy reliance on expatriates, however, does not necessarily reflect an initial stage of development. Most of the island states in the Caribbean and the Pacific, for example, have relatively high living standards in terms of GDP per capita and education levels. However, countries with populations of fewer than one million employ, on average, 61 agricultural researchers per million population, which is substantially more than the average less-developed country, which employs 29 researchers per million. Moreover, countries with small populations are often unable to support their own comprehensive university systems, requiring that students study abroad. This may lower the chances of their own nationals pursuing a formal professional education.

In the majority of Latin American countries, the share of expatriate researchers has never been especially high because most of these countries were independent prior to the growth in formal agricultural research in the late 19th century. European immigrants, however, did play an important role in the establishment of the first agricultural research stations in Latin American.

NARSs in less-developed countries are presently at many different stages of national independence in terms of their reliance on expatriate researchers. Some, especially the smaller countries, still depend heavily on expatriate researchers, while the larger countries have more or less completed the replacement of expatriates by nationals. Expatriate researchers within the larger NARSs hold mainly advisory functions, while in the smaller NARSs they often still hold management-related positions. Many of the former are associated with development contracts from the aid agencies of the more-developed countries. Sometimes these retain links with colonial times, but increasingly the donors of such resources are nontraditional in this sense, such as the Scandinavian countries and the US. Whether these development contracts are traditional or not, there is a problem that they tend to be not only time-bound but short. This means that many expatriate staff are not involved in the NARSs for the typical gestation periods required for many applied agricultural research projects. Accordingly, the effectiveness of the research endeavor is often compromised, and this may be a key issue in any research policy analysis of the efficiency of contemporary NARSs.

8.4   SIZE , SCOPE, AND SPILLOVERS

Size may "not matter" in every aspect of life, but in the context of a country, size is clearly associated with many aspects of development. Development assistance, for instance, has long been disproportionately channeled to small countries, at least as measured by population magnitude (de Vries 1973). The claims for preferential treatment on behalf of relatively small states (e.g., Jalan 1982) seems both articulate and effective, and their alleged special problems of vulnerability, remoteness, high unit costs in general, and capital in particular may lead to a persistence of past trends in development assistance. This is notwithstanding the finding of Srinivasan (1986b) that such difficulties are neither peculiar nor special.

Diversity within both a NARS and the agricultural sector it serves also carries with it many policy and management concerns. Choices must be made concerning the commodity, agroecological, technological, and problem orientation of a system's research program. And, it is these decisions concerning the scope of a NARS that have a substantial bearing on the cost structures of national research programs.

Our present purpose is not to contribute to the wider debate about the size and diversity of a country and development in general, but rather to address issues of size, scope, and spillovers as they relate more directly to NARSs and the costs of doing research. We offer some broad observations supported by comparative data across countries and NARSs of varying size and give some special attention to those issues that are particularly pertinent, but not necessarily unique, to small NARSs.

8.4.1   Size

Over the two decades after 1961-65, the average size of less-developed NARSs increased from 150 to 600 full-time equivalent (fte) researchers, while average research expenditures (expressed in constant 1980 PPP dollars) grew from $8.4 million to $27.9 million per system.[17] The number of less-developed micro systems (i.e., those with fewer than 25 fte researchers) dropped by nearly one-half from 74 systems in 1961-65 to 39 systems in

---

[17] Corresponding figures across all (i.e., both more- and less-developed) NARSs are presented in section 7.1.3.

1981-85, although 95 of 130 of the world's less-developed NARSs still employ fewer than 200 fte researchers. By way of contrast, there are now 14 less-developed NARSs that employ more than 1,000 researchers. These data point to a considerable diversity among NARSs in their scale of operation, growth, and development over the past 25 years, and their dynamics are surely worth closer scrutiny. Here we consider in more detail some of the key policy issues arising from the (changing) size of NARSs and the agricultural sectors in which they operate.

There is, unfortunately, a dearth of empirical evidence concerning the economies of scale, or probably more appropriately size, of the research process in general and the agricultural research process in particular.[18] Much of the difficulty in generating evidence on size economies stems from specifying just what is being measured or counted as research output.[19] The options range from using tangible measures of the research process per se through to the realized or potential economic benefits of research. In the former case analysts have sought to estimate research production functions that relate inventive output, as proxied by patent levels or intensity (measuring, for example, patents per billion dollars of sales), scientific publications, and the like to relevant research inputs (Griliches 1984; Pardey 1989). In summarizing these types of studies Kamien and Schwartz (1982, p. 66) concluded that the innovation process itself does not appear to exhibit economies of scale and that constant or even diminishing returns are more likely the case. If constant returns to scale do indeed prevail, then there is no particular unit cost advantage or disadvantage to bigness or smallness and so there is no single, optimal-sized research operation. But one needs to be circumspect in drawing such conclusions from these findings.

For one thing, the temporal dimension of the research process can act to confound measured research-input to research-output relationships. Mansfield et al. (1971), for instance, found that total R&D costs for a given project seemed to increase at an increasing rate as the project period was shortened. This phenomenon may be especially relevant for agricultural research where seasonality constraints to biological processes (at least for some crops) can only be overcome by investing in expensive greenhouses and accelerated growth facilities, multilocational trials, or even more modern biotechnology procedures, which can all serve to increase overheads substantially and hence also average costs.

In a related vein, slotting new research personnel and facilities into an ongoing, longer-run research program usually incurs additional organizational costs. Lucas (1967) and Prescott and Visscher (1980) have argued that firms incur adjustment costs that are an

---

[18] Beattie and Taylor (1985, pp. 52-3) reaffirm that returns to scale in its traditional (single-output) sense has to do with the expansion of output in response to a proportionate expansion of all factors, while returns to size refers to the behavior of costs as the production level of output changes. We usually have in mind the latter, and economically more meaningful, concept when contemplating NARS growth or optimum size, but (in keeping with the literature being cited) we use economies of scale and size in an interchangeable and not always entirely rigorous fashion throughout this section.

[19] There are also difficulties of dealing meaningfully with stocks versus flows on the input side. For a treatment of these issues in the context of agricultural research see section 8.2.

increasing function of the rate of adjustment or growth. It generally takes some time for new personnel to "learn the ropes," while newly installed buildings and equipment usually require a shakedown period before reaching their productive potential. To the extent that NARSs must invest productive resources in the organizational capital required for growth and change, and this level of investment is an increasing function of the speed of adjustment, then rapidly growing or changing NARSs will have higher average cost structures than slower growing systems.[20]

At an even more fundamental level, it is not always clear which scale or size economies are being (or ought to be) assessed. The unit of analysis can range from an individual research project, through a research program or facility, up to the level of an entire research system. Certainly the research effort at the facility or system level involves aggregating across a heterogeneous group of individual research projects or programs that vary in their commodity, type-of-technology, site, and problem focus, to name but a few of the characteristics that are noteworthy in this regard. In the case of agricultural research, for instance, the capital or overhead costs required to support a crop breeding program as compared with a crop protection or management program depend on many factors, such as the particular commodity focus — a perennial crop versus an annual crop (or, in the case of livestock research, a large versus a small ruminant program) — and the particular set of research problems under study.[21]

A further and related difficulty, especially from a policy perspective, is that economies of size and scale of inventive processes or institutions per se represent only part of the story. It is often the *multibenefit economies of scale*[22] relating overall research investments to the eventual economic benefits flowing from the research that is of ultimate concern. It is in this context that size as well as the technological, agroecological, and socioeconomic diversity of the agricultural sector itself, and the research system served by it, comes more directly into play. It is to these diversity issues that we now turn.

8.4.2 Scope

Agricultural research produces new materials, knowledge, processes, and ultimately economic benefits that have commodity, technology, and site-specific characteristics. A new rice variety may have waterlogging tolerance bred into it, making it suitable for swampy areas, while another breeding exercise may isolate the genetic basis for brown planthopper

---

[20] Although there are commonly expressed concerns that precedent rather than opportunity plays too large a role in research funding decisions, it is no doubt the case that, in some instances, the cost of organizational capital includes the foregone benefits from experiments or programs that are prematurely terminated or redirected.

[21] See Ruttan (1982, pp. 166-71) for additional discussion on these issues.

[22] This is analogous to the concept of multiproduct scale economies, but here the product-specific benefits arising from research rather than research products per se constitute the output measure. These issues are considered in more detail in the following subsection.

resistance, making it possible to incorporate this trait into rice varieties suitable for various rainfed and irrigated regimes. Still other research may develop management techniques aimed at increasing rice yields through improved fertilizer placement, timing, and application rates. In this sense a NARS may profitably be viewed as a multiproduct operation where the products of its research are differentiated according to commodity, site, technology, and other characteristics. Under this scheme a new swamp-rice variety could be classified (at least conceptually if not in practice) as a product different from a new rice variety suited to rainfed conditions, in exactly the same manner that a new wheat variety is more readily seen as a product different from a new breed of chicken.

By classifying NARSs (or their components) as multiproduct operations in this way, it is possible to draw directly from the literature on economies of scope to gain more complete and operationally relevant insights into the cost structure of a NARS.[23] Put simply, positive economies of scope exist when a research operation (be it a research unit, program, station, or even an entire system) can produce a given bundle of products more cheaply than a combination of separate operations, each producing a single product at the same general level. Such scope economies arise from the sharing or joint utilization of inputs. More specific examples for the case in point include the joint utilization of indivisible assets (such as greenhouses, mainframe computers, electric generators, etc.) to perform different lines of research that lead to differentiable research products, the economies of networking within a research operation to avoid or minimize unnecessary duplication of effort, the reuse of an input (such as using the same pool of parent breeding material to produce both swamp and upland rice varieties), and the sharing of intangible assets or know-how across different lines of research.

The more familiar single-product concept of economies of scale quite naturally broadens into a notion of *multiproduct economies of scale* where it turns out that the overall degree of scale economies (or, indeed, diseconomies) is a weighted average of the product-specific degrees of economies of scale, magnified by the degree of economies of scope. A particularly important implication of this result is that sufficiently strong scope economies can confer scale economies on the entire product set, even if there are constant returns or, up to a point, diseconomies of scale in the separate products.

This provides a powerful conceptual, and perhaps operationally useful, framework for jointly assessing size and diversity issues within a NARS. For example, there are likely (at least over certain levels of output) to be strong and positive economies of scope across some product lines (e.g., rainfed, irrigated, and swamp-rice technologies) and either weakly positive, constant, or even negative economies of scope across other product lines (e.g., rice breeding versus chicken breeding). Characterizing these potential scope and size econo-

---

[23] The notions of economies of scope (and multiproduct economies of scale) discussed here owe much to the work of Panzar and Willig (1981). Bailey and Friedlaender (1982) give a useful overview of the pertinent issues while Pardey (1986) represents an early application to agricultural research of a perspective on economies of scope.

mies and, most important, their interactions to yield multiproduct economies of scale requires a substantial, but perhaps not insurmountable, amount of product-specific cost and output data. While this task is challenging in the context of the multiproduct firm, it is even more difficult, and perhaps not entirely relevant, for a research organization with multiple lines of research. The difficulty, as described earlier, is coming up with meaningful measures of research output. While direct measures of research output are possible, it may be more appropriate for research policy purposes to identify the overall "economies of scale" linking research inputs to research benefits rather than to identify the scale characteristics of the research-input to research-output relationship. There are certainly several research evaluation methods that could be used to assess the realized or potential economic benefits arising from research. These methods could well make it feasible to investigate the multibenefit economies of scale arising from a joint program of research.

8.4.3 Spillovers

Agricultural research impacts can spill over well beyond their target location, commodity, or even market level. Analogous spillovers are found to occur between firms and industries in the nonagricultural sector (Griliches 1979; Jaffe 1986, 1989). But in contrast to industrial-sector research, much agricultural research exhibits a degree of site specificity. This limits the extent to which the results or products of agricultural research targeted towards one region (or, more broadly, production environment) are applicable or freely available to another.

Accurately assessing the degree to which research results are transferable is a tricky business requiring, at a minimum, disaggregated information at a commodity or even subcommodity level (e.g., differentiating between irrigated versus rainfed rice) concerning the types of technology and their agroecological sensitivity. While generalizations on this score are inevitably false in some respects, it is often the case that new plant varieties (or some specific traits therein) exhibit a higher degree of locational specificity than do crop protection or crop and soil management technologies.

The notion of research spillovers within the context of a particular research operation or system is, in essence, economies of scope under a different guise. For example, locational spillovers are conceptualized as one research product or technology that is applicable to several locations, while a perspective on economies of scope would distinguish between multiple products that are specific to a single location. System-to-system spillovers are more conventionally thought of as an external benefit (i.e., an un- or under-priced technology) being made available to one system through the efforts of another. While there is no real substitute for disaggregated information when it comes to assessing the magnitude of these spillovers, there are other means by which the spillover potential between NARSs can be characterized. One of these involves assessing the technological or *research proximity* of NARSs, where this measure refers to the degree by which the research interests of one NARS overlaps with that of another.[24] The idea here is that potential spillovers are more likely between proximate NARSs than between those whose research interest are far apart. Proximity in this sense can be captured by the commodity focus, research discipline and speciality, agroecological focus, etc.

8.4.4 Some Empirical Observations

At the outset, the point must be made that, for NARSs in the aggregate sense, size does indeed matter. The differences apparent in table 8.11, where data are averaged according to a NARS-size classification, reveal systematic trends with respect to the number of research workers employed in NARSs at the beginning of the 1960s. What were then the smallest systems have grown rapidly in terms of total expenditures and, especially, numbers of research personnel.

Much of the literature on the effects of a country's size has taken size to be best represented by total population, usually for the sake of simplicity. In order to provide

Table 8.11: *Features and Growth of NARSs According to Size of Research System*

Size of NARS[b]	Research expenditures[a]			Average annual growth between 1961-65 and 1981-85	
	as a % of AgGDP	per unit labor[c]	per unit land[d]	Research expenditures	Researchers
	%	*(1980 PPP dollars)*		%	%
Less-Developed Countries					
Micro (63)[e]	0.43	3.69	0.48	6.5	9.4
Small-medium (43)	0.48	6.78	1.31	5.6	6.2
Medium-large (9)	0.39	3.06	1.96	6.6	7.3
More-Developed Countries					
Small-medium (7)	0.64	47.99	5.06	3.5	2.7
Medium-large (15)	2.15	240.98	4.10	4.0	1.6

[a] Observations represent 1981-85 weighted averages.

[b] The classification of NARSs into micro, small-medium, and medium-large was based on the number of researchers employed by NARSs in 1961-65. A NARS was classified as micro if it employed fewer than 25 researchers, small-medium if it employed between 25 and 400, and medium-large if it employed more than 400.

[c] Labor is measured as the economically active agricultural population.

[d] Land is measured as arable land and permanent crops plus permanent pastures.

[e] Bracketed figures represent number of countries included.

---

[24] Jaffe (1986, 1989) used a distance metric based on patent data to quantify the proximity or "closeness" of firms doing R&D in technological space, while Pardey (1986) used a similar metric to measure the proximity of US agricultural experiment stations in terms of their disciplinary mix.

convenient links between that literature and the present discussion of NARSs, a second presentation of descriptive data is offered in table 8.12. Several contrasts can be drawn from these data. Overwhelmingly, the differences between less- and more-developed countries that have been described previously stand out again. With one exception, when reviewed in this tabular manner, size seems to have little influence on the measures of research intensity in more-developed countries. This is the anomalously low measure of intensity per unit of agricultural land, arising, presumably, from the livestock/pastoral orientation of several of the medium-sized more-developed countries.

Table 8.12: *The Link between Population Size and Research Intensities, 1981-85 Average*

Population size[a]	Research expenditures			Researchers		
	as a % of AgGDP	per unit labor[b]	per ha land[c]	per billion AgGDP	per million labor units[b]	per million ha land[c]
	%	(1980 PPP dollars)		(full-time equivalents)		
Less-Developed Countries						
Micro (26)[d]	1.74	23.82	5.02	248	339	72
Small (31)	0.91	11.31	0.66	113	140	8
Small-medium (29)	0.63	6.53	0.83	73	76	10
Medium-large (15)	0.47	6.44	0.92	82	113	16
Large (14)	0.37	3.20	2.03	91	79	50
More-Developed Countries						
Small (5)	2.09	247.21	7.47	317	3,755	113
Medium (11)	1.91	182.40	1.84	227	2,165	22
Large (6)	2.04	226.68	6.73	234	2,599	77

Source: AgGDP data primarily taken from World Bank (1989), population data from FAO (1987b), and land data from *FAO Production Yearbooks*.

Note: Observations represent 1981-85 weighted averages.

[a] Size of population has been classified as follows: < 1 million as micro; 1-5 million as small; 5-15 million as small-medium; 15-40 million as medium-large; > 40 million as large.

[b] Labor is measured as the economically active agricultural population.

[c] Land is measured as arable land and permanent crops plus permanent pastures.

[d] Bracketed figures represent number of countries included.

In contrast, most measures of research intensity diminish progressively with increasing size of country. Perhaps the most surprising average of those reported in table 8.12 is the especially high research-intensity ratio (measuring agricultural research expenditures as a percentage of AgGDP) of 1.7% for the micro less-developed countries in 1981-85. In this limited sense, such countries superficially appear to be like more-developed countries. The explanation, however, rests in the expatriate domination of such small NARSs, as elaborated in section 8.3.2, along with an inability to exploit size economies and probably a reluctance to specialize selectively among commodities.

The political economy forces discussed in chapter 1 are, as always, pervasive in their influence on public expenditures for agricultural research. In this case, international "willingness to pay" seemingly helps the smallest countries enjoy (or suffer) a relatively intensive research investment. This, in turn, doubtless reflects the continuing success of the already-noted special claims on behalf of small countries and the political importance of maintaining historical (often ex-colonial) links and responsibilities, in spite of the typically high costs of supporting expatriate rather than national research staff.

It might reasonably be hypothesized that, for many reasons, small NARSs are less likely than large ones to be "mature" systems, as indicated by stability of staffing and management, adequacy of research infrastructure, refinement of research policy, etc. In turn, it can be further hypothesized that less mature systems are less likely than others to be able effectively to borrow research findings from elsewhere. There are frequent complaints, for example, that small NARSs are unable to deal with their charged responsibilities in international testing networks such as those coordinated by several IARCs (Anderson, Herdt, and Scobie 1988).

In an ideal world, the economies of size and scope as they apply to NARS organizations might be studied by examining the varying historical experiences of countries around the world. For instance, regression models of the following type might be postulated:

$$
\begin{bmatrix} \text{Cost per unit} \\ \text{of research} \\ \text{output} \end{bmatrix} = F \begin{bmatrix} \text{Size of the NARS,} \\ \text{the agricultural sec-} \\ \text{tor, or the country} \end{bmatrix}, \begin{bmatrix} \text{Diversity of} \\ \text{national} \\ \text{agriculture} \end{bmatrix}, \begin{bmatrix} \text{Strength of} \\ \text{support for} \\ \text{research} \end{bmatrix}, \begin{bmatrix} \text{Opportunities} \\ \text{for spill-in} \\ \text{and free-riding} \end{bmatrix} \quad (8.1)
$$

There are, predictably, considerable conceptual and empirical difficulties with even these few variables. We leave to others, or to a later occasion, attempts to flesh out such an important relationship, but, by way of making a start, a few observations can be made and some preliminary analyses entertained.

The left-side variable itself has several implicit problems. For the numerator, these concern temporal aspects of accounting, including (a) handling the depreciation of physical and human capital and (b) appropriate alignment of costs with research products, given the variable lags in research productivity. Analogously, the denominator has all the measurement challenges documented elsewhere in this volume and, in order to be useful in research planning, the output measure should, minimally, be one of expected research benefits — but this makes assessment awkward.

The explanatory variables present their own difficulties. Some of those pertaining to size have been noted above. The diversity of the agricultural sector is not straightforward either. Perhaps something like the number of major agroecological regions in a country raised to the power of the average number of major commodities per region (say, those exceeding 10% of output in each respective region) could capture the phenomenon adequately. The nature of support for the system might be measured by political economy variables of the type discussed in chapter 1. The final variable concerning "free-riding" and the efficient exploitation of opportunities for "spilling in" technical innovations, etc., from elsewhere might be forged from the rather subjective data embodied in the spill-in matrices of Davis, Oram, and Ryan (1987) and Ryan and Davis (1990) or by using proxies for the research proximity of NARSs as described in section 8.4.3. But this will inevitably be tricky and controversial.

Until these problems can be resolved, we must be content with much more simplistic and partial quantitative insights. By way of illustration, we take agricultural research intensity (representing agricultural research expenditures as a percentage of AgGDP, i.e., *ARI*) as an indicator of the costliness of research endeavor and we relate this to a size variable (the magnitude of *AgGDP*) and a richness-of-country variable (gross domestic product per capita, *GDP/CAP*). The data are for individual countries recorded as the quinquennial averages reported elsewhere in this volume.[25] The results for a double-log specification are reported in equation 8.2:

$$
\begin{aligned}
\ln\ (ARI)\ =\ &2.11 - 0.18 \ln\ (AgGDP)\ + 0.42 \ln\ (GDP/CAP) \\
&\quad\ \ (0.02) \qquad\qquad\quad\ (0.05) \\[6pt]
&+ 0.57\ SSA\ + 0.24\ ASP\ + 0.20\ WANA\ + 0.72\ MDC \qquad\qquad (8.2)\\
&\quad (0.10)\qquad\ (0.11)\qquad\ (0.11)\qquad\quad (0.12) \\[6pt]
&- 0.57\ T_1\ - 0.35\ T_2\ - 0.25\ T_3\ - 0.16\ T_4 \\
&\quad (0.10)\quad\ (0.10)\quad\ (0.09)\quad\ (0.09)
\end{aligned}
$$

$n = 616,\ \ R^2 = 0.42,\ \ SEE = 0.76$

where the second row reports coefficients for regional intercept dummy variables[26] with Latin America & Caribbean as the base class, the third reports counterparts for the first four sample periods with the 1981-85 period as the base class, and respective errors are reported in brackets.

This empirical association further supports the conclusion in chapter 7 that richer countries do indeed invest in agricultural research more intensively (with an elasticity of 0.42 with respect to average income) but it is suggestive of significant "size economies" with an elasticity of −0.18 for intensity with respect to size of AgGDP. We note, however, that when the sample is partitioned into less- and more-developed countries, this size effect is not so readily apparent in the case of more-developed countries but it is even more pronounced than reported here for less-developed countries (section 1.3). After controlling for sectoral size and per capita income differentials, the pattern of coefficients across the

---

[25] The sample size used here is smaller than the 760 observations for which we have research expenditure data (see appendix to this volume). This is largely because of missing GDP and AgGDP observations.

[26] SSA = sub-Saharan Africa, ASP = Asia & Pacific, WANA = West Asia & North Africa, and MDC = More-Developed Countries.

four time dummies (normalized on the 1981-85 period) points to a continuous, but not clearly accelerating or decelerating, increase in agricultural research-intensity ratios. The positive and significant coefficients on the regional dummies suggest that all countries appear to spend more on agricultural research as a share of AgGDP than do those in Latin America & Caribbean, again after sectoral size and income differentials have been accounted for. The apparent size, scope, and spillover issues implicit in the formulation of equation 8.1 surely deserve careful empirical investigation to guide future decisions. For the moment, we are obliged to return to more speculative, a priori commentary.

Policy Implications

On the face of it, small countries are, ceteris paribus, more likely to have agroecologically similar neighbors and near neighbors, and to have fewer major agricultural commodities and major agroecological zones than other countries. The existence of small but highly diverse countries and isolated island nations surely complicates the picture, but these generalizations seem incontestable from simple logic and geography. To the extent that the research programs of these small states are proximate to others in the sense described above, it can be argued that they should give top priority to organizing an effective capacity for exploiting research conducted elsewhere. In short, they need to develop a well-targeted, carefully specialized "spill-in" capability rather than to seek to develop a conventional, broadly based, across-the-commodity-board research system.

If this argument is valid, it has implications for many aspects of NARS strategy and management, too few of which appear to have been the subject of cogent investigation. Some of the more obvious implications include the following:

1. Recruit research staff experienced in closely similar ecology and commodity emphases, even if such people are not nationals.
2. Select staff with an established ability to borrow and adapt new technology rather than necessarily being able to invent it.
3. Send young national staff members to work and learn in the best NARSs in closely similar agroecologies.
4. Budget adequately for international travel to research institutions working on similar commodities and in similar agroecologies.
5. Reward productive staff members adequately to discourage departures to other NARSs pursuing similar policies.
6. Enter into agreements with appropriate near-neighbor NARSs that will facilitate regional cooperation, particularly through (3) above.
7. Maintain a flexible and responsive stance to emerging opportunities, such as may be identified through (4) above.

On the last-mentioned point, it is generally understood that small-country NARSs are particularly susceptible to the vagaries of national funding and to boom/bust cycles. An extremely small but extremely unstable example is documented by Treadgold (1984). The challenge for research administrators is to recognize the reality of market instability and natural variability and thus to be prepared to abandon existing plans and to seize new possibilities — all of which is so much easier said than done.

Albeit through a circuitous approach, this brings us to the question of the scope of agricultural research programs, particularly for small NARSs. Demonstrably, small systems cannot do everything that may warrant consideration in a research program, and a high degree of selectivity is clearly required. Selectivity, in turn, requires specialization according to commodity, discipline, and function. As for consideration of flexibility and responsiveness, much judgment is called for in planning resource allocations for research. There are difficult trade-offs between the flexibility that comes within short-term (often two-year) employment contracts and the turmoil and low productivity that come with rapid staff turnover. One guiding principle is to exploit (but pay fairly for) human capital in its broadest sense. In the language of Schultz (1975), the essence of human capital is the ability to deal with disequilibria. Buying the skills of bright, well-trained, motivated people is likely to be the most worthwhile staffing policy for all NARSs, but this is especially the case for small-country NARSs. While diversification of the research portfolio at any given time may be highly circumscribed, an adaptive and responsive system can handle desired diversification over time, even given the often long leads and lags that make such dynamic optimization so demanding of judgment and luck.

8.5   SOURCES AND STRUCTURE OF SUPPORT

Rather than pursue a prescriptive approach concerning financial support for public agricultural research, the discussion and evidence in chapter 1 sought to foster an understanding of the political economy dimensions of this support. In this subsection we explore briefly some extensions to this positivist approach that may have practical public-policy implications related to investing in NARSs. The relatively modest research component of official development assistance, which nevertheless constitutes a significant share of total funds available for the NARSs in many less-developed countries, is then quantified.

8.5.1   A Transactions Perspective

Quantitative evidence and analysis at the aggregate *level* of the costs and benefits of public agricultural research is quite widely available. But, while the *incidence* of actual and potential benefits from agricultural research is subject to an increasing amount of analysis and policy debate, there has been comparatively little attention paid to the incidence of the overall cost of research under existing or alternative funding scenarios.

To be sure, there have been increasingly sophisticated attempts to discern the incidence of research benefits among different producing and consuming groups stratified by income class, by location (e.g., international versus domestic), by ownership of factors of production (e.g., labor or land), or vertically across stages in a multimarket production system (e.g., on- versus off-farm). One of the primary motivations for producing information of this type is the desire of NARS managers and others to improve the accountability of their research systems, which is demanded by the various interest groups on whose support they ultimately depend. Identifying distributional consequences with regard to the actual or potential benefits of agricultural research is seen as a useful means of giving public research systems an opportunity to address some of these accountability concerns. Unfortunately, when working through their distributional calculus, many of these studies explicitly omit the level of investments and, more important, the sources of investments in public agricultural research. Net as opposed to gross benefits from research need not be coincident. As Alston and Mullen (1989) note, there are disparities between the benefits from research and the costs of research investments. These disparities generate, for any particular sector of an industry, incentives for research investments that may differ from those that would maximize benefits for the industry as a whole (chapter 1).

Over the past few decades, for many but not all NARSs, a goodly portion of public-sector research budgets has been financed from general taxpayer revenues. More recently, an increasing number of systems have begun to move (or at least contemplate moving) the burden of support away from the general taxpayer and closer to the agents (be they farmers, large commercial estate-crop operations, or private input and processing companies) who are among the direct beneficiaries of research. These tendencies are not only a response to greater budgetary pressures arising from tighter fiscal policies but are also viewed as a means to achieve a more complete and seemingly equitable correspondence between the incidence of research benefits and the sources of support for publicly executed research.

A potentially constructive framework within which to assess the public policy options surrounding agricultural research is to classify public policy instruments in a transactions context (Tisdell 1986; Bollard, Harper, and Theron 1987). Governments can carry out *internal transactions* to secure the direct public provision of R&D goods and services, or they may undertake, facilitate, or influence *market transactions* in order to enhance the market provision of new agricultural technologies. The latter set of policy instruments includes macroeconomic policies (e.g., interest and exchange rate policies) that influence the nature and level of private investment in R&D activities, trade policies that influence the international transfer of technologies, public funding of privately or jointly executed research, and the enactment or enforcement of intellectual property-rights legislation (e.g., patents or plant variety protection acts).[27] In contrast with a market failure rationale for public intervention in agricultural R&D, this transactions approach may well provide more operational guidelines for government involvement. It could do this through a quantitative assessment of the expected relative efficiency as well as distributive merits

---

[27] See Evenson and Putnam (1990, table 25.1) for a listing of patent and plant variety protection acts in force in selected less-developed countries.

of securing R&D services either by internal, market, or joint (i.e., essentially public/private) transactions.

For those services provided directly by research agencies within the public sector, there still remains a policy concern over how these activities are to be financed. In addition to funding agricultural research out of general revenues (involving taxpayers to both domestic and donor governments), there are many cost-recovery mechanisms that may be (and indeed, are being) used.[28] Alternative sources of revenue for public research systems include taxes or legislatively sanctioned check-off schemes on agricultural output or exports, fee-for-service (i.e., contract) research, license fees related to third-party use of publicly provided research output, or even the proceeds from state-run football pools as in the case of the Norwegian system.

To augment the transactions perspective on public interventions, it may be worth-while contemplating the degree by which such interventions enhance the *contestability* of agricultural R&D markets.[29] One of the major insights from the contestability literature is that potential entry in a market, not just the presence of current rivals, can act to discipline market behavior and promote the public interest. This is so even in the presence of economies of size and scope that rule out the concept of perfect competition as a guiding principle for policy formulation. A more complete understanding of the (potential) inci-dence of research benefits is most useful when designing and developing alternative mechanisms for allocating funds and recovering costs that embody contestability princi-ples. For those R&D activities generating clearly identifiable benefits that accrue to agents considered in no need of special assistance, fee-for-service schemes (or some variant thereof) may well be appropriate. But even for those cases in which the public funding of R&D services is deemed appropriate (due to appropriability difficulties, distributional concerns, or whatever), minimizing the transactions costs of providing such services, through due regard to contestability principles, may still be a guiding principle (Sandrey and Reynolds 1990). For instance, changing the mechanisms by which public funds are allocated within the public research system (e.g., switching at least a portion of disburse-ments from core or formula funding to competitive or project-based funding schemes) can raise the threat of entry, be it from "competing" laboratories or colleagues within the public system or elsewhere, thereby generating increased efficiencies within the public system itself.

---

[28] For example, out of a total 1988/89 budget of NZ$ 110 million (with $80 million spent on 'basic' research), New Zealand's MAF Technology collected some $25 million in revenue — up from $15 million the previous year and $2-3 million in 1985/86 (Sandrey and Reynolds 1990, p. 103).

[29] Baumol and Lee (1991) give an introductory exposition of contestability; for a more complete treatment see Baumol, Panzar, and Willig (1988).

8.5.2 Donor-Sourced Funding[30]

Foreign development assistance takes many forms: financial, technical, and food. This aid may be transferred through projects or programs and may represent grants or concessional loans. If official flows from one country to another are aimed at economic development or welfare improvements and if they have at least a 25% grant element, they are called *official development assistance* (ODA). Foreign assistance to agriculture is a portion of total ODA and includes such diverse components as agricultural research and extension, irrigation projects, rural roads, agricultural education and training, flood control projects, health improvement programs, integrated rural development projects, and agricultural policy assistance.

The rationale for foreign aid in general as well as for aid to agriculture rests on humanitarian (moral or ethical), political (strategic self-interest), and economic self-interest grounds (Krueger 1986).[31] Ruttan (1989b) notes that several variants of the humanitarian argument have been made on the basis of compensation for past injustices, uneven distribution of global natural resources, and a moral obligation to help the least-advantaged members of society. The political self-interest rationale rests on the notion that aid will strengthen the political commitment of the recipient to the donor(s). A quick glance at the distribution by country of US foreign assistance makes it clear that strategic political considerations have been a major motivation for aid, regardless or whether the intended results have been achieved. One line of argument, emanating from certain (often farmer) groups within donor communities, suggests that aid may generate foreign competition, while the contrary perspective argues that agricultural growth stimulates income growth, which in turn increases less-developed country imports of agricultural products from donor countries (see section 2.2.6).

So it is with that small, but significant, part of ODA used to promote and fund agricultural research. In the remainder of this section we review several recent attempts to measure the level and incidence of this component of ODA.

Notes on Measurement

Information on how much donors are contributing to less-developed NARSs is difficult to get and often of poor quality. Donors tend to have their own ways of classifying contributions, which causes serious comparability problems. Sometimes rough assumptions have to be made about whether a budget item, or a part of it, contributes to agricultural research. Also, at the receiving end, information about donor contributions is often incomplete. Some NARSs keep donor contributions completely outside their financial reporting, while in others it cannot be identified as such because the donor contributions have not been made

---

[30] The introduction to this section draws heavily on Norton, Ortiz, and Pardey *(forthcoming)*.

[31] Precedence and, in particular, old colonial relationships also play a role here.

directly to the research organization, but rather through a ministry of finance, for example. Another reporting difficulty concerns in-kind contributions, such as equipment or expatriate staff, that are paid for directly by the donor. Expatriate salaries especially can constitute a large part of a donor contribution, but in most instances they do not appear in the financial reporting of the NARSs. It can, therefore, be assumed that donor-sourced support reported by the NARSs generally understates the full extent of donor contributions.

Two sources, Oram and Bindlish (1981) and Lewis (1987), compiled much of what little information is available on the research component of ODA from the donors' side. Oram and Bindlish provide information on donor contributions to NARSs for 1976 and 1980, and Lewis provides similar information for 1970, 1975, 1980, and 1984. In those instances where the data sets are nominally comparable (namely 1975/76 and 1980), there are some considerable differences that we were unable to resolve entirely. While Lewis's sample coverage improves over time and apparently accounts for about 80% of ODA in 1984 (Lewis 1987, p. 19), the Oram and Bindlish data appear to have somewhat more complete coverage. Consequently, the 1984 Lewis total of US $590.73 million (Lewis 1987, table 5) was first rescaled[32] and the donor contributions to the CGIAR were then deducted in an attempt to improve its comparability with the Oram and Bindlish estimates. We also applied appropriate deflators and convertors to recalibrate the data from both sources and to express them in comparable 1980 dollar units.

Some Estimates

A decline in donor contributions to less-developed NARSs since 1980, as depicted in table 8.13, seems quite plausible. Oram and Bindlish (1981, p. 65) reported that among 12 donors for which they had year-by-year data for the period 1976-80, eight showed decreased levels of funding in 1980 compared with 1979, and four were lower in 1979 than in 1978. It is always possible that our rescaling procedure did not fully redress the possible bias inherent in Lewis's (1987, p. 12) admittedly "harder core concept of research" vis-à-vis Oram and Bindlish's 1980 estimate. As a result, measurement difficulties may well account, at least in part, for the lower 1984 estimate. But, in the broader context of stagnating total ODA expenditures throughout the 1980s (OECD 1989), it is not unlikely that a decline of donor contributions to less-developed NARSs has occured.

Recent data on World Bank supported agricultural research in less-developed countries (table 8.14) corroborates the stagnant if not contractionary pattern of donor support evident in table 8.13. In addition to its substantial commitment to the CGIAR system (see table 9.3, chapter 9), the Bank accounts for around one-quarter of all donor funds (grants and loans) in support of agricultural research in less-developed countries. In real terms,

---

[32] The rescaling increased Lewis's 1984 total by the inverse of 0.76, representing the proportion of Oram and Bindlish's 1980 research expenditure total that corresponds to the more restrictive institutional coverage in Lewis's 1984 sample.

Table 8.13: *Donor Contributions to NARSs in Less-Developed Countries as Reported by Donors*

Year	Donor contributions to NARSs[a]	Official development assistance		
		Total	Research share	
	(millions of 1980 US dollars)	*(millions of 1980 US dollars)*	*(%)*	
1980	691	33,780	2.0	
1984	624	34,172	1.8	

Source: Donor contributions to less-developed NARSs adapted from Oram and Bindlish (1981) and Lewis (1987). Total ODA expenditures adapted from OECD (1989).

[a] The US was the major single donor in both years providing around 27% of all donor contributions followed by the World Bank representing around 23%.

World Bank support for country-specific agricultural research initiatives trended downward over the 1981 to 1987 period. Commitments to "free standing" projects — i.e., those projects designed specifically to strengthen NARSs and to develop their linkages to extension systems — tended to be quite lumpy. Nearly two-thirds of the US$817 million committed by the Bank over the 1981-87 period specifically to strengthen NARSs are accounted for by just six projects. The research share of agricultural and rural development projects generally fell in the 2% to 3% range. The polar years of the time series in table 8.14

Table 8.14: *World Bank Contributions to Agricultural Research in Less-Developed Countries*

Fiscal Year[a]	Research component of			Research share of ARD[c] projects
	"Free standing" research projects[b]	ARD[c] projects	Total[d]	
	(millions 1980 US dollars)			*(%)*
1981	156.7	173.8	330.5	5.1
1982	118.1	53.1	171.3	2.0
1983	120.5	53.6	174.2	2.3
1984	54.9	51.4	106.3	2.3
1985	84.0	73.3	157.4	2.8
1986	70.0	48.0	118.0	2.1
1987	66.0	88.9	154.9	4.4

Source: Compiled from data reported in Pritchard (1990), particularly tables 4, 5, and annex I tables 1 to 7.

Note: Data are reported in nominal US dollars and were deflated by the GDP implicit price deflator for the US. Data may not add up exactly because of rounding.

[a] The World Bank fiscal year runs from July 1 to June 30.

[b] "Free standing" projects include those projects designed specifically to strengthen NARSs.

[c] ARD = Agricultural and Rural Development

[d] Excludes both the non-Bank component of those projects cofinanced by the World Bank and other donors as well as World Bank grants to the CGIAR system.

are the exceptions, with the 1981 percentage buoyed by a particularly large research component (17%) of a sizable Mexican project and the 1987 figure similarly inflated by larger than average research shares of projects executed in Brazil and Côte d'Ivoire.

As far as we are aware, there has to date been no systematic compilation of donor contributions to NARSs in less-developed countries from a recipient perspective. On the basis of the information collected in constructing the Indicator Series, we derived estimates of the donor component in total expenditures as reported by the NARSs for 58 of the 130 less-developed countries. Shares of external funding differ considerably between countries. The range is from zero in Venezuela and South Korea, for example, to 85% in Tuvalu. The regional averages also differ greatly. Sub-Saharan Africa tops the list with 35% of agricultural research expenditure financed by donors. It is followed by Asia & Pacific with 26%. Substantially lower average percentages, namely 7% and 11%, were reported by the NARSs of Latin America & Caribbean and West Asia & North Africa, respectively. In the case of Latin America & Caribbean, the sample was fairly complete (92% of total regional expenditures), so the estimate of external funding in this case should be quite accurate.

These regional averages mask a good deal of intraregional diversity in the levels of donor support. To gain some quantitative perspectives on this issue, countries were first ranked within a region in terms of the proportion of total NARS expenditures accounted for by donor-sourced funds. Weighted average donor shares were then formed for those countries most intensively supported by donors and, in turn, accounting for one-third of a region's externally sourced expenditures. A similar average was formed across those countries receiving the least intensive donor support in each region. As demonstrated in figure 8.4, there are wide intraregional disparities in the donor component of NARS expenditures, particularly within sub-Saharan Africa and Asia & Pacific. In the latter region, a substantial portion of NARS expenditures is internally financed, while in sub-Saharan Africa, a sizable proportion of locally-sourced research expenditures is matched by a similar proportion of donor-sourced funds.

Multiplying total regional expenditures by the ratio of shares of external funding gives an estimate of the donor contribution to the NARSs in each of the less-developed regions.[33] Aggregating the donor contributions to each of the regions gives a total that should be comparable with the data collected by Oram and Bindlish and Lewis at the donor end. An average of the 1980 and 1984 estimates of the donor contributions to less-developed NARSs based on donors' information should thus be comparable with the 1981-85 estimate of donor contributions based on NARS information.

Annual donor contributions to less-developed country NARSs during 1981-85 averaged $658 million, based on donor information (table 8.13), and $357 million, based on information from NARSs (table 8.15). It is quite likely that the discrepancy between these

---

[33] To enhance the comparability of these figures with those of Lewis and Oram and Bindlish, we substituted the 1980 annual average exchange rates used in table 8.13 for the 1980 PPP indices used more generally throughout this volume to translate agricultural research expenditures.

Figure 8.4: *Donor component of less-developed country NARS expenditures*

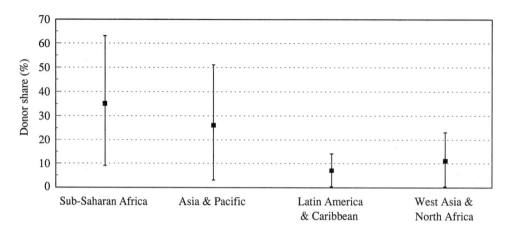

Note : The share of external funding in NARS expenditures by region has been calculated on the basis of the following samples: sub-Saharan Africa — 22 countries (representing 51% of total regional expenditures); Asia & Pacific — 13 countries (42%); Latin America & Caribbean — 15 countries (92%); West Asia & North Africa — 8 countries (26%).
The boxes are centered on the regional averages weighted by NARS expenditures, while the lines indicate the regional disparity in donor share as measured by the range between the averages for the one-thirds of data expenditures in each region that are, respectively, most and least heavily reliant on donor-sourced funds.

figures is due in large part to a systematic understatement of the donor contributions by the recipients. As mentioned above, expatriate salaries and other contributions in kind are in most cases not known and not included in the financial reports of the NARSs. In other words, nearly one-half ($301 million) of the donor contribution does not get reported in the financial statements of the NARSs. This interpretation assumes that the donor-specified figures do not overestimate their contribution to the NARSs of less-developed countries. It may be, however, that some such overstatement is inevitable in the procedures used in reporting ODA expenditures. To take one obvious case for illustrative purposes, consider the handling of air travel to NARSs by donor-agency personnel supervising disbursement of agricultural research project funds. This cost is certainly real to the donor and appropriately categorized as ODA, yet the reluctance of a NARS accountant to report this as received aid (or even to know the extent of such costs) can readily be understood.

Another seeming inconsistency between the data reported by NARSs and by donors is the regional distribution of the donor contributions. Based on data from NARSs, 37% of the donor contributions goes to sub-Saharan Africa, 44% to Asia & Pacific (including China), 9% to Latin America & Caribbean, and 11% to West Asia & North Africa. Based on 1980 data from donors reported by Oram and Bindlish (1981), these percentages are 40% to sub-Saharan Africa, 26% to Asia & Pacific (including China), 23% to Latin America & Caribbean, and 11% to West Asia & North Africa. The differences are perhaps

Table 8.15: *Share of Donor Funding in Total Agricultural Research Expenditures as Reported by Recipient NARSs, 1981-85 Average*

Regions[a]	Donor funding as a share of total	Public research expenditures	
		Total	Donor component
	(%)	(millions 1980 US dollars)[b]	
Sub-Saharan Africa	35	373	131
China[c]	4	497	20
Asia & Pacific, excl. China	26	520	136
Latin America & Caribbean	7	480	33
West Asia & North Africa	11	342	38
Less-Developed Countries	16	2214	357

Note: Data may not add up exactly because of rounding.

[a] Details of regional sample coverage given in note to table 8.4.

[b] Research expenditures are expressed here in millions of constant 1980 US dollars using annual average exchange rates rather than the purchasing power parities used elsewhere in this volume.

[c] During 1981-85 China received an average of $800 million ODA per annum. We assume that about 2.5% of the assistance to China is given to agricultural research. This is slightly more than then 1.8% to 2.0% for all less-developed countries (table 8.12), so that the externally funded component of Chinese agricultural research amounts to 4%.

not too surprising given the crudeness of the estimates. In the case of Latin America & Caribbean, however, the incongruence is more striking, given the high coverage of the sample in this region. NARS data for Latin America & Caribbean annually report around $33 million during 1981-85, while an estimate based on donor data would amount to an annual average of $151 million during the same period.

The coverage of donor contributions in the measurement of NARS expenditures, based on NARS data, is thus rather incomplete. A major reason is that contributions in kind are difficult for NARSs to measure. As noted, however, there are analogous difficulties with donor-sourced data, and thus the problems of making an accurate assessment of what are considerable flows of research resources are endemic and unlikely to be easily resolved without further detailed case-by-case study. The rewards to such refinement of data are probably few, however, and for the purpose of policy analysis, the choice of data source will usually be set by whether the analysis pertains to either the donor or the recipient.

8.6   CONCLUDING COMMENT

Our discussion of these selected topics of importance to national agricultural research policy has revealed some serious gaps in information. The difficulties begin with fundamental inadequacies in the data available on NARSs, their investment patterns, and personnel details. These difficulties are compounded at more disaggregated levels of examination. There are conceptual uncertainties at all levels, and perhaps this helps to explain why there has thus far been much little effort towards investigating and resolving these problems, which present substantive questions to research administrators concerning efficiency in public research systems. The situation may not be much better in private research, the topic that is covered in part IV, which follows the international dimensions of agricultural research considered in the next chapter. In making the start that we have in this chapter, however, we believe areas for investigation have been identified that will be helpful in boosting efficiency in research, and indications given of the types of analyses that can be and must be made.

Chapter 9

International Agricultural Research

Guido Gryseels and Jock R. Anderson[1]

Since the early days of exploration and subsequent colonialism, there has always been an international dimension to the knowledge base concerning agricultural production in what is today's less-developed world. Often, this was rather informal, especially prior to the establishment of the science-based agriculture that has largely been a feature of the 20th century, and invariably it was what in contemporary terms would be described as applied or adaptive research. There was much direct transfer of European farming techniques, practices, and materials and, of course, acquisition and exploitation of novel species and varieties for enriching agricultural technology in Europe and the New World as well as the tropics.

Because of time and space limitations the authors do not attempt to do justice to the rich experience embodied in these antecedents of contemporary institutional arrangements. Rather, they concentrate on the past three decades and spotlight developments related to the most significant international initiative of the period, namely, the Consultative Group on International Agricultural Research (CGIAR).

9.1 THE CGIAR

During the past three decades, major efforts have been undertaken to develop a number of international agricultural research centers (IARCs) to assist less-developed countries in developing technology to increase agricultural production and in building research capacity. The first four institutes were established during the 1960s through the joint efforts of the Ford and Rockefeller Foundations. Beginning in the mid-1940s and accelerating in the 1950s, the Rockefeller Foundation had articulated its concerns for the perceived forthcom-

1 Many people assisted in assembling this compact discussion of complex and rapidly evolving phenomena, including colleagues in TAC and in the TAC and CGIAR Secretariats. Without implicating any of our helpful colleagues, we especially wish to record our gratitude to our friends at ISNAR, Wilhelmina Eveleens, Philip Pardey, and Johannes Roseboom, for providing such material assistance so unstintingly.

ing world food problem by placing its agricultural staff to work alongside national scientists in the agricultural research organizations of several less-developed countries. Out of these efforts grew the fledgling programs on the major staple foods that became the initial research centers, coordinated support for which evolved to become the CGIAR, or CG, as it is often referred to for brevity (Baum 1986). The CGIAR is an informal association of more than 40 countries, international and regional organizations, private foundations, and representatives from national research systems in the less-developed world, formed to guide and support a system of international research centers.

9.1.1   Institutional Background

The CGIAR was established in 1971 and, up to 1990, sponsored the work of 13 centers, of which 10 had headquarters located in a less-developed country. The goal for the system, adopted in 1986, reads: "Through international agricultural research and related activities, to contribute to increasing sustainable food production in developing countries in such a way that the nutritional level and general economic well-being of low-income people are improved" (TAC/CGIAR 1987). Subsequent analysis and reflection led to a suggestion that this goal of increasing food production should be modified to incorporate the concept of achieving food self-reliance in the less-developed world (food self-reliance being defined as the capacity of a nation to provide a sufficient and stable food supply to all of its inhabitants, either from domestic production or from production of exportable goods to enable food imports). It implies that a country produces those things it is best able to do and, where necessary, trades them for required food. A goal of self-reliance has several implications. Although the range of commodities that are potentially in a production system is likely to be large, this does not commit the CGIAR to working on all commodities, but it does commit the CGIAR to taking account of diverse production systems and their capacities to produce income and employment as well as marketable commodities.

From this new perspective, the CG at its October 1990 meeting accepted a revised mission statement (TAC Secretariat 1990b, p. 87): "Through international research and related activities, and in partnership with national research systems, to contribute to sustainable improvements in the productivity of agriculture, forestry, and fisheries in developing countries in ways that enhance nutrition and well-being, especially among low-income people."

CGIAR research activities cover a broad spectrum of crop and livestock production, plant breeding, farming systems, natural resource conservation and management, animal diseases, plant protection, post-harvest technology, and food policy. Ten centers (CIAT, CIMMYT, CIP, ICARDA, ICRISAT, IITA, ILCA, ILRAD, IRRI, and WARDA) have mandates that cover either food commodities, agroecological zones, or both. CIMMYT, CIP, and IRRI have a global mandate for particular crops. One other center, IBPGR, is devoted to the collection, conservation, and utilization of plant germplasm. Of the two remaining centers, IFPRI deals with food policy issues while ISNAR deals with strengthening national agricultural research systems in the areas of research policy, organization, and management. Three centers have headquarters located in Latin America, two in Asia, one in the Near East, four in sub-Saharan Africa, two in Western Europe, and one in the US. The location of each of the 13 centers, the year of establishment, and an overview of formal and operational mandates is presented in appendix A9.1.

The key characteristics of the concept of the international center for the CG system have been defined recently, in what is characteristic prose, as the following (TAC/CGIAR 1987):

(a) the *global perspective* of mandates and programs, which facilitates a clear focus on problems lending themselves to an international solution;
(b) the *international status* of centers and their governance, staffing, program design, and resource support, which protect their mandates and programs from undue political pressures and from purely national or regional influences;
(c) the *international mobility* of germplasm, center staff, and knowledge;
(d) the *principle of universality*, which ensures accessibility of research results to all interested parties and openness of centers to all partners seeking collaboration.

The major thrusts of the CG can now, in the early 1990s, be encapsulated as enhancing sustainable agriculture through resource conservation and management, increasing the productivity of commodity production systems, improving the policy environment, and strengthening national research capacities. The primary functions of the centers are research and technology development, training and institution building, germplasm conservation, catalyzation and coordination of research on specific topics or commodities, and facilitation of linkages among national systems and advanced institutions. The specific role of each center is shaped largely by the scale of the effort of its partners in the global agricultural research system, particularly those of national systems. The CGIAR has played many vital roles, some of which might be categorized as gap-filling and bridging in agricultural research. Centers fill many gaps in research and technology generation that cannot be effectively addressed by national research systems in less-developed countries, and they provide a bridge to basic and strategic institutions, wherever these are to be found.

The operational approach of the typical CG center is characterized by a problem-oriented, multidisciplinary, commodity approach, with a critical mass of committed scientists, relative freedom from political constraints, and an ability to maintain a continuity of effort. An important consideration in the planning and prioritizing of international research relates to the nature and size of likely spillover effects that will result from a research activity, i.e., the benefits of research undertaken in one region but applicable to other regions, especially in those with similar agricultural environments. Phrasing these contemporary generalizations about the nature of the CG centers so succinctly and bluntly is possible through the consensus reached by the many actors on the CG stage during the many reflective deliberations of the 1980s. The CG system wasn't always this way and wasn't always perceived to have been so intended — but that is another story told variously by Baum (1986) and in the studies alluded to in section 9.1.4 below.

The CG is assisted in its work by a Technical Advisory Committee (TAC), which now has 18 members and a chairperson, and which meets three times a year. TAC proposes priorities and strategies for the group, reviews the quality and relevance of center programs, and makes recommendations on resource allocation among centers. TAC's work is supported by a secretariat located at FAO headquarters in Rome. The CGIAR also has a secretariat that deals primarily with donor relations, financial matters, and management issues and which is based at the World Bank in Washington, D.C. The TAC and CGIAR secretariats respectively commission external program and management reviews that, once every five years or so, evaluate the work of each center. Donors accept these review processes as assuring accountability of the CG supported activities, and thus these reviews are at the heart of the donors' continuing ability to provide unrestricted core funding.

As is clear from this volume, the CGIAR is just one of the participants in the global agricultural research system and commands only a fraction of its resources. In terms of global public-sector spending in agricultural research, the share handled by the CG in 1981-85 was a modest 1.8%; in relation to the expenditures by and for less-developed countries, its share amounted to only 4.3%.[2] The CGIAR as a whole has been reviewed twice (CGIAR 1977, 1981). A major effort has also been made to assess its impact and achievements. Amongst many things, it was found that its activities have indeed had a substantial impact on food production as well as on institution building in the less-developed world (Anderson 1985b; Anderson, Herdt, and Scobie 1988).

At the start of its operations in 1971, the CGIAR had 20 donors and a total annual budget of US$ 20 million. By 1990, the number of donors had grown to 40 and the total budget to approximately US$ 280 million.

9.1.2   Evolving CGIAR Priorities and Strategies

In its early years, the CG gave highest priority to research on the food staples of the poor, particularly cereals. About two-thirds of CG resources were allocated to research on rice, maize, and wheat. High priority was also given to improving the quality of diets through research on food legumes and ruminant livestock. Research on starchy food sources, including roots and tubers, was also prioritized, because of their dietary significance in less-developed countries, their potential for producing high outputs of energy on a unit-area basis, and the prospects for yield increases.

CGIAR priorities and strategies have been reviewed and revised at regular intervals. The most recent and most in-depth review of priorities and strategies took place in 1985 (TAC/CGIAR 1987). It reaffirmed the prevalent CGIAR view of the international center

---

2 These precentages are based on the global agricultural research expenditure series presented in chapter 7.
   The major omissions from the series are the USSR and Eastern Europe, Mongolia, North Korea, Vietnam,
   Cambodia, Cuba, and South Africa.

concept as being sound and relevant, as well as the earlier decision that factor-oriented research should be sought through collaboration with other specialized institutions, rather than through factor-based institutions within the system. Gradually, the scope of CGIAR objectives shifted from the narrow focus of increasing food production to the broader aim of improving agricultural productivity while sustaining the natural resource base. It was slowly acknowledged that increased food production was a necessary but not sufficient condition to alleviate malnutrition, and that increased cash incomes could also provide the means to improve access to food.

After the 1985 review it was contended that the CG should maintain its focus on food crops, rather than expand its coverage to include others. Because of the impact achieved and the growing strength of national programs, TAC also urged a reduction in the relative allocations to wheat and rice. It suggested increasing relative allocations to sorghum, millet, and maize. Funding for research on food legumes was to be maintained, while research on lentils, faba beans, and cocoyams was to be phased out.

It was also proposed that, after these shifts were accomplished, the CG should consider engaging in three new ventures if additional funding were to become available. In order of priority, the three commodity groups recommended were tropical vegetables, vegetable oils (in particular coconut), and aquaculture. The key considerations in these recommendations were the potential of these ventures for income and employment generation and the opportunity for filling nutritional gaps and dietary inadequacies.

It was, of course, recognized from an early stage that socioeconomic issues created major constraints to the adoption of new technology, and centers were encouraged to give greater attention to social science research. It was also perceived that the need for factor-oriented research was best met through a commodity approach. The need for centers to build national research capacity was repeatedly stressed, as well as the need for developing truly collaborative relations with national programs as well as with advanced institutions.

The centers were urged to move their scientific focus "upstream," through more strategic research away from location-specific applied and adaptive research, which was seen as the responsibility of national programs. It was also acknowledged that the urgency of food problems had shifted from Asia to sub-Saharan Africa and it was recommended that the group further increases its effort in sub-Saharan Africa.

In 1978, the CG acknowledged the crucial importance of food policy issues by bringing in the International Food Policy Research Institute (IFPRI). As of 1990, the last center to join was the International Service for National Agricultural Research (ISNAR), established in 1980 to assist less-developed countries in the areas of research policy, organization, and management. Decisions taken at the October 1990 meeting of the CGIAR will, however, result in the addition of several more centers to the system, including ICRAF, IIMI, INIBAP, eventually AVRDC, and probably also ICLARM.

Although reviewing CG priorities and strategies is a continuing activity, TAC is committed to presenting to the group every five years or so a major report on priorities and strategies. The next such report is due in 1991 and, among other initiatives, new strategic approaches such as the development of "eco-regional" centers will be appraised (TAC Secretariat 1990b). New methods of analysis to inform the research priority setting process, such as elaborated by Ryan and Davis (1990), may also be applied.

9.1.3   Trends in CGIAR Resource Allocation

Expenditures

In CG-center budgets, a distinction is made between expenditures for core or essential programs, which are central to a center's mandate, and expenditures for special projects. Special projects are usually specific to a country or topic for which funding is being provided directly by a donor (not necessarily a member of the group) without being channelled through CG mechanisms. Table 9.1 provides an overview of the evolution of expenditures by centers between 1960 and 1989, both for core and special-project activities.

Figure 9.1:   *Breakdown of total CGIAR expenditures by core operating, core capital, and special-project expenditures, percentage shares by year*

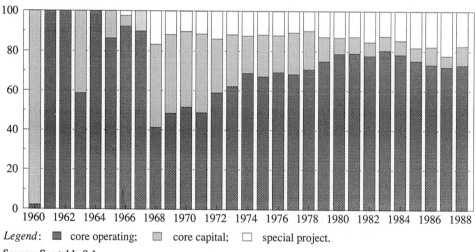

Legend: ■ core operating;   ▨ core capital;   □ special project.
Source: See table 9.1.

As depicted in figure 9.1 and table 9.2, the core component of expenditures has gradually dropped from 88% during 1971-75 to 81% during 1986-88, while "special-project" or "complementary-activity" funding has increased in relative importance. Within core funds, a distinction is made between those used for operating expenditures and those for capital. During the first five years of the CGIAR (1971-75), 25% of total core and special funds was allocated to capital expenditures, as the number of centers was increasing rapidly. Between 1976 and 1980, the rate of expansion slowed and the two centers that then joined the system, IFPRI and ISNAR, did not require construction capital for headquarters. This

Table 9.1: *Total CGIAR Core and Special-Project Expenditures by Center, in Millions of Current US Dollars*

	IRRI	CIMMYT	IITA	CIAT	CIP	ICRISAT	ILRAD	ILCA	WARDA	IBPGR	ICARDA	IFPRI	ISNAR	CGIAR
1960	7.4	—	—	—	—	—	—	—	—	—	—	—	—	7.4
1961	0.2	—	—	—	—	—	—	—	—	—	—	—	—	0.2
1962	0.4	—	—	—	—	—	—	—	—	—	—	—	—	0.4
1963	0.9	—	—	—	—	—	—	—	—	—	—	—	—	0.9
1964	0.6	—	—	—	—	—	—	—	—	—	—	—	—	0.6
1965	1.0	—	0.3	—	—	—	—	—	—	—	—	—	—	1.3
1966	1.1	0.5	0.4	—	—	—	—	—	—	—	—	—	—	2.0
1967	1.2	1.2	1.0	—	—	—	—	—	—	—	—	—	—	3.4
1968	2.4	1.8	2.6	0.2	—	—	—	—	—	—	—	—	—	7.0
1969	2.5	3.0	4.7	1.4	—	—	—	—	—	—	—	—	—	11.6
1970	2.9	5.1	4.4	2.3	—	—	—	—	—	—	—	—	—	14.8
1971	3.7	6.1	6.8	3.6	—	—	—	—	—	—	—	—	—	20.2
1972	4.4	6.5	6.4	4.5	0.5	0.3	—	—	—	—	—	—	—	22.7
1973	4.6	7.7	6.4	6.4	1.3	2.7	—	—	—	—	—	—	—	29.1
1974	7.8	7.5	7.2	6.1	2.3	3.8	0.7	0.3	0.5	—	—	—	—	36.2
1975	10.6	9.1	9.8	6.7	2.9	6.2	2.1	1.7	0.6	0.5	—	0.3	—	50.4
1976	12.3	10.7	11.1	7.0	4.7	7.3	4.7	4.3	0.8	0.9	1.5	0.8	—	66.0
1977	15.4	11.4	12.8	10.2	5.9	11.2	5.4	6.7	1.3	1.3	4.6	1.2	—	87.4
1978	15.8	13.9	17.4	13.0	5.8	14.1	7.9	7.5	1.9	1.7	7.6	1.6	—	108.3
1979	18.7	16.8	19.5	15.2	7.4	13.5	7.4	9.0	1.7	2.4	10.6	1.9	—	125.2
1980	21.1	18.3	20.0	17.3	8.2	14.4	9.1	10.0	2.9	3.0	13.1	2.5	1.1	141.5
1981	22.4	20.4	22.8	18.9	9.6	15.7	9.9	10.5	3.3	3.6	15.8	3.2	1.6	158.5
1982	25.0	21.1	27.2	21.6	9.9	19.5	8.9	10.5	4.1	3.1	15.6	4.2	2.9	174.8
1983	24.6	20.6	25.9	23.1	11.9	20.9	9.3	12.0	5.3	4.5	20.6	5.0	4.1	188.8
1984	26.9	24.9	27.9	24.0	11.7	21.1	9.1	14.6	6.4	4.2	20.8	5.9	4.3	201.4
1985	31.3	24.9	33.5	23.3	11.1	24.5	9.6	16.3	5.9	4.5	21.9	6.5	4.7	216.9
1986	29.8	27.2	36.4	24.3	13.8	31.6	10.7	18.5	4.9	5.0	21.8	7.2	6.1	239.3
1987	32.7	28.4	35.9	29.6	15.3	41.0	12.2	16.3	6.9	5.3	23.6	8.0	7.1	261.1
1988	31.0	32.4	41.3	29.7	18.2	39.3	13.3	18.5	6.8	6.9	24.0	8.8	8.8	279.0
1989	33.9	34.1	32.1	32.6	21.9	36.3	14.1	20.6	6.3	7.6	22.6	10.9	9.8	282.8

Source: 1960-82 – CGIAR Secretariat (1983a); 1983-84 – CGIAR Secretariat (1988a); 1985-86 – CGIAR Secretariat (1988c, 1989b); 1987 – CGIAR Secretariat (1988c, 1989b), IBPGR (1988) & ISNAR (1988); 1988 – CGIAR Secretariat (1988c, 1989c), CIMMYT (1989), IBPGR (1989) & ISNAR (1988); 1989 – CGIAR Secretariat (1990b).

Table 9.2:   *Total CGIAR Expenditures by Type and Source*

Expenditure category	1961-65	1966-70	1971-75	1976-80	1981-85	1986-88
	(millions 1980 US dollars)					
Core operating	1.5	9.2	32.7	88.7	121.6	137.4
Core capital	0.3	5.7	13.2	18.4	10.9	15.3
Total core	*1.7*	*14.8*	*45.9*	*107.1*	*132.5*	*152.8*
Special projects	—	1.9	6.4	14.4	23.1	36.2
Total core & special projects	1.7	16.8	52.3	121.5	155.6	188.9
	%	*%*	*%*	*%*	*%*	*%*
Core operating	84	55	63	73	78	73
Core capital	16	34	25	15	7	8
Total core	*100*	*89*	*88*	*88*	*85*	*81*
Special projects	—	11	12	12	15	19
Total core & special projects	100	100	100	100	100	100

Source: See Table 9.1.

Note: "Core" and "special projects" categories were reclassified as "essential" and "desirable" categories after 1987 and the latter as "complementary" after 1990. Totals may not sum due to rounding.

resulted in overall capital expenditures declining to just over 15% of total core and special funding. Since 1980, capital expenditures have only accounted for between 7-8% of total expenditures, assuming that capital components of special-project expenditures are negligible.

Until 1980, the four centers that provided the founding basis of the system had the largest budgets. During the mid-eighties, and because of the rapid expansion of its activities in sub-Saharan Africa, ICRISAT has grown to the size of the "big four." During 1987 and 1989 it had the largest of all center budgets. WARDA and IBPGR are the smallest centers although, with the construction of a new headquarters site in Côte d'Ivoire, WARDA will make expanded claims on the system's budget.

Between 1971-75 and 1986-88, CG core funds increased annually by 9% and special-project funds by 13.1%. During the 1970s, when the number of centers was expanding, total funds increased by 18% annually, slowing to 5% in the 1980s. The growth of core funds was more constrained than that of special projects, which had a high growth rate because of the crisis perceived in sub-Saharan Africa.

Centers' expenditures are largely financed from donor grants. Tadvalkar (1989) has analyzed trends of and prospects for CGIAR funding and noted that the composition of donors has changed considerably over time. In the beginning, the Rockefeller and Ford Foundations jointly provided 26% of core funds, but this declined to 3.8% between 1976 and 1980, and to less than 2% during the 1980s (table 9.3). European countries increased

Table 9.3: *Donor Contribution to Core Programs of the CGIAR*

	Average annual contribution to centers' core programs				Share of core budget			
	1972-75	76-80	81-85	86-87	1972-75	76-80	81-85	86-87
	(millions 1980 US dollars)				%	%	%	%
EEC	8.3	23.7	25.8	37.0	16	23	20	25
Non-EEC European	2.9	6.9	8.1	13.1	6	7	6	9
Subtotal Europe	*11.2*	*30.6*	*33.9*	*50.1*	*22*	*30*	*26*	*34*
Japan	0.5	4.3	7.8	12.5	1	4	6	9
United States	10.7	24.9	34.9	32.0	21	24	27	22
Other	5.6	12.5	11.2	11.1	11	12	9	8
More-developed countries	*27.9*	*72.3*	*87.8*	*105.7*	*55*	*70*	*68*	*73*
Less-developed countries	*0.2*	*1.2*	*2.3*	*1.1*	*0*	*1*	*2*	*1*
World Bank	3.9	10.5	16.8	21.6	8	10	13	15
Regional development banks	2.4	7.3	6.8	7.8	5	7	5	5
Other international and regional organizations	3.2	7.5	13.6	8.0	6	7	10	5
Foundations	13.2	3.9	2.2	1.6	26	4	2	1
Total	50.9	102.6	129.5	145.7	100	100	100	100

Source: CGIAR Secretariat (1983a, 1988a, 1989b).

Note: Donor contributions only include those to core activities because of incomplete data on special-project funding. Totals may not sum due to rounding.

their share of core funds from 22% during 1972-75 to 34% during 1986-87.[3] The USA was the largest individual donor, contributing about one-quarter of core funds, but Japan has also become important. The World Bank acts as a balancing "donor of last resort," allocating its funds after other donor intentions are known. It accounts for about 15% of total contributions. Funding procedures were considerably improved in 1984 when the CGIAR Secretariat established a stabilization fund to protect centers from exchange-rate and inflation risks.

Staffing

The number of senior professional staff in the centers increased from about 250 in 1974 to nearly 750 in 1989 (table 9.4). Those with the largest professional staffs are ICRISAT and

---

[3] The preconverted data that were available to us meant that donor contributions were first converted and then deflated to a 1980 base. Thus the highly valued US dollar (vis-à-vis many European currencies) that prevailed during the early 1980s may well account for the "apparent" decline in the European donor share during the 1981-85 period, as shown in table 9.3.

CIMMYT, followed by CIAT, IITA, and IRRI. WARDA, IBPGR, and ISNAR have had the smallest staffing levels. Since 1983, the annual increase in the number of senior staff positions has averaged 2.8%. IBPGR and IFPRI increased their senior staff by more than 10% per annum, while CIP, ILCA, and ICRISAT reported virtually no growth in senior staff levels in this period.

These data, which were not available for the period prior to 1974, should be treated with some caution. Some centers label all internationally recruited staff as senior professionals, whereas others only include designated senior scientific staff, thus excluding those in administrative or post-doctoral positions. However, since 1987 when a new resource allocation process was introduced, there has been a systematic attempt by the CGIAR Secretariat to standardize staff reporting by centers. Before then, data on staffing levels were reported only for core programs, where the number of senior staff positions is subject to TAC approval. Centers that fall short of resources can, however, overcome a temporary need by recruiting staff on a consultancy basis. The actual input in staff-year equivalents is thus likely to have been well above the reported staffing levels.

Senior staff are supported in their work by supervisory and support staff, as well as post-doctoral and other visiting scientists. In 1988, the unit cost, including cost of support staff, per senior staff member in CG centers amounted to an average of US$ 179,000 per annum, ranging from $257,000 at CIP to $125,000 at WARDA. The variation between centers in cost of senior staff can be attributed largely to the number of staff that centers assign to support scientists. Personnel costs account for approximately 60% of the centers recurrent expenditures, supplies and services for 28%, travel for 8%, and replacement of equipment for 4% (CGIAR Secretariat 1989a).

Allocation by Program and Commodity

Needless to say, research has traditionally received the largest share of core expenditures allocated to operations, stepping up from close to 50% during the early set-up phase to over 60% recently (table 9.5). The balance of core expenditures is divided between training and conferences; library, documentation, and information; and administration.

A breakdown of the research component of the CG core operating expenditures by "commodity" orientation is given in table 9.6. It clearly shows that, with the expansion of the system, the share allocated to cereals research declined steadily to about 40%. Rice is the crop that still receives the largest share of resources, although this has declined to a stable 17%. Research on maize was the second largest recipient during the initial years but has been reduced steadily to about 7%. Throughout the CG's history, allocation to research on wheat and barley declined only gradually. Research on triticale is probably being phased out. Although in relative terms the share of resources allocated to research on cereals declined, in real terms it more than tripled from 10.1 million constant 1980 US dollars during 1971-75 to 32.5 million in 1986-88.

About one-third of the funds for livestock research is allocated to research on animal

Table 9.4: *Senior Professional Staff by CGIAR Center*

	IRRI	CIMMYT	IITA	CIAT	CIP	ICRISAT	ILRAD	ILCA	WARDA	IBPGR	ICARDA	IFPRI	ISNAR	CGIAR
1974	42	53	57	45	19	21	1	12	na	—	—	—	—	250
1975	50	55	74	46	19	26	8	22	na	na	—	na	—	300
1976	49	56	73	48	20	31	15	32	na	na	na	na	—	324
1977	54	60	61	54	21	34	21	54	na	na	20	na	—	379
1978	57	69	65	61	29	44	31	64	na	na	22	na	—	442
1979	58	77	65	61	29	41	41	74	26	na	24	25	—	521
1980	59	83[a]	55	62	30	57	51	81	30	2	26	25	7	578
1981	59	81[a]	61	63	30	60	51	52	33	5	29	25	13	556
1982	61	79[a]	52	61	30	62	51	54	33	9	37	20	26	584
1983	66	77	57	60	30	77	45	54	45	16	41	21	23	607
1984	66	79	61	60	30	73	48	57	45	21	41	21	23	621
1985	66	82	61	65	30	73	50	59	45	23	42	21	25	642
1986	72	84	69	65	31	67	48	52	45	23	49	24	25	654
1987	71	85	69	70	31	72	59	53	28	25	54	26	25	668
1988	71	87	69	73	31	81	60	56	26	26	54	35	28	697
1989	71	85	69[a]	83	31	91	62	63	27	30	54	39	31	736

Source: 1974-79 — CGIAR Secretariat (1982); 1980-81 — CGIAR Secretariat (1982, 1983b); 1982 — CGIAR Secretariat (1983b); 1983-88 — CGIAR Secretariat (1989a); 1989 — CGIAR Secretariat (1990c).

[a] The number of senior professional staff at CIMMYT for 1981 and 1982 are estimates, as is the 1989 figure for IITA.

Table 9.5: *Functional Breakdown of Core Operating Expenditures*

	1971-75	1976-80	1981-85	1986-88
	%	%	%	%
Research	53.5	58.4	60.0	61.2
Training & conferences	8.5	7.1	6.9	8.4
Library, documentation, and information	4.9	5.9	5.4	5.6
Administration and general operations	33.1	28.6	27.7	24.8
Total	100.0	100.0	100.0	100.0

Source: Adapted from CGIAR Secretariat (1983a, 1986, 1989d).

diseases, and the balance is allocated for animal production. Research on food policy has increased steadily from 0.3% at its introduction as a CG activity in 1975 to 3.7% of core resources in 1986-88. Research on genetic resources has followed a similar path. Farming systems research (FSR) has been an important activity in most centers since the inception of the CGIAR, accounting for about 12% of the system's core research resources during

Table 9.6: *"Commodity" Orientation of CGIAR Core Research Operating Expenditures*

	1971-75	1976-80	1981-85	1986-88
	%	%	%	%
Rice	21.5	17.2	17.3	17.2
Wheat, barley & triticale	13.8	10.9	10.3	9.1
Maize	19.5	9.3	7.2	7.3
Sorghum & millet	3.1	3.3	4.8	5.0
Subtotal, cereals	*57.9*	*40.6*	*39.6*	*38.7*
Potatoes	4.6	7.0	6.1	6.8
Other roots & tubers	6.8	5.4	4.8	4.5
Legumes	8.1	11.4	11.2	12.9
Subtotal, crop research	*77.4*	*64.4*	*61.7*	*62.9*
Livestock	10.2	19.8	19.1	19.7
Subtotal, commodity research	*87.6*	*84.2*	*80.8*	*82.6*
Farming systems	12.2	11.7	9.9	8.5
Food policy	0.1	2.0	3.1	3.7
Genetic resources	0.1	2.0	4.2	2.8
NARS capacity building	0.0	0.0	1.9	2.4
Subtotal, other research/activity	*12.4*	*15.8*	*19.2*	*17.4*
Total	100.0	100.0	100.0	100.0

Source: Adapted from CGIAR Secretariat (1983a, 1986, 1989d).

Note: The 1971-85 shares are based on core operating research expenditures exclusive of an administrative component. This administrative component was included, apparently on a prorated basis, in the 1986-88 data. Totals may not sum due to rounding.

1971-75, but it has since then gradually declined. There are always definitional questions surrounding FSR (Simmonds 1985) and, with changing donor enthusiasm for work in this area, it may well be that both the early emphasis and the subsequent fall in FSR efforts have been overstated in such data.

Regional Allocations

Since 1983, when data first became available on center operating expenditures by region, the major share of resources has been directed to sub-Saharan Africa. In 1986-88, on average 39% of expenditures were directed towards sub-Saharan Africa, 26% to Asia, 21% to Latin America, and 14% to West Asia & North Africa (table 9.7). Regional allocations

Table 9.7: *CGIAR Core Operating Expenditures by Category, Apportioned by Geographic Region, 1986-88 Average*

	Sub-Saharan Africa	Asia & Pacific[a]	Latin America & Caribbean	West Asia & North Africa
	%	%	%	%
Research activities				
Rice	28	63	8	0
Wheat, barley & triticale	21	14	20	44
Maize	43	18	34	6
Sorghum & millet	53	42	5	0
Subtotal, cereals research	*33*	*40*	*16*	*11*
Potatoes	30	15	45	10
Other roots & tubers	45	0	55	0
Legumes	18	30	27	25
Subtotal, crop research	*30*	*33*	*24*	*13*
Livestock	68	0	21	11
Subtotal, commodity research	*39*	*25*	*23*	*13*
Farming systems	43	28	0	29
Food policy	42	55	2	1
Genetic resources	25	25	25	25
NARS capacity building	25	25	25	25
Subtotal, other research/activity	*38*	*33*	*8*	*22*
Nonresearch activities				
Information, communication, library, and documentation	47	22	18	13
Training and conferences	40	30	21	9
Total operating expenditure	39	26	21	14

Source: Adapted from CGIAR Secretariat (1989d).

Note: Totals may not add up because of rounding.

[a] Includes China.

among commodities and research activities vary considerably, however.

Research on cereals is focused on Asia, while research on food legumes appears to be relatively equally balanced between the four major less-developed regions. CG-sponsored activities on roots and tubers are predominantly focused on sub-Saharan Africa and Latin America, while almost 70% of its investment in livestock research is concentrated in sub-Saharan Africa. Research on genetic resources is equally divided between the four regions, while research on farming systems is largely concentrated in sub-Saharan Africa and West Asia & North Africa, while now receiving little attention in Latin America. Most of the food policy research is concerned with Asia and sub-Saharan Africa. Training efforts mirror the overall allocation of operating expenditures.

9.1.4 Issues Related to CGIAR Policy Choices

The CGIAR is primarily a donor's club for concerted action in agricultural research. The individual donors are subject to many influences from the diversity of groups that seek to express interest in this field of development work. The CG operates through the formation of a consensus on each of its policy choices and it is thus inevitable that, on some matters, some donors (and the interest groups whose influence they may serve) may be less than perfectly content with decisions taken. This surely explains much of the enthusiasm that has been shown for special projects and the implied individual emphasis. It also underpins the importance of TAC's role in analyzing options and presenting reasoned recommendations. There is, of course, no shortage of critics of CGIAR policy both from within the group itself and certainly from many points beyond it. It is not our present purpose to review such criticism — for one such attempt see Anderson, Herdt, and Scobie (1988) — or, indeed, to attempt a balanced analysis of major policy issues facing the group. Rather, we here offer some remarks on four broad policy matters that both TAC and the CG have anguished over and have dealt with in an evolutionary manner, namely (a) regional focus, (b) focus on food supply, (c) modus operandi of the CGIAR, and (d) sustainability issues.

Regional Focus

The distribution of CG resources by region, although not a perfect reflection of the degree of effort, provides an indication of regional emphasis within the system. The first center, IRRI, was established largely in response to the immensity of the food needs of Asia (Baum 1986). The creation of the CG itself was also motivated largely by the problems of food supply, malnutrition, and poverty that prevailed in Asia. During the early 1980s, attention shifted from Asia to Africa as the region of greatest concern with respect to needed growth in food production. Between 1970 and 1985, per capita agricultural production in sub-Saharan Africa declined by 1.3%, while in Asia it increased by 0.2%, in Latin America by 1.7%, and in West Asia & North Africa by 7% (FAO 1987a). Center activities in sub-Saharan Africa expanded rapidly, facilitated by the strong donor response to the food crisis in the region.

A survey carried out in 1986 (ISNAR 1986) showed that more than 42% of the centers' activities in sub-Saharan Africa were supported through special-project funding. It was noted that special-project funds gave centers the flexibility to respond quickly to identified problem areas, while the stability of core funding provided them with the long-term sustained commitment required for agricultural research. On the negative side, special-project funding can also be the tail that wags the center dog!

As indicated in table 9.7, in 1986-88 approximately 39% of the CG core operating expenditures were allocated to sub-Saharan Africa. Data on regional allocations of special projects are not available but it is likely that the share of sub-Saharan Africa is well above that going to core operating expenditures. Unfortunately, data on regional allocation of such expenditures are not available for the period prior to 1983. It has thus not been possible to estimate the size of the shift in CGIAR resource allocations in favor of sub-Saharan Africa and the extent to which resources have been diverted from other regions. Given the increase in funding available, it is likely that the shift has to some extent been financed from additional funding sources. The emphasis on Africa at the expense of Asia is revealed by contrasting the final row of table 9.7 (or, equivalently, the final column of table 9.8) with corresponding NARS 1981-85 expenditure shares taken from table 7.1 (chapter 7), namely 10% for sub-Saharan Africa, 57% for Asia & Pacific (including China), 20% for the Latin America & Caribbean, and 13% for West Asia and North Africa. It thus can be hypothesized, especially taking into account the prospects as opposed to the needs for success, that the Group has "gone overboard" on sub-Saharan Africa, perhaps to Asia's ultimate cost.

Progress in achieving farm-level increases in food production through CGIAR-sponsored research has been relatively slow, and the green revolution of Asia has not been repeated in sub-Saharan Africa (Anderson, Herdt, and Scobie 1988). Conditions for rapid technological progress in the agriculture of sub-Saharan Africa seem more difficult than in Asia, and therein lies a major challenge for agricultural research, both national and international. The prevalence of dryland farming and the absence of large-scale irrigation potential has inhibited the adoption of technologies developed elsewhere under more favorable conditions (TAC/CGIAR 1987). In addition, national research capacity was generally weak and the CGIAR had to make substantial investments in training and institution building. Many of Africa's environments are fragile and many areas face severe problems of resource degradation. Given the wide range in ecological conditions, breeding strategies are to a large extent location-specific. Finally, sub-Saharan Africa is deficient in basic infrastructure and effective marketing systems and, in contrast to Asia, the policy environment has generally not been conducive to smallholders boosting food production.

While during the 1980s sub-Saharan Africa may have had the most urgent need for CG support for enhancing research capacity and for the development of improved technology for increasing food production, it is appropriate to reconsider the regional allocation of CGIAR resources. Despite the rapid increases in food production in Asia during the past two decades, problems of poverty and malnutrition remain acute throughout the region, particularly in South Asia. On the basis of the size of population, the number of poor, and the value of agricultural production, Asia appears to be significantly underfunded in comparison to other regions, as illustrated in table 9.8.

Table 9.8:   *Distribution of Population, the Poor, Agricultural GDP, and CGIAR Operating Expenditures among Less-Developed Regions*

	Less-Developed Countries			
Region	Population 1985	The poor[a] 1985	AgGDP 1981-85	CGIAR expenditures 1986-88
	%	%	%	%
Sub-Saharan Africa	12	16	8	39
China	29	19	26	26
Asia & Pacific, excl. China	40	53	41	
Latin America & Caribbean	11	6	15	21
West Asia & North Africa	7	5	9	14

Source: Population data extracted from FAO (1987b), data on poverty adapted from *World Development Report 1990* (table 2.1), value of agricultural production primarily taken from World Bank (1989b), and CGIAR expenditures from table 9.7.

Note: Totals may not add up because of rounding.

[a] The poverty line in 1985 PPP dollars used in *World Development Report 1990* is $370 per capita per year.

In addition, recent evidence has suggested that the effect of each of the technologies that contributed to the green revolution in Asia, namely, modern varieties, irrigation water, and fertilizers, may be reaching a plateau and that further "breakthroughs" in raising yield potential are not expected in the short run (Byerlee 1989). However, projections by the World Bank on poverty in the less-developed world in the year 2000 show an absolute decrease of the number of poor in Asia & Pacific, China, and Latin America & Caribbean, a stagnation of this number in West Asia & North Africa, and about a 50% increase in the number of poor in sub-Saharan Africa (*World Development Report 1990*, p. 5). The issue of regional "balance" must thus be carefully considered in the continuing reviews of CGIAR priorities and strategies.

Focus on Food Supply

CG priorities and strategies were traditionally linked to the role of agriculture in food supply. Increases in food production will, however, be a necessary but not sufficient condition for sustained economic and agricultural development. Many people are simply too poor to buy available food. Alleviation of poverty and malnutrition and a reversal of environmental degradation will require efforts on a much broader front. This will require that the CG give much greater emphasis to the role of agriculture in generating income and employment, and in management and conservation of natural resources, in addition to its orientation to food production.

As noted in section 9.1.1, TAC has recognized this need for a change in emphasis by proposing that the goal of increasing food production be modified to incorporate the concept of achieving food self-reliance in the less-developed world. (TAC/CGIAR 1989). The direct implications of this change in goal are that the CG should not automatically rule out support for any commodity that is not a food staple but that contributes to food self-reliance, such as cash crops.

It is arguable that analysis of this research policy issue by TAC and others is yet far from complete. Those concerned with equity issues in general and with poverty-oriented agricultural research in particular, such as Lipton with Longhurst (1989) for one eloquent exposition, will regard the implied de-emphasis of poor peoples' crops as a grave dereliction of responsibility. However, for food self-reliance to be sustainable, increased attention will have to be given to management and conservation of natural resources, as well as to research on labor-intensive, income-elastic commodities that can meet the growing demand resulting from higher incomes and employment creation. Others, more concerned with seeking economic growth through the release of labor for nonfarm employment through labor-saving innovations, will see these issues through different eyes. The actual change in absolute factor use associated with agricultural research investment is, of course, a rather challenging econometric research issue in itself. Notwithstanding the absence of empirical resolution of such questions bearing on commodity coverage, TAC and the CGIAR are formulating proposals on how to incorporate research on forestry into the system, and ICRAF is likely to be invited to enter the CGIAR system as a center for agroforestry research. TAC has recommended incorporating research on vegetables and aquaculture as CGIAR-sponsored activities and AVRDC and ICLARM will probably join the system early in the 1990s.

The revision of the CGIAR's goal statement has also renewed debate on the issue on how the Group should balance its activities between those that favor areas of high potential and those that favor areas marginal to crop production.[4] In its early years, the CGIAR gave highest priority to basic food crops that were grown in relatively favorable environments, for example, where irrigation was available and where good responses could be obtained from the application of fertilizers and improved cultivars. It was assumed that the marketable surplus produced in these areas would help to feed the urban population, the majority of which was poor. The CGIAR Impact Study (Anderson, Herdt, and Scobie 1988) and the several other studies that it synthesized supported the validity of this assumption and demonstrated that both rural and urban poor, particularly in Asia, have benefited substantially from the green revolution.

With the establishment of ICRISAT and ICARDA, the CG gave explicit recognition to

---

4 See, e.g., Baum (1986), Anderson, Herdt, and Scobie (1988), TAC/CGIAR (1987, 1989), and chapter 3.

the research needs of less-favored, "marginal" areas with poor soils and low or erratic rainfall. For these areas, there is a need to increase the productivity of the indigenous subsistence crops without creating a demand for purchased inputs that would be beyond the reach of the target farmer. Although achieving any production impact in these areas is much slower and more difficult than in favored environments, the CG has allocated substantial resources for research on marginal areas. For the period 1983-86, for example, research on millet, sorghum, pigeonpeas, chickpeas, cowpeas, food legumes, and groundnuts as well as on farming systems at ICARDA and ICRISAT together accounted for 16.8% of core resources. If it is further assumed that, during this period, 20% of the research on rice, wheat and barley, maize, potatoes, and cassava, 30% of the research on livestock and beans, and 20% of the research on food policy, genetic resources, and institution building was allocated specifically to the needs of marginal areas, then an additional 13.5% of core resources were being so mobilized. A total of more than 30% of core resources would thus appear to have been allocated to research on marginal areas, which is probably about congruent with the share of these areas in the number of rural poor. Congruence is, of course, much less than half of the story and it is the conditioning multipliers of real research opportunity that must guide any informed research policy analysis of this issue. The definitive analysis is not yet to hand but it is clear that minimally one has to come to terms with the marginal land hypothesis (chapter 3) and to assess realistically the opportunity costs of taking scarce research resources from somewhat less marginal areas.

Mellor (1988) has argued that most of the poor live in areas of high potential and emphasized the need for economic growth as a means of radically reducing poverty. In such areas the CG should focus on the development of high-yielding agricultural technology, while governments should give high priority to the development of roads and infrastructure to allow the technology to spread and to allow employment multipliers to work. Increasing the agricultural capacity in favored environments would also help to relieve the pressure on more fragile environments. The balance of CG efforts between favored and marginal areas is an important policy issue and must be reviewed at regular intervals. The size of these efforts will be determined by evolving CGIAR priorities and their underlying perceptions, including those of developments in national systems. Part of the continuing difficulty of good decision making about the issue is its inherent complexity and the much less than satisfactory situation of detailed data, from meteorological and edaphic through demographic and social. Prospective advances in geographical information systems may permit better interpolation and organization of data and thus more penetrating analysis of such research priorities.

Modus Operandi of the CGIAR

Along the continuum of types of agricultural research — basic, strategic, applied, and adaptive — centers have been involved mainly in applied research, creating new technology that potentially can be transferred across countries (CGIAR 1981; Anderson, Herdt, and

Scobie 1988). Most centers have been also involved in basic and strategic research when gaps in knowledge required such action, and in adaptive research when national systems lacked such capability. All this emphasis on such formal types of research categories is not to deny the fundamental importance of informal activities such as those deriving from innovative farmers. Difficulties of access, data, and documentation oblige us not to give such work the attention it surely deserves.

Over time, the development in national systems of a capacity for formal research will increasingly allow these systems to take a leading role in generating technology and will gradually change the demands for CG activities. National systems will gradually assume the responsibility for applied and adaptive research, thereby enabling the centers to allocate a greater share of their resources to strategic and basic research, and to exploit the technological opportunities from advances such as those in biotechnology. TAC has strongly encouraged this shift in emphasis and this process of "devolution" (TAC/CGIAR 1987). This will change the nature of the collaboration between the CGIAR and national systems, as the latter will be involved to a much larger extent than at present in contract research, in the organization and management of research networks, and in training.

There is little quantitative information on the extent to which CG centers presently allocate resources to the various types of formal research. The CGIAR Impact Study, as summarized in Anderson (1985b), attempted to portion by center the allocation of internationally recruited staff to four broad categories of activities, i.e., strategic research, crop improvement research, other applied research, and finally, training, research support, and administration.

Nine of 13 centers were regarded as having an involvement in strategic research. For seven of these the involvement was estimated at between 10% and 15%, while for one center it was estimated at 6%. The only center with the major share of its resources involved in strategic research was ILRAD, which was estimated to allocate 83% of its senior staff time to this category.

The main category of work of senior staff in the centers was crop improvement. Crop-oriented centers devoted 25% to 40% of senior staff resources to applied crop improvement research and genetic resource conservation. Other applied research constituted 15% to 35% of senior staff resources, except for ILCA (livestock production) and IFPRI (food policy research) where the corresponding shares were 59% and 93%, respectively. The proportion of staff time allocated to training, research support, and administration averaged 15% to 30% in most of the centers. Unfortunately, there has been no follow-up to these estimates, which date from 1984, and there are thus no quantitative indicators available that clearly show shifts in overall resource allocation that may have taken place since that time.

Most relevant centers appear to have increased their involvement in strategic research, particularly in molecular biology. Given the rapid increase in special projects in the CG centers in sub-Saharan Africa since 1984, there is also likely to be a greater involvement in adaptive and applied research. In order to address this research policy issue more effectively, information mechanisms should be developed that allow for the systematic monitoring of ongoing efforts.

As centers move "upstream" and increase their involvement in basic and strategic research, there should be greater spillover effects to other regions and agroecological zones. These spillovers can either be commodity-specific or across commodities. Any assessment of resource allocation between regions and environments as discussed in the two previous sections will thus become increasingly difficult.

There is as yet little clarity, at least apparent to us, as to how the process of devolution and the hand-over of responsibilities to national systems will occur and in what time frame. Although collaborative networks are proving useful in facilitating the process (Plucknett, Smith, and Ozgediz 1990), many national systems remain weak and do not yet have even sufficient capacity to organize adaptive research on their basic food crops (Ruttan 1987b). A systematic attempt to assess comprehensively the strength of national systems at the program level has not been made. Few quantitative indicators for assessing the quality of national research efforts are available, other than general ones on staffing and budget levels. Within one national system, differential strengths exist by commodity, discipline, and program, while scientific quality may sometimes be compromised by poor management. Some national research systems are already partly involved in priority setting both within the CGIAR and at wider levels, and in principle, these systems could take greater responsibility for charting a course for devolution. Stronger national systems could possibly assume regional responsibilities and assist weaker systems in developing research capacity, perhaps in part through using donor funds. Some centers already actively support such a process. CIP, for example, organizes its training courses largely through national programs.

Research on Resource Management and Conservation

The need for rapid increases in food production during the coming decades will inevitably lead to greater pressures on natural resources. Soil erosion, deforestation, desertification, and salinization have led to widespread land degradation, and the need for agricultural research to have a sustainability perspective is now widely recognized (TAC/CGIAR 1989).

The CG has already incorporated forestry into its mandate and is evaluating institutional options for how forestry can best be integrated into the research agenda. Further, there is a renewed debate on whether factor-oriented research (e.g., on soil and water management) should be conducted by institutions within or outside the group. TAC's position has consistently been that factor-oriented research in the CGIAR can best be carried out as part of its multidisciplinary commodity-oriented approach (Crawford 1977; TAC/CGIAR 1987). Specialized, factor-oriented institutions should remain outside the CGIAR but CG centers should maintain active collaborative relationships with these institutions. Some centers have organized some research on soil, fertilizer, and water management within their commodity programs or have closely collaborated with specialist institutions for this purpose. The rapid growth in the activities of some of the other centers dealing with such factor-oriented research (IFDC, IIMI, and IBSRAM) indicates the growing demand for such activities. Although the importance of research on natural resource issues is widely acknowledged, many of these issues are rather location-specific and may thus be addressed most effectively by national programs. The 1990 decision to incorporate IIMI within the system is at least a step towards recognizing CG responsibilities and opportunities in one resource field.

9.2 OTHER INTERNATIONAL AGRICULTURAL RESEARCH

Stemming to a large degree from its well-recognized contribution to the success of the green revolution, the CGIAR is one of the more visible actors in the field of international agricultural research. There are, however, many other agricultural research institutes with mandates that transcend national boundaries. Some of them are truly international in the sense that they have global research mandates and are supported by a broad base of donors. Others have regional mandates and are largely funded by local contributors. A broad spectrum of different setups is to be found between these extremes. The historical pattern of establishing these centers is depicted in table 9.9, where the frenzied formational activity of the 1970s is clearly revealed.

As noted, toward the end of the 1980s, the notion within the CGIAR that food security could best be achieved solely through food production was replaced by one that also considers the income-generating possibilities of farmers so that they can buy food. Additionally, developments outside the CGIAR forced the group to reconsider its position. Whereas there were only a few international institutes that were not included in the CGIAR in the 1970s, during the 1980s the number of such institutes with a global or regional mandate and funded by largely the same donors as support the CGIAR increased rapidly. By the end of the 1980s the affiliation of some 10 other international agricultural research institutes was placed on the agenda of the CG. These institutes are discussed below as the CG-kindred centers.

9.2.1 CG-Kindred Centers

Both the first and second reviews of the system (CGIAR 1977, 1981) recommended that the group forego new additions not only so that existing activities could be consolidated but also because of financial constraints. In recent years, several new entities, patterned after the CG centers and, in seeming contradiction, supported by many of the same donors, have been established. At the mid-term meeting in Berlin in May 1988, members of the CG requested that the TAC undertake an examination of a possible expansion of the CGIAR to incorporate some or all of these entities.[5] The themes and institutions considered for

---

5 Centers considered for inclusion were initially labelled "non-associated centers" (TAC Secretariat 1988). Since these centers have indeed been actively associated with the CG centers in collaborative activities and

Table 9.9:   *Grouping of Dates of Foundation of Multilateral Agricultural Research (-Supporting) Centers*

Date of establishment	Affiliation	Mandate				
		Global	SSA[a]	A&P[b]	LAC[c]	WANA[d]
Before 1961	CGIAR	0	0	1	0	0
	CG-kindred	0	0	0	0	0
	Other	1	6	1	7	1
	Total	*1*	*6*	*2*	*7*	*1*
1961-70	CGIAR	1	0	0	1	0
	CG-kindred	1	0	0	0	0
	Other	0	3	3	3	1
	Total	*2*	*3*	*3*	*4*	*1*
1971-80	CGIAR	4	4	1	0	1
	CG-kindred	3	0	1	0	0
	Other	1	11	4	3	1
	Total	*8*	*15*	*6*	*3*	*2*
1981-85	CGIAR	0	0	0	0	0
	CG-kindred	3	1	0	0	0
	Other	0	1	4	1	0
	Total	*3*	*2*	*4*	*1*	*0*
All	CGIAR	5	4	2	1	1
	CG-kindred	7	1	1	0	0
	Other	2	21	12	14	3
	Grand total	14	26	15	15	4

Source: Compiled from data in IDRC (1986).

[a] SSA = Sub-Saharan Africa; [b]A&P = Asia & Pacific; [c]LAC = Latin America & Caribbean; [d]WANA = West Asia & North Africa.

possible inclusion, include the following: banana and plantain, INIBAP; vegetables, AVRDC; fisheries, ICLARM; research related to livestock diseases in sub-Saharan Africa, ITC and (with insect research in general) ICIPE; natural resource conservation and management, IBSRAM, IFDC, IIMI, and (with tropical agro-forestry, in particular) ICRAF. In addition, an organization not reported in the following tabulations but the subject of on-going attention from TAC in the implementation of forestry research within the group is

have common donor support, the term seems to us an unfortunate misnomer and here we opt for the term "CG-kindred" for referring to them.

the Special Program for Developing Countries of the International Union of Forestry Research Organizations (IUFRO/SPDC). Four of the CG-kindred centers are located in Asia, three in Africa, and one in Europe. An overview of the location, date of establishment, and mandate of the centers is presented in appendix A9.2.

In assessing the activities of these centers, TAC developed a set of criteria for receiving CGIAR support (TAC Secretariat 1990a). The activities under consideration must be of direct relevance to the mission and goals of the Group, and they must be regarded as research or directly related to research. A substantial part of these research activities must be strategic and applied in nature, and international in character. The CG system must gain a comparative advantage in undertaking the candidate activity. The activities of the candidate center must be complementary to those of other research organizations. A high quality of work is also an essential criterion for CG support. Finally, mandates and governance must conform to agreed guidelines. Some decisions were taken at International Centers Week in October 1990 namely, to include ICRAF, IIMI, and INIBAP forthwith; AVRDC, once some political issues are settled; and ICLARM, subject to a further review process.

Trends in Resource Allocation[6]

Between 1986 and 1990, annual expenditures by the CG-kindred centers increased from $33 million to $65 million. Annual growth rates in total expenditures amounted to 19% in nominal and 13% in real terms, compared with an average annual growth of 6% in nominal and 1% in real terms of corresponding expenditures in the CG centers. The distribution of expenditures between operations and capital, respectively, amounted to 94% and 6% in 1990, compared with 93% and 7% for the CG centers. Similarly, the composition of operating expenditures between research, development of research capacity, and general adminstration and management differs little between both groups. Approximately 57% was allocated to research and research support, 21% to development of research capacity, and 22% to general administration and management of the CG-kindred centers compared with 54%, 21%, and 25%, respectively, for the CG centers in 1989. As with the CG centers, the deployment of operational expenditures varies widely.

The number of senior staff increased by 50%, from 181 to 272, between 1986 and 1990. This rapid growth is due to the fact that centers such as IBSRAM, IIMI, and INIBAP were in an establishment phase during this period, while ICRAF was also substantially expanding its activities. ICIPE, ICRAF, and IFDC have the largest numbers of senior staff, accounting for 56% of the total staff in the CG-kindred centers. INIBAP has the smallest staff, as it is basically a network operation with no in-house research facilities.

Most CG-kindred centers rely heavily on a small number of donors. Six of these centers rely on three donors for 55% to 66% of their funding. For seven of the nine centers, the three major donors are members of the CGIAR. The centers depend heavily on

---

6 The data presented in this section are taken primarily from CGIAR Secretariat (1990a).

special-project funding, which has contributed about 60% of their revenue, compared to 19% for the CG centers. This limited extent of unrestricted funding may hamper the continuity of their core research activities.

9.2.2   Other Multilateral Agencies

In addition to the 13 CGIAR centers and nine CG-kindred centers, some 52 other multilateral agricultural research and research-complementing institutions based in less-developed countries were identified in 1985 (IDRC 1986). Of these 52 only 17 were directly involved in conducting agricultural research. The remaining 35 filled a research-support role by providing training, coordination, information, and extension services in addition to providing financial and professional support to agricultural research agencies. Financial institutions such as the World Bank and the regional development banks have been excluded from this tally even though they have also supplied some support other than directly through their lending programs.

   Table 9.10 indicates that only two of these institutions work in more than one region. Of note is the high concentration of multilateral agricultural research and research-complementing institutions in sub-Saharan Africa. The majority of these have a research-support function with only three of the 21 in the region actually conducting agricultural research. The region with the largest number of multilateral institutions actually conducting agricultural research is Latin America & Caribbean, including some subregional agencies such as CARDI in the Caribbean and CATIE in Central America. Only three multilateral institutions are reported for the West Asia & North Africa region.

Table 9.10: *Other Multilateral Agricultural Research and Research-Supporting Institutes Based in Less-Developed Countires*

Mandate	Total number	Executing research	Funding sources		
			Local	International	Mixed
Global	2	1	0	0	2
Sub-Saharan Africa	21	3	5	6	10
Asia & Pacific	12	3	2	2	8
Latin America & Caribbean	14	9	3	0	11
West Asia & North Africa	3	1	3	0	0
Total	52	17	13	8	31

Source: Adapted from IDRC (1986).

   We defined an institution as being internationally funded if most of its funds are provided by donors outside the region, "mixed" if both local governments as well as donors from outside the region provide funding, and local if most of the funds are provided by local governments. As shown in table 9.10, only one-quarter of the multilateral institutions are funded solely by their client governments. In most cases, however, the existence of the multilateral institutions in less-developed countries is heavily dependent on funding from outside the region in which they are operating. In many cases, Western donors have been the driving force behind the creation of these regional agricultural research (-supporting) institutions, and their existence often has little to do with regional cooperation. In particular, the explosion of regional institutions established during the 1970s in sub-Saharan Africa appears to have been donor-driven, with only one of 15 institutions (CGIAR, CG-kindred, and other) established during 1971-80 classified as locally funded. Regional cooperation is often impeded by political instability, strong nationalistic tendencies, and a prevalence of national institutions that are too weak to participate. Future regional approaches to agricultural research in sub-Saharan Africa are presently being discussed within the CGIAR, the World Bank, and SPAAR. New initiatives should be seen early in the 1990s.

9.3   BILATERAL AGRICULTURAL RESEARCH

Another layer of agricultural research activities that transcend national boundaries is bilateral agricultural research. The discussion here will focus only on tropical agricultural research. As described in chapter 7, most European colonial powers had built up considerable colonial agricultural research systems by the time their colonies achieved political independence. In some cases, independence meant an abrupt cutoff from the colonizing country and a disintegration of institutional arrangements for agricultural research, while in other cases, strong bilateral relationships were continued. In general, however, there has been a strong shift from institutionalized program support to project-oriented donor activities in agricultural research. This has seen a changing cast of actors, with the expertise once found primarily in colonial services now residing variously in universities, specialist research organizations, and increasingly in private consulting firms that manage competitively won projects. Quantitative data on these trends are not available and we thus resort to brief descriptions of the approaches that have been taken by a couple of major ex-colonial powers.

France, for example, continued her overseas agricultural research operations on a bilateral basis for many years and has considerably expanded activities in (sub)tropical agricultural research since her former colonies gained independence. Presently there are two major French organizations, ORSTOM and CIRAD, with extensive headquarters in France and additional research locations spread around the world. ORSTOM conducts research of a more basic character, with only a small part of its research activities directly applicable to agriculture. CIRAD, however, focuses exclusively on agricultural production issues. France spent a total of about $100 million on (sub)tropical agricultural research in 1985 through its CIRAD and ORSTOM operations (ORSTOM 1986; CIRAD 1987), thus amounting to nearly one-half of the total CGIAR budget in that year.

Great Britain has also continued some of its tropical agricultural research activities but took a quite different and less involved approach than did France. At the time of independence, Great Britain transferred all of its colonial agricultural research structures to the newly established governments. For some years after independence, budgetary aid was provided to finance agricultural research in the former colonies. In the late 1960s, this transitional form of support was phased out and increasingly replaced by project aid (ODA 1979). Although the British colonial agricultural research system was never heavily centralized, several rather specialized agricultural research institutes were headquartered in Great Britain. With the decline of the empire, some major reorganizations and mergers of these institutes took place. The objectives and mode of operation of these institutes were gradually redefined. By the early 1970s, only three institutes, operating with the financial support of the Overseas Development Administration (ODA), had survived, namely, the Center for Overseas Pest Research, the Tropical Products Institute, and the Land Resources Development Centre (Central Office of Information 1972). In 1983 the Center for Overseas Pest Research and the Tropical Products Institute were amalgamated into the Tropical Development and Research Institute. In 1987 the Land Resources Development Centre was then merged into this new institute, which was renamed the Overseas Development Natural Resources Institute. In addition to these institutes under the direct administrative control of ODA, there are others linked to British universities or independent agricultural research institutes that receive core support from ODA. The most important are the Centre for Tropical Veterinary Medicine, the Tsetse Research Laboratory Bristol, the National Institute of Agricultural Engineering, and the Oxford Forestry Institute.

Total ODA R&D expenditures averaged some 37.5 million 1980 PPP dollars per annum during the early 1980s, of which nearly 60% could be identified as being spent on tropical agricultural research (ODA 1981 to 1987). On average about 33% of this $22 million for tropical agricultural research was spent on the UK-based research institutes and another 30% through the IARCs. The remaining 37% was allocated to research projects conducted mainly by UK universities and institutes in collaboration with counterparts in the tropics (20%) and through other bilateral country support (17%).

Although strictly speaking, the Commonwealth Agricultural Bureau International (CABI) entities are appropriately classified together as a multilateral institute, they represent an additional and important component of British tropical agricultural research that continued throughout the post-colonial period. Founded in 1929, CABI originally functioned as an institute that provided backstopping services for tropical agriculture throughout the British Commonwealth. It has subsequently evolved into an organization that provides services to the world agricultural science community in three main areas: (a) It acts as a clearing house for the collection, collation, and dissemination of published material in every branch of agricultural science and related aspects of applied biology, sociology, and economics. It publishes 43 abstracting journals and maintains the largest agricultural science reference database in the world; (b) It provides an identification service for insects, fungal and bacterial diseases of plants, helminth pests of animals and man, and plant-parasitic nematodes through its institutes of entomology, parasitology, and mycology; and (c) The Institute of Biological Control provides a biocontrol service to manage animal and plant pests using biotic agents. CABI employs some 185 scientists and 195 support staff spread over 10 bureaux and four institutes, and it is financed by the Commonwealth member countries and through payments for the services it provides.

In singling out the French and British institutions involved in bilateral agricultural research for special attention in this section, we are conscious that this does not do justice to the analogous endeavors of several other nations such as our own. We have in mind particularly the non-CG research activities of, for example, USAID that are executed indirectly under the umbrella of the NARSs and thus are not picked up in the data of chapter 7. This also includes the similar activities conducted by Belgian, Dutch, German, Scandinavian, Swiss, and other agencies working alongside the NARSs. Other research and research-support activities are conducted by Australian and Canadian agencies, namely, ACIAR and IDRC, respectively. All these activities provide something of a difficulty for the quantitative purview of this section since data that describe operations outside those of national programs per se and yet largely involve research in less-developed countries are not readily available. In many cases, such as those involving ACIAR and IDRC, the mode of working is deliberately collaborative and thus budgetary dissection is especially difficult, although usually a majority of the incremental funding comes from the donor agency. In closing this section, we thus do so with a plea for further quantitative analysis in this field to document more comprehensively and in a functionally more detailed manner the many different arrangements for bilateral international agricultural research.

9.4   FINAL COMMENTS

Although the CGIAR accounts for only a minor share of resources available for agricultural research, it has become an important actor in the global agricultural research system. As investigations such as the Impact Study (Anderson 1985b; Anderson, Herdt, and Scobie 1988) have shown, the CG centers led to a major increase in national research capacity and in generating improvements in agricultural productivity.

This chapter has provided an overview of trends in resource allocation in international agricultural research and a brief discussion of some of the major policy issues faced by the CGIAR. With an enlarged CG research system, these issues will become even more complex. Already questions are being raised as to whether the present organizational structure of the system is the most appropriate one to respond to the future demands for international agricultural research (McCalla 1988). New concepts for executing research more effectively, such as "ecoregional centers" are being proposed and evaluated. The continuing review of priorities and strategies will surely be the subject of active attention and discussion among the many concerned parties in both the less- and more-developed countries. Deliberations and decisions will, it is hoped, be facilitated and improved by the availability of data and material such as have been assembled in this volume.

Table A9.1: *Key Features of the CGIAR Centers*

CENTER		Commodity/activity	MANDATE[a]
Name	Headquarters location (year established)		Region/agroecological zone
CIAT: International Center for Tropical Agriculture	Cali, Colombia (1968)	Phaseolus bean, cassava Rice Tropical pastures	World Latin America Latin America/lowland tropics
CIMMYT: International Center for Improvement of Maize and Wheat	Mexico City, Mexico (1966)	Wheat, maize, triticale Barley	World Latin America
CIP: International Potato Center	Lima, Peru (1971)	Potato, sweet potato, other root crops	World
IBPGR: International Board for Plant Genetic Resources	Rome, Italy (1974)	Promote activities to further collection, conservation, evolution, and utilization of germplasm	World
ICARDA: International Center for Agricultural Research in Dry Areas	Aleppo, Syria (1976)	Farming systems Barley, lentils, faba beans Wheat, kabuli chickpeas	North Africa/Near East World North Africa/Near East (concentration on nonirrigated agriculture)
ICRISAT: International Crops Research Institute for the Semi-Arid Tropics	Hyderabad, India (1972)	Farming systems Sorghum, millet, pigeonpeas, chickpeas, groundnuts	Semi-arid tropics (Asia, Africa) World
IFPRI: International Food Policy Research Institute	Washington, DC, USA (1975)	Identify and analyze national and international strategies and policies for reducing hunger and malnutrition	World, with primary emphasis on low-income countries and groups

Table A9.1:   *Key Features of the CGIAR Centers (Contd.)*

CENTER		Commodity/activity	MANDATE[a]
Name	Headquarters location (year established)[a]		Region/agroecological zone
IITA: International Institute of Tropical Agriculture	Ibadan, Nigeria (1967)	Farming systems Rice, maize, cassava, cocoyams, soybeans Sweet potatoes, yams, cowpeas	Humid & sub-humid tropics (Africa) Africa **World**
ILCA: International Livestock Center for Africa	Addis Ababa, Ethiopia (1974)	Livestock production systems	Africa
ILRAD: International Laboratory for Research on Animal Diseases	Nairobi, Kenya (1973)	Trypanosomiasis, theileriosis, other diseases	Africa
IRRI: International Rice Research Institute	Los Baños, Philippines (1960)	Rice, rice-based cropping systems	World, with emphasis on Asia
ISNAR: International Service for National Agricultural Research	The Hague, Netherlands (1980)	Strengthen national agricultural systems	World
WARDA: West African Rice Development Association	Bouaké, Côte d'Ivoire[b] (1971)	**Rice**	West Africa

Source: TAC/CGIAR (1987).

[a] Represents current operational mandate of centers.
[b] Relocated from Monrovia, Liberia, in 1989.

Table A9.2:   *Key Features of the Nine CG-Kindred Centers under Consideration for Entry to the CGIAR*

Center, location (year established)	Mandate	Programs	Commodities
AVRDC: Asian Vegetable Research & Development Center, Tainan, Taiwan (1971)	Improve nutritional quality and production potential of vegetables in humid and sub-humid tropics.	Crop improvement, production systems development, training	Tomatoes, chinese cabbage, soybeans, sweet potatoes
IBSRAM: International Board for Soil Research & Management, Bangkok, Thailand (1985)	Promote improved and sustainable soil management technologies to reduce soil constraints to food and agriculture production.	Networks for research on soils	—
ICIPE: International Center of Insect Physiology and Ecology, Nairobi, Kenya (1970, reconstituted in 1986)	Undertake research in aspects of insect life for the control of major crop and livestock pests and insect vectors responsible for tropical disease.	Crop pests, livestock ticks, tsetse flies, plant resistance, medical vectors, training	—
ICLARM: International Center for Living Aquatic Resources Management, Manila, Philippines (1977)	Conduct and stimulate research on fisheries and other living aquatic resources to assist developing countries' nutritive, economic, and social needs.	Aquaculture, resource assessment and management, coastal development, networks, training, information	Fish
ICRAF: International Council for Research in Agro-Forestry, Nairobi, Kenya (1978)	Improve nutritional, economic, and social well-being of people in developing countries by promoting agroforestry systems non-deterrent to environment	Agroforestry, networks, training, information	—
IFDC: International Fertilizer Development Center, Muscle Shoals, AL, USA (1977)	Research, development, and transfer of appropriate fertilizer technology to developing countries at lowest possible cost.	Fertilizers, nutrients and technology (nitrogen, phosphorus, sulfur, potassium), economics, technical assistance	—

Table A9.2: *Key Features of the Nine CG-Kindred Centers under Consideration for Entry to the CGIAR (Contd.)*

Center, location (year established)	Mandate	Program	Commodities
IIMI: International Irrigation Management Institute, Colombo, Sri Lanka (1984)	Strengthen national efforts to improve and sustain irrigation system performance through development and dissemination of management innovations.	Irrigation research, training, information	—
INIBAP: International Network for the Improvement of Banana and Plantain, Montpellier, France (1984)	Coordinate and stimulate research on improvement of bananas and plantains.	Networks	Bananas and plantains
ITC: International Trypanotolerance Center, Banjul, The Gambia (1982)	Research seeking to understand and utilize the natural resistance exhibited by West African livestock breeds to infection from trypanosomiasis.	Trypanosomiasis, tsetse flies, helminthiasis, livestock development	Livestock

Source: CGIAR Secretariat (1988b) and TAC Secretariat (1988).

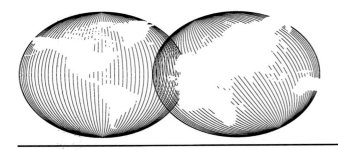

PART IV

P<small>RIVATE</small>–S<small>ECTOR</small>
A<small>GRICULTURAL</small> R<small>ESEARCH</small>

Chapter 10

Private-Sector Agricultural Research in Less-Developed Countries

Carl E. Pray and Ruben G. Echeverría

The institutional environment in which agricultural research is conducted is changing rapidly. The role of market forces and the reduction of government intervention has received considerable attention in recent years, especially in the more-developed countries. Similar thinking is reflected in the foreign policies of more-developed countries as well as in the domestic policies of many less-developed countries.

Although the main focus of privatization in less-developed countries has been on the industrial sector, attention has also been given to reducing the role of government in agriculture. The main targets for privatization have been public agencies supplying inputs and marketing outputs (Maddock 1987). This is important, for example, for the seed industry in less-developed countries where plant breeding, seed multiplication, and distribution are often public-sector activities.

Research conducted by private firms is a growing source of agricultural technology in less-developed countries. Firms sometimes also fund research in public institutions but in this chapter we concentrate on research *conducted* rather than *sponsored* by the private sector. Unfortunately, there are few quantitative data on how much private agricultural research is being conducted and still less information about its determinants or its impact on public research, farmers, and consumers. This lack of information may lead public research administrators to set research priorities and allocate resources inefficiently.

With the exception of Griliches' (1957a, 1958) studies, the private sector has until recently been largely ignored in the agricultural economics literature pertaining to technological change in production. For the industrial sector, Mansfield (1977, 1984) quantified the social and private rates of return to industrial innovations developed by the private sector during the 1970s. Griliches (1980) analyzed returns to research expenditures in the private sector but concentrated his attention on large (1000-plus employees) US manufacturing companies that engage in research and development (R&D) activities.[1] Caves (1982) has reviewed the R&D behavior of multinational corporations, but not those involved in agriculture such as the large chemical, seed, and farm machinery companies that are based in the more-developed countries and tend to operate globally.

During the past two decades, various studies[2] have analyzed the key role of new technologies in increasing agricultural productivity and the role of agricultural research in the generation of those technologies. They have also shown that research increases agricultural productivity by improving input and output quality, lowering production costs, creating new products, and improving production processes.[3] Several have demonstrated that technical change is induced largely by changes in demand and shifts in resource availability.

The literature on technical change in agriculture has been concentrated on public-sector research, rather than on the totality of the research system. The contribution of the private sector has not been closely considered in most studies analyzing the role of research in agricultural development. This omission is not only a source of bias for studies on the rates of return to agricultural research, but it also implies that science and technology policy is synonymous with public-sector R&D activities. Despite the lack of studies focusing on private agricultural R&D in the past, the subject at last received some due attention in the 1980s.[4]

By ignoring private R&D transfer, agricultural technology policies may have unexpected consequences. Moreover, the understanding of the roles of the public and private sectors in R&D and their potential complementarities through institutional innovation is also limited. This is a major issue on the agricultural development agenda in the 1990s and it deserves further research.

Some of the relevant questions related to the involvement of the private sector in generating and transferring agricultural technology are the following: Should the private sector play a larger role in developing and transferring technology? Should public research concentrate on the crops and more difficult agroclimatic areas where private investment is less likely to occur? How should the public sector account for private R&D activities when setting research priorities? In what instances does the private sector substitute for or complement public research?

---

[1] Griliches (1984) contains a representative group of studies on private-sector research in more-developed countries.

[2] See, for example, Evenson (1967), Peterson and Hayami (1977), Binswanger, Ruttan et al. (1978), and Hayami and Ruttan (1985).

[3] See Echeverría (1990b) for a summary of estimates of rates of return to investment in agricultural research and extension from 1958 to 1990.

[4] See Trigo and Piñeiro (1981), Ruttan (1982), Evenson and Evenson (1983), Pray (1983), Piñeiro (1985, 1986), Trigo (1988), Echeverría (1988a), Pray and Echeverría (1988, 1989b), and Sarles (1990).

In this chapter we have two objectives: (a) to examine the determinants of private-sector investments in agricultural R&D and (b) to characterize and, to the extent currently possible, quantify private-sector research activities in less-developed countries.

10.1 RELATIONSHIPS BETWEEN PUBLIC- AND PRIVATE-SECTOR AGRICULTURAL RESEARCH

We focus our discussion of the relationships between public and private agricultural R&D on the different types and potential sources of agricultural technologies and on the public-good nature of much agricultural research output. The other sections of this chapter examine the determinants and the nature and scope of private-sector research in less-developed countries.

10.1.1 Types and Sources of Agricultural Technologies

Agricultural technologies aimed at the primary production sector can be broadly classified into four categories that are not necessarily mutually exclusive:

(a) *managerial:* crop and livestock management techniques and other managerial practices;
(b) *biological:* crop cultivars, animal breeds, hormones, vaccines, and other living organisms;
(c) *chemical:* growth regulators, fertilizers, fungicides, insecticides, and herbicides;
(d) *mechanical:* tractors, harvesters, and other farm equipment.

There are also other types of technologies (either products or processes) related to agriculture, such as post-harvest and food processing technologies. In this chapter we focus primarily on the four types identified above.

Private research activities in more-developed countries have concentrated on developing mechanical and chemical technology, and less on biological and managerial technology. However, with advances in biotechnology, private research on biological technology is increasing (Persley 1990). Patent protection has been a factor influencing the type of technology being developed. Although the legal framework in many countries offers certain rights to research organizations for appropriating some of the potential benefits of new technology, effective protection of those rights is quite difficult to achieve. Once the technology is made available, the costs of replication may be negligible compared with the costs of initial discovery. An exception is hybrid seed, which cannot be reproduced from its own seed.

In addition to variations in the type of agricultural technology, there are also its different sources. It is the combination of technology types and sources that provides a framework to understand the relationships between public and private research.

Table 10.1 identifies four sources of agricultural technology: domestic public and private research and foreign public and private research. The distinction between foreign and domestic and between public and private research is relevant when analyzing each sector's activities.[5]

Table 10.1:   *Potential Sources of Agricultural Technologies*

	Institutional location	
	Domestic	Foreign
Public	Ministry	Ministry
	Research institute	Research institute
	Research council	University
	University	IARC
Private	Cooperative	National company
	Foundation	Multinational
	Commodity institute	Cooperative
	Plantation	
	Processing company	
	Input company	

Research institutes, ministries of agriculture, and universities, among others, are usually domestic, public sources of technology. The research done by these organizations can be directed to adapting technologies developed elsewhere or to creating new ones. International agricultural research centers and national agricultural research systems in other countries are the main foreign, public sources of agricultural technology.

Private-sector R&D is conducted by agricultural processing, production, and input industries. Agricultural processing industries develop technology for farmers so that the product purchased by the industry will be cheaper or of better quality. Agricultural production industry, i.e., farms and plantations, develops technology to reduce costs and to increase the demand for its products. The agricultural input industry produces technologies that are intended to increase farmers' productivity. Within these industries, research is conducted mainly by two types of private institutions: individual companies and groups of firms or farmers. In a few countries, nonprofit corporations and nongovernmental organizations also conduct agricultural research but are excluded from further consideration in this chapter.

---

[5] The distinction between an organization funding versus executing research also has important policy implications. In this chapter we concentrate on public- and private-sector organizations executing research.

Foreign sources of technology consist of multinational companies that transfer their technology to local subsidiaries, and foreign companies that export technology directly. Foreign sources tend to be firms, but some cooperatives are important too.

The distinction between public and private research and between domestic and foreign sources of technology is less clear-cut than portrayed in table 10.1. In fact, there is a continuum of institutional arrangements from a broadly based public research institute to a private company dealing with a single input. For instance, many organizations that play an important role in developing and transferring technology in less-developed countries, such as nongovernment organizations, cooperatives, foundations, and joint ventures (between public and private organizations, domestic and foreign) fall between these extremes. All these different entities constitute what we broadly refer to as the private sector.[6]

10.1.2 The Public-Good Nature of Agricultural Research Output

Much new knowledge produced from research has the nonrivalness and nonexcludability characteristics of a public good. *Nonrivalness* means that the research output is available to everybody at zero marginal cost. A purely "rival" (private) good or factor is one such that the use of a unit by any agent precludes its use entirely by anyone else. Knowledge in this sense is a pure public good, i.e., one for which the use by any agent has no effect on the amount available for use by others. Consider, for instance, the development of a new crop-rotation pattern that improves crop production and reduces soil erosion. The use of this information by a particular farmer does not prevent the adoption of the same practice by other farmers.

The second attribute, *nonexcludability*, implies the infeasibility (or high cost) of denying use to those who do not pay for it so that a "free rider" problem is present. For a nonrival (public) good, exclusion does not have the same importance as for private goods. Since the marginal social cost of a new user is zero, it is not socially optimal to set prices that will exclude anyone who benefits from the public good, i.e., exclusion is economically inefficient. A common aspect of the products of agricultural research is that many are nonexcludable.

Private firms usually do not produce goods that are nonrival or nonexcludable (like most public goods) because they would be unable to capture benefits to cover the costs resulting from their research activities. Farmers seldom conduct formal research because farms are small and capture only a small part of the benefit of an innovation.[7] A socially

---

[6] When referring to private R&D, we use the National Science Foundation (NSF 1989, p. 2) definition of R&D as "... basic and applied research in the sciences and in engineering and the design and development of prototypes and processes." Most of the R&D information presented in this chapter is based on companies' definition of R&D. This may bias that information upwards for large companies that believe R&D adds to their prestige and downwards for smaller companies that do not have an explicit R&D budget and in general do not consider "the design and development of prototypes and processes" to constitute R&D.

[7] Many farmers do, however, conduct their own trials of new technology.

optimal level of public good will, therefore, not be supplied if its production is left to private firms. Since information is not perfectly appropriable by its discoverer, the excess of the social over the private value of new technological knowledge leads to underinvestment in inventive activity.[8] Consider, for example, the development of an open-pollinated variety of a crop in a country with no plant variety protection. After it is released, it can spread among farmers without benefiting the inventor. Hence, private firms alone would typically produce nothing or at least suboptimal quantities of such a technology.

Peterson (1976) has shown that if research is carried out by private firms and if the new technology is adopted, social returns to private research must be greater than private returns. Griliches (1958) has also argued that the difference between social and private rates of return is a necessary but not a sufficient reason for public intervention.[9]

Private underinvestment in research is a strong argument in favor of government intervention in the supply of new technology. The most common types of intervention are government funding of research and legislation on intellectual property rights such as patents, which endeavor to ameliorate the nonexcludability attribute.

Rationales other than the public-good argument, have been advanced to justify public involvement in agricultural research (chapter 1). Ruttan (1982) lists two reasons in addition to low levels of private investment: the complementarity of research and education and the preservation or enhancement of competition. Another argument is that the direction of private-sector endeavor could be biased. It would concentrate on producing knowledge that could be embodied in private goods such as agricultural machines and pesticides, rather than new crop rotations or biological pest control, which may be more valuable to society. Even joint ventures between private and public research can be biased against the interests of society (Ulrich, Furtan, and Schmitz 1986).

On the basis of case studies of US public and private research, Ruttan (1982) concluded that mechanical technology will remain a low priority for US public research. More public resources will most likely be devoted to chemical technologies in the areas of new pest control methods that use fewer chemicals. In biological technology, public resources will probably be reallocated from plant breeding per se to more basic supporting areas such as genetics, physiology, and pathology.

10.1.3 Is Private Research a Substitute for Public Research?

With an expanding role for private agricultural research, arguments for less funding for public research may be common in the future, the more so given widespread pressures to reduce government budget deficits. Such an argument presumes that public and private research are substitutes for each other. In many instances, however, they are complementary

---

8 See Hirshleifer (1971) for an analysis of the private and social values of inventive activity.

9 Griliches (1958) argues that it is not a sufficient reason because private returns may still be high enough to induce firms to invest in research.

activities. Basic public research provides opportunities for firms to profit from R&D and to accelerate the spread of publicly produced technology by adapting it to the needs of farmers. Whether a specific public research program substitutes for or complements private R&D is an empirical question.

In general, the public and private sectors are not simply direct substitutes because they are doing different types of research to produce different types of technology. An exception is in some biological research where there is more potential competition. Private agricultural research tends to be more applied than public research, and it concentrates more on mechanical and chemical technology. The public sector does most basic research, and it is more involved in biological and agronomic technology. It is also a major contributor to human capital, the supply of which is a necessary condition for the conduct of research in any sector.

Chemical technologies typically have a short economic life span, and benefits are relatively appropriable by the innovator. In more-developed countries, mechanical technologies are usually patentable, and innovators' rights are enforced. In most less-developed countries, where innovators' rights are often not enforced, private firms have fewer incentives to invest in research to develop new products. In the case of mechanical and chemical technologies then, a mixed public and private effort is common in the more basic stages, but it is the private sector that undertakes much of the applied research work.

Given the difficulty of capturing benefits from managerial and biological innovations, unless covered by patents or a plant variety protection act, public research has a key role to play in supporting both the generation and diffusion of those technologies.[10]

Evenson's (1983) argument for private-public complementarity is based on a classification of the output of research into *pre-technology, prototype technology,* and *usable technology.* He argues that private R&D focuses mainly on the development of usable technology, with some effort on prototype technology and very little in pre-technology. This is because the private incentive system usually stimulates the invention of usable technology but does not provide protection to pre-technology research. Public research activities are, therefore, important not only in pre-technology development (by the public-good argument) but also in (a) prototype technology development when markets or firms are small, and (b) usable technology to enhance technological competition.

10.2 DETERMINANTS OF PRIVATE INVESTMENTS IN AGRICULTURAL RESEARCH

Private research is growing in a number of countries, in spite of the fact that much of the output of research has the characteristics of a public good. This section examines the reasons why firms invest in agricultural research.

---

[10]See Lesser and Masson (1983) for an analysis of the Plant Variety Protection Act of the United States.

10.2.1 Theoretical Concepts

According to neoclassical economic theory, firms seek to maximize expected profits. This objective can be translated into three main determinants of private investment in R&D: (a) market factors, (b) a firms' ability to appropriate economic gains from R&D, and (c) the technological opportunities for innovation. These main determinants are detailed in table 10.2 and classified into two groups: economic and technical, and government policies.

Table 10.2: *Factors Influencing the Level of Private Agricultural R&D*

Main determinants of private agricultural R&D expenditure	Factors influencing determinants	
	Economic and technical	Government policies
Market factors		
Expected demand	Income growth Income elasticities Export demand Demand elasticity	Agricultural price policies Import/export policies
Input prices	Level of industrialization Supply and demand of inputs	Input price controls Credit policies Government supplies Input import policies Industrial policies
Appropriability	Nature of technology Market structure	Public R&D effort Anti-trust policy Patents and plant breeders' rights legislation Enforcement of rights
Technological opportunity	Private local R&D	Public R&D IARC research
	Foreign technological developments	Policies on multinationals Technology import policies
	Quality and cost of scientific inputs	Output of universities Subsidies on R&D costs Imports of R&D equipment

Market Factors

The key market factors are expected demand and input prices. Food and agricultural processing industries will not conduct research unless they expect a profitable level of demand for processed goods. The role of demand in inducing seed firms to innovate was demonstrated by Griliches (1957a) in his study of the spread of US hybrid maize. Schmookler (1966) also emphasized the role of demand in his work on industrial patenting.

Anecdotal evidence suggests that expected demand is an important factor in companies' decisions to invest in R&D in less- as well as more-developed countries. Several international seed firms (Pray 1986) cited the importance of demand-side policies. One example is government price support for yellow maize and subsidies for farmers who purchase hybrid maize seed, in a firm's decision to invest in hybrid maize research in the Philippines, which thus emphasizes public-private interactions.

Most economics literature emphasizes the importance of relative input prices in determining the direction of research and thus of technical change.[11] An example of this is Malaysian plantation research to find technologies that will reduce production costs. Input prices also influence the level of private research. An example of a processing firm reacting to input price changes is the Bangladesh Tobacco Company. When Bangladesh separated from Pakistan, the source of cigarette tobacco was cut off, tobacco prices increased, and induced the company to initiate a research program to develop local supplies.

Appropriability

Investment in R&D also depends on the firm's ability to appropriate the gains from innovation. This ability depends on four main factors. The first is the structure of the industry. Schumpeter (1950) argued that large monopolistic firms would have higher rates of technical change than small competitive firms. Scherer summarizes the findings of subsequent research on Schumpeter's theory: "A bit of monopoly power in the form of structural concentration is conducive to invention and innovation, particularly when the advances in the relevant knowledge base occur slowly. But very high concentration has a favorable effect only in rare cases, and more often it is apt to retard progress by restricting the number of independent sources of initiative and by dampening firms' incentive to gain market position through accelerated research and development" (Scherer 1980, p. 438).

The second factor is the nature of the technology. Some innovations, by their technical nature, are more appropriable than others. For example, hybrid maize gives its inventor a monopoly if the inbred lines required to produce the hybrid can be kept secret. In Asia and Latin America, private breeding is almost entirely devoted to hybrids of cross-pollinated crops. Hybrids give their developers the ability to preclude others from easy duplication and help to ensure a market because farmers must buy seed every year to get maximum yields (chapter 11). Companies, therefore, can charge more for seed if they develop a higher-yield-ing hybrid, and they can then sell it every year. The third factor is lead time. If a firm can keep improving its product or developing new products more rapidly than its competitors,

---

[11] The importance of relative factor prices in inducing technical change was first discussed by Hicks (1932). Induced innovation resurfaced in the 1960s when the work of Fellner (1961) stimulated research on the importance of factor prices in technical change. Hayami and Ruttan (1971) incorporated both demand and input prices into a model of induced innovation, which they applied to agricultural development. Binswanger (1974) integrated these ideas into a more rigorous microeconomic theory of induced technical change.

it can charge higher prices and can thus profit from the research underlying the product.

The fourth factor is the existence and enforcement of patents and plant breeders' rights legislation. A large body of literature has developed around (a) the value of patents as an incentive to R&D and (b) the costs of patents to society (Scherer 1980; Griliches 1984). Surveys of firms in more-developed countries find that firms feel patents are an important stimulus to research. As mentioned in section 10.1.2, patents allow firms to exclude others from using an invention. There is, however, surprisingly little empirical evidence on the impact of patents on agricultural research or productivity in more-developed countries. Some studies of the impact of the US Plant Variety Protection Act found that it had a positive impact on private plant-breeding research (Perrin, Hunnings, and Ihnen 1983; Butler and Marion 1985).

Technological Opportunity

Expenditures on R&D are also influenced by the potential for development of new technology. The relevant dimension of this potential from the perspective of a private firm is the cost of producing an innovation relative to expected profits. Technological opportunity can be divided into two components, a physical component related to the technical efficiency of the R&D process, which depends on the state of knowledge and R&D management, and a price component that depends on the supply and demand of R&D inputs.

Research by other firms can lead to new opportunities. One purpose of patents is to ensure that the technology embodied in an innovation can be made public for other firms to use in making further innovations. Another way to learn about other firms' R&D is through reverse engineering. A third way by which knowledge is frequently transferred is by hiring another firm's scientists and engineers.

Yet another source of technological opportunity is local adaptation of foreign ideas and innovations. For example, the development of a disease-resistant variety in, say, an international agricultural research center, can lead to matching plant breeding in less-developed countries that will incorporate that characteristic into local varieties. Yet again, the synthesis of new chemicals in more-developed countries will require field testing of that new material in less-developed countries, which, if the chemical is proved effective against local pests, may stimulate further R&D to develop suitable application methods. Farming implements such as power tillers quickly become models for local tillers, which are modified to meet local conditions.

Judiciously targeted public-sector agricultural research can also increase technological opportunities for private research. For example, the private hybrid maize breeding programs in Southeast Asia are based on genetic material that confers resistance to downy mildew, identified by the Kasetsart University-Rockefeller Foundation maize program in Thailand.

10.2.2 Recent Evidence

Levin and Reiss (1984) and Levin, Cohen, and Mowery (1985) have attempted to quantify the determinants of industrial R&D through incorporating demand, appropriability, and technological opportunity variables in empirical models. Pray and Neumeyer (1989) tested the importance of appropriability and technological opportunity on US agricultural input industries and found that technological opportunity had an important influence on R&D, while appropriability did not. Hayami and Ruttan (1985) showed that, in aggregate, technical change in agriculture in less-developed countries does respond to relative factor scarcity; i.e., presumably both private and public innovators respond to relative prices.

Table 10.3 summarizes evidence of the importance of these different determinants of private R&D in the agricultural input industries and in all industrial R&D in less-developed countries.

Determinants of R&D in Agricultural Input Industries

Mikkelsen (1984) reported a survey of 56 Philippine agricultural machinery firms in 1981 in which it was found that the amount of research effort was positively related to firm size but that the increase in R&D was less than proportionate with firm size. The research and extension effort of the International Rice Research Institute (IRRI) was the only technological opportunity variable in his analysis. Cooperation with IRRI was found to increase the productivity of private research.

Some 49 Brazilian agricultural machinery firms were surveyed in 1981 by Dahab (1985), who found that the research investments of these firms increased more than proportionately with firm size up to a certain point and then declined. Holding firm size constant, firms with smaller market shares in their segment of the industry conducted more R&D than those with larger shares. The demand-side variable was the share of industry sales exported, and it was positive and significant. Technological opportunity variables included foreign ownership and the number of Brazilian majority-owned joint ventures. Both were negatively related to R&D but only the latter was significant.

Data for 24 Indian seed firms in 1987 were analyzed by Ribeiro (1989). Among the 17 firms that then conducted R&D, firm size was positively and significantly related to research expenditure. At first, R&D effort grew faster than firm size, but the converse held after a certain firm size was reached. The technological opportunity variables were contacts with the International Crops Research Institute for the Semi-Arid Tropics (ICRISAT) and with government research programs. The effect of such contacts was found to be positive but not statistically significant. Although the technological opportunity variables in Ribeiro's study were also not statistically significant, most private pearl millet and sorghum hybrids were bred using ICRISAT inbred lines, which indicates that public-sector research did, in fact, play an important role in providing technical opportunities (Pray et al. 1989).

Table 10.3:   *Determinants of Private Agricultural R&D in Less-Developed Countries*

	Agricultural input industries			Industries		
Source	Mikkelsen (1984)	Dahab (1985)	Ribeiro (1989)	Mikkelsen (1984)	Kumar (1987)	Deolalikar & Evenson (1988)
Country	Philippines	Brazil	India	Philippines	Brazil	India
Period of analysis	1981	1981	1987	1965-66 & 1979-80	1978-79 & 1980-81	1960-70
Dependent variable	Machinery R&D	Machinery R&D	Seed R&D	R&D expenditure	R&D intensity	Patents
Demand						
Price						a
Exports		+				
Differentiated product					b	
Appropriability						
Firm size	+	+	+	+		–
Concentration		–		–	–	
Opportunity						
Technology		+		c	+	
Public R&D	+	+				
Foreign R&D		–		+		+
Imports				+	+	–d
Multinational links		–		–	–	

a Higher wages in light industry decreased demand for innovation but higher fuel costs increased demand for innovation

b Estimated coefficient was positive for advertising and negative for consumer industries.

c Industrial dummies were significant but no interpretation was given.

d Negative and significant for light industry only.

Determinants of R&D in Nonagricultural Industries

Given the scarcity of studies focusing on determinants of R&D investment in agriculture, this section reviews selected studies relating to other industries in less-developed countries. Mikkelsen (1984) estimated the determinants of industrial R&D in the Philippines using firm-level data from two periods: 1965-66 and 1979-80. He found that research expenditures by Philippine firms were negatively related to concentration and, as firm size grew, R&D expenditure increased. He also found that firms reacted positively to the technological opportunities provided by international technology.

Indian manufacturing was the focus of Kumar's (1987), study. He analyzed data on 1143 firms for the period 1978-79 to 1980-81. The dependent variable in his single-equation model is research intensity (research expenditure over sales) rather than the absolute level of research expenditure used by Mikkelsen. He included expenditures on advertising and a dummy variable for consumer goods to capture demand. Consumer industries or industries with more advertising would presumably face higher elasticities of demand and thus do more research. Advertising was positively related to R&D but consumer goods were negatively related. Using the four-firm concentration ratio as an appropriability variable, a weak negative relationship between research intensity and concentration was found.[12]

Kumar included a number of technological-opportunity variables in his analysis. He argued that imported technology can either increase technological opportunity or decrease the incentive to do R&D, depending on how it enters the country. If it comes from a multinational corporation to its local affiliate, R&D at the company's headquarters can be a ready substitute for local research. If the technology is imported in return for a royalty to a nonaffiliated company, then the local company will probably do more adaptive research. His results support this hypothesis: the share of foreign-controlled firms in an industry is negatively related to R&D intensity and the royalty fees as a percent of industry sales are positively related to R&D intensity.

A third study of the determinants of industrial R&D (Deolalikar and Evenson 1988) takes a different theoretical and econometric approach by using duality theory and estimating cost-minimizing input-demand functions with Indian industry data from 1960 to 1970. This study treats domestic technology and technology imports (measured by patents) as variable inputs. This is one of the few studies that actually measures the demand for technology. For the light industry group, the local level of patenting was responsive to changes in wages and the price of fuel. For the chemical and engineering industries, there was no significant response to the level of local patents to any of the price variables.

The only variable in the Deolalikar and Evenson study that is related to appropriability is firm size. This had a negative impact on local patenting and technology imports. The

---

[12] The Indian patent law excludes food- or health-related products from patent protection, while other products have quite strong patent protection. This provides the yet-untried possibility of estimating the impact of patents on R&D expenditure.

negative impact on patenting is unexpected since most studies in more-developed countries show a positive impact of firm size on patenting.

Determinants of the Location of Multinational R&D Companies

Multinational companies are an important component of agricultural research in less-developed countries. Their location decisions are made on the basis of worldwide (expected) profit maximization and of country-specific factors. Behrman and Fischer (1980) mention three factors encouraging firms to locate their R&D activities in a country: (a) the existence of a profitable affiliate, (b) a growing and sophisticated market, and (c) an adequate scientific and technical infrastructure for doing research. They also mention as obstacles the scale economies of centralized R&D at headquarters and the difficulties of assembling adequate R&D staff in less-developed countries.

If the existence of a profitable subsidiary is the main determinant of R&D, then the question shifts to why subsidiaries locate in a specific country. In one of the few empirical studies on this issue, Nankani (1979) found that the existence of tariff and nontariff barriers to imports had a positive impact on foreign investment in less-developed countries. In addition, his results indicate that there is more investment by countries in their former colonies than in other countries. Davidson (1980) shows that the US invests most of its resources in countries that it "knows well," such as Canada, the UK, and Mexico. Caves (1982) interprets this and other studies as showing that information and transaction costs are the main determinants of the location of the investment of multinational companies.

Summary of Empirical Studies

Most studies did not directly test market factors, but those that did (Dahab 1985; Deolalikar and Evenson 1988) found evidence that greater demand for technology does stimulate R&D. Firm size and concentration were the only variables used to test the impact of appropriability. R&D grows with firm size but is negatively related to concentration, which runs counter to the Schumpeterian hypothesis that large monopolistic firms are the main innovators. No studies were found that tested the impact of patents on R&D in less-developed countries. However, the potential importance of intellectual property rights on plant R&D is indicated by the concentration of private plant breeding research on hybrid crops. The primary findings from the technological opportunity variables were that public agricultural research and imports stimulate R&D, but links with multinational corporations depress a firm's local R&D efforts. Finally, the location decision concerning research by multinational corporations was examined, since multinational corporations account for an important share of research activities in less-developed countries. It was found that growing markets, a profitable affiliate, and local availability of scientists were key factors in the decision to locate.

10.3 THE NATURE AND SCOPE OF PRIVATE AGRICULTURAL R&D

On the basis of the scattered information assembled, we will try, in this section, to examine the agricultural research activities undertaken by the private sector. First we will focus on the type of technology under development, be it biological, chemical, mechanical, or managerial, followed by a more quantitative regional assessment.

10.3.1 Private Research Activities by Type of Technology

Biological Technology

Research by private seed companies consists almost entirely of breeding hybrids based on inbreds developed by multinational corporations or by public research programs such as local institutes, (US) agricultural universities, and international agricultural research centers. Both multinational corporations and local companies have active breeding programs. Most private plant breeding worldwide is conducted on maize (chapter 11), probably followed by sorghum and sunflowers. In some less-developed countries, there is also work on hybrid pearl millet, hybrid cotton, hybrid rice, and hybrid wheat. Many companies also breed horticultural seeds of which hybrid tomatoes are probably the most important.

Agricultural research conducted by processing industries and by plantations includes both plant breeding and management. Plant breeding and selection is done, for example, by oilpalm, rubber, pineapple, and tobacco companies and plantations.

The main focus of livestock research includes breeding and aspects of animal nutrition, primarily pasture and feed research. The goals of feed research are improving the quality of feed and reducing its cost by using inexpensive sources of protein and energy. In Latin America there is private research in animal breeding (beef, dairy, and sheep) and some research on improved pastures and management of large livestock/crop operations. There is a limited amount of private research on veterinary pharmaceuticals, mainly through local affiliates of multinational companies.

In addition, there is a substantial amount of private research in poultry breeding. Most of this research is conducted at the headquarters of multinational corporations in the more-developed countries, and its results are directly transferred to local affiliates or joint ventures in less-developed countries.

Chemical Technology

Research by agricultural chemical companies is centralized in the more-developed countries. Some of the research at headquarters is done on less-developed country issues, but most of the technology is developed for markets in the US, Europe, or Japan. If the technology is deemed suitable for less-developed countries, it is then tested and perhaps modified. Most major multinational corporations have a few research farms situated in less-developed countries for early screening of new products. They also conduct research on different formulations, on the ecological impact of new pesticides, and to meet registration requirements.

Agricultural chemical companies in less-developed countries undertake little or no research that leads to the synthesis of new chemicals. Taiwan, which does not have a strong patent system, has a sophisticated chemical industry that reverse-engineers products developed elsewhere. In countries such as India and Brazil where the process for producing a pesticide can be patented but the pesticide itself cannot, much agricultural chemical R&D is oriented toward developing new process technology.

Mechanical Technology

Two types of agricultural machinery R&D are carried out by local companies. The first consists of minor modifications of existing machines. This is usually not done under the auspices of a formal research program. It is, however, quite important in terms of the actual amount of innovative activity (Mikkelsen 1984). The second type is more basic R&D, involving the adaptation of engines, transmissions, and brakes of agricultural machinery to less-developed country conditions. This is particularly being done in India where several companies are investing in tractor and irrigation pump research.

Managerial Technology

Plantations invest most of their research resources in developing improved management procedures. Malaysian plantations, for example, focus most of their research on ways of reducing fertilizer and pesticide costs. Private plantations probably invest more than chemical companies in integrated pest management research in less-developed countries. Banana plantations also allocate most of their research resources in managerial technology to reduce input costs.

Consulting firms in the southern part of Latin America (Chile, Argentina, Uruguay, Rio Grande do Sul-Brasil) conduct applied research on cultural practices such as fertilizer application and pastoral management for ranches that specialize in livestock and crops, and transfer information from public research stations. Also, farmers' organizations hire experts to provide technical advice on farm management and conduct applied research on managerial technology.

10.3.2 Private Research Activities by Region

The currently available estimates of private food and agricultural R&D expenditures in more- and less-developed countries are summarized in table 10.4. Private research is concentrated in more-developed countries. The United States has the largest amount. France and the United Kingdom have roughly one-tenth and one-fifth the private research expen-

Table 10.4: *A Summary of Private-Sector Food and Agricultural R&D Expenditure Estimates in the 1980s*

Country	Year of estimate	Coverage	R&D expenditures
More-Developed Countries			*(millions US $)*
Australia	1986-87	Agriculture	7.2
France	1985	Food & Agriculture	270
	1986	Agriculture	122
UK	1987	Food & Agriculture	530
		Agriculture	370
US	1984	Food & Agriculture	2400
		Agriculture	1400
Less-Developed Countries			
India	1985	Agriculture	16.7
Indonesia	1985	Agriculture	2.0
Malaysia	1985	Agriculture	10.0
Pakistan	1985	Agriculture	0.8
Philippines	1985	Agriculture	4.4
Thailand	1985	Agriculture	4.3
Chile	1984	Seed	0.2
Argentina	1989	Seed	10.0

Sources: Australia — Kerin and Cook (1989); France — Conesa and Casas (1986); UK — Thirtle et al. (1991); US — Crosby (1987), Pray and Neumeyer (1990); India, Indonesia, Malaysia, Pakistan, Philippines, Thailand — Pray (1987b); Chile — Venezian (1987); Argentina — Pray and Echeverría (1989b).

diture of the US. Large chemical corporations and manufacturers of agricultural machinery have large research programs on agricultural inputs in Germany, Switzerland, and Japan, but no national totals seem to be available.

Data on the trends in private agricultural research are even more limited than data on levels of current research. Private R&D appears to have grown by about 2% annually since 1960 in the US (Pray and Neumeyer 1989), which is slower than the 2.7% rate of growth in public R&D expenditures (Pardey, Eveleens, and Hallaway 1991). In the 1980s private research in Europe has grown as have investments in public research,[13] although public-sector growth during the 1980s is somewhat slower than in earlier periods.

Because there are no estimates of private R&D in Africa, and estimates for only two countries in Latin America (table 10.4), we surveyed multinational corporations to get an alternative perspective on the location of private agricultural R&D. Table 10.5 shows the amount and location of agricultural research conducted by the major agribusiness multina-

---

[13]Pardey, Eveleens, and Hallaway (1991) estimate a 3.1% rate of growth in public agricultural research expenditures by EEC countries over the 1976-80 to 1981-85 period.

Table 10.5: *An Estimate of Multinational R&D Expenditures in Less-Developed Countries, 1985-90 Average*

	Sub-Saharan Africa[a]	Asia & Pacific	Latin America & Caribbean	Total
	(thousands of 1985 US dollars per year)			
Plant breeding[b]				
no. of stations	5	14	36	55
expenditure ($206)[c]	*1,030*	*2,884*	*7,416*	*11,330*
Poultry breeding[d]				
no. of programs	0	2	0	2
expenditure ($470)	*0*	*940*	*0*	*920*
Veterinary pharmaceuticals[e]				
no. of programs	0	1	0	1
expenditure ($140)	*0*	*140*	*0*	*140*
Pesticides[f]				
no. of programs	4	10	12	26
expenditure ($240)	*960*	*2,400*	*2,880*	*6,240*
Agricultural machinery[g]				
no. of programs	0	0	2	2
expenditure ($2000)	*0*	*0*	*4,000*	*4,000*
Feed production[h]				
no. of programs	0	2	1	2
expenditure ($38)	*0*	*76*	*38*	*76*
Food processing[i]				
no. of programs	2	9	6	17
expenditure ($146)	*292*	*1,314*	*876*	*2,482*
Plantations[j]				
no. of programs	5	10	4	19
expenditure ($667)	*3,335*	*6,670*	*2,668*	*12,673*
Total	*5,325*	*13,636*	*17,002*	*35,963*

[a] Plant breeding and pesticide research also includes Egypt. [b] Number of plant breeding stations of the multinational corporations and their subsidiaries in Philippines, Thailand, Indonesia, Malaysia, India, Pakistan, and Bangladesh. Based on the following companies; Pioneer, Cargill, DeKalb, Continental, Asgrow, Agrigenetics, and Ciba-Geigy/Funk. The average expenditure per station in the seven Asian countries is extrapolated to Africa and Latin America. [c] Expenditures are averages per station or program in thousands of 1985 US dollars and include all operational and personnel research costs. [d] Based on the following companies: Arbor Acres, Cobb, and Babcock. [e] Based on the following companies: American Cyanamid and Monsanto. [f] Includes only research locations that conduct screening and field tests of new compounds. Based on the following companies: ICI, Bayer, Hoechst, Ciba-Geigy, Monsanto, Du Pont, American Cyanamid, Chevron, and Shell. [g] Using the average R&D expenditures of large Indian tractor companies. Based on the following companies: Deere, Massey-Ferguson, and International Harvester. [h] Based on the following companies: Cargill and Ralston Purina. [i] Based on the following companies: BAT Industries, Philip Morris, DelMonte, Unilever/Brooke Bond, Dole, and United Fruit.

tionals in less-developed countries. The locations of research programs of 27 multinational companies were identified through interviews of company personnel and from annual reports. To gain a rough estimate of R&D expenditures, the number of stations is multiplied by the average cost per station for each type of research — seed, tractors, plantation, and poultry breeding. The R&D expenditures per station were calculated from the multinationals surveyed in seven less-developed countries in Asia. This may bias our estimates of private research expenditures downwards because research inputs in Asian countries are less expensive than in Latin America or Africa (chapter 7).

These estimates only include the R&D activities and expenditures of the multinationals listed in table 10.5. The level of domestic private R&D, plus other foreign research not accounted for in table 10.5, may be substantial and is seen clearly to be an area that needs more research effort. To supplement this sparse empirical evidence, some informed, but nevertheless impressionistic discussion of the nature and level of private research activity on a regional basis is presented below.

Asia

Most of the estimates of total private research expenditures in less-developed countries in table 10.4 are from surveys conducted by Pray in Asia in 1985 and 1986. Expenditures ranged from $17 million in India to negligible amounts in Bangladesh. Table 10.5 shows that multinational corporations in Asia concentrate their research on seed, agricultural chemicals, and plantations. There is seed-industry R&D in India, Thailand, and the Philippines. Asian agricultural chemical research is concentrated in the Philippines and Thailand. The only poultry breeding programs done by multinational corporations outside the US and Europe are two joint ventures between Indian and French companies. Research by multinational companies on animal feed or veterinary pharmaceuticals is very limited in less-developed countries. Most private plantation research is in Malaysia, with some important research programs in Papua New Guinea and Indonesia.

In South and Southeast Asia, multinational corporations conduct between 35% and 40% of the total private agricultural R&D effort and the rest is undertaken by locally owned companies.[14] Multinational corporations spend more than local firms on plant breeding and pesticide investigations, about an equal amount on plantation R&D, and almost nothing on agricultural machinery and livestock research.[15] The Indian hybrid seed industry is largely locally owned. Seventeen companies had research programs in 1987, but only two of those programs were controlled by multinational corporations (Pray et al. 1989).

---

[14]These percentages are based on unpublished data collected by Pray in 1985 and 1986 through surveys of approximately 100 private firms in the Philippines, Thailand, Indonesia, Malaysia, India, Bangladesh, and Pakistan.

[15]This information is doubtless biased against local firms because many of the local companies in India and the many small agricultural machinery firms in other countries were not interviewed.

Latin America

In Latin America more private research is conducted in Brazil than elsewhere but unfortunately estimates of total private R&D are not available. In Argentina and Chile, most private research is conducted by seed companies. Plant breeding R&D is concentrated in countries with large areas planted to maize, such as Argentina, Brazil, and Mexico. There is some research on agricultural machinery by multinational corporations in Brazil but little elsewhere in the less-developed world. Plantation research is carried out in a significant way in Central America and Colombia.

In Latin America the ratio of R&D carried out by multinationals compared to domestic agricultural R&D seems likely to be higher than in the other regions. Unfortunately, there are few data with which to test this hypothesis. In Argentina, private hybrid maize, sorghum, and sunflower breeding is almost entirely done by multinational corporations, and they have a strong position in those crops in Brazil. However, several Argentine companies have wheat research programs, and the largest Brazilian seed firm is locally owned (Jacobs and Gutierrez 1986). In Chile all hybrid maize seed comes from the US with very little local R&D input, but wheat and rice breeding is carried out by local private companies and farmers' groups (Venezian 1987). In Argentina none of the tractor companies, multinational or local, has a formal R&D program (Huici 1984). Dahab (1985), in a detailed study of the Brazilian agricultural machinery industry, found that only 11 of 49 firms conducting research were owned primarily by foreigners. Echeverría (1990a) surveyed maize research in Mexico and Guatemala. He found that 25 companies in Mexico spent a total of US$ 1.7 million on R&D in 1987. Of this, US$ 1.3 million was spent by four large multinationals and the rest by local companies.

Africa

Multinational companies do much less research in Africa than in Asia or Latin America. Agricultural chemical companies have at least four research stations in Africa. The only company with active plantation research in Africa seems to be Unilever, which undertakes oilpalm research in Cameroon and Zaire and tea research in Kenya. In Kenya BAT Industries conducts some applied research on tobacco and reforestation and Del Monte has some research on pineapples. In addition, during the past five years, Pioneer began conducting maize research in Cote d'Ivoire.

Little is known about R&D by local companies in Africa, but the presumption is that it is very scarce. Local maize breeding in Zimbabwe has been successful. About half of Zimbabwe's total agricultural research budget is financed and performed by commodity organizations (Billing 1985). In Kenya most R&D by private companies seems to be by multinational corporations and their affiliates. There are also strong R&D programs by the tea and coffee producer organizations. Commodity organizations undertake considerable research in francophone Africa — especially in Côte d'Ivoire.

In summary, agricultural research by the private sector is largely concentrated in the more-developed countries. Among less-developed regions, Latin America appears to have more private R&D research than Asia, whereas Africa is lagging far behind. Agricultural chemical, veterinary pharmaceutical, and poultry breeding R&D are almost entirely conducted by multinational corporations. They also undertake much of the maize and sorghum breeding, but local companies are important in some countries. Agricultural machinery is the one industry where private local companies conduct most of the R&D.

10.4 SUMMARY AND CONCLUSIONS

Most of the literature on technical change in agriculture has focused on public-sector research. Agricultural technology is, of course, not produced only by government research institutes. As sketched above, private-sector organizations, play an important role in generating and transferring new materials and methods. It must be emphasized that the distinction between public and private sector research is not clear-cut. There is a complex, almost a continuum, of institutions conducting research, which extends from government research institutes to private agricultural input and processing companies.

Three groups of factors influence the nature and the level of private agricultural R&D investments: (a) market factors, such as the expected growth in demand for agricultural products, derived demand for modern agricultural inputs, and factor prices facing farmers and agribusiness; (b) the ability of firms to appropriate the benefits from new technology; and (c) the technological opportunities for producing profitable products.

In two of the studies reviewed, demand and factor-price issues played an important role in research investment decisions by private firms. The role of appropriability is revealed by the propensity of seed firms to concentrate their research efforts on breeding hybrids rather than open-pollinated varieties. Technological opportunity measured by patents, technology imports, and local public-sector R&D were found to have had positive impacts on the level of local R&D investment in the Philippines and India. Unfortunately, these few studies fail to provide a solid basis for future technology policy in this area.

Despite the paucity of information on private R&D, some generalizations are possible. Private R&D is located primarily in Latin America and Asia, and it is concentrated in a few large countries such as Brazil, Mexico, Argentina, and India. Research conducted by local companies seems to be more important in Asia than in Latin America. Private research expenditures in the seed and machinery industries is growing. Agricultural chemical research seems to be growing in Asia, but the evidence is less clear for less-developed countries as a whole. There are too few data on plantation and processing R&D to be confident about the trends. The amount of private food and agricultural R&D is low in low-income countries and generally grows with per capita GDP.

Private research expenditure is still low relative to public R&D and to agricultural output in most less-developed countries. For instance, our estimates of annual R&D expenditures by multinationals on agriculturally related technologies are about US$ 36 million, while annual public-sector research expenditures by less-developed countries during 1981-85 were about US$ 3.6 billion (chapter 7). The studies reviewed in this chapter show that private research generally has a positive impact on agricultural productivity. This suggests that many less-developed country governments could increase national welfare by releasing constraints on private R&D activities.

Governments have a number of policy instruments with which to influence private R&D. Public-sector research can foster private-sector research by providing (or selling) research results and by training the personnel needed by private companies to conduct research. Patents and plant variety protection laws, if they are well designed and enforced, can create the necessary incentives for private companies to invest in R&D. Technology imports can stimulate local R&D, so more liberalized technological trade could also increase private-sector R&D activities.

More accurate data on private R&D expenditures and their impact and further research on potential public-private research interactions would help policymakers and public administrators identify the potential areas of conflict versus complementarities between public and private research endeavors. In addition, empirical studies of the impact of the various technology-policy instruments and of alternative public-private institutional arrangements for research could help policy-makers use them more effectively.

Chapter 11

Impact of Research and Seed Trade on Maize Productivity

Ruben G. Echeverría

The primary determinants of private-sector investments in agricultural R&D were examined in chapter 10 along with the scope of these activities in less-developed countries. This chapter builds on that analysis in an endeavor to assess the economic impact of public and private research and the seed trade on maize productivity. As mentioned in chapter 10, the private R&D sector has been largely ignored in the agricultural economics literature, particularly in regard to its impact on crop productivity. One notable exception is Griliches' (1957a) pioneering study of hybrid maize.[1]

In terms of the quantity produced, maize is the second most important cereal crop after wheat.[2] Maize production involves important technological and institutional policy issues, many of which are of relevance to other crops. These issues include the wide diversity of technology types utilized by farmers (namely, own seed, improved varieties, or hybrids) and the sources of these technologies, ranging from public through private research. There is virtually a continuum of maize seed types, from farmers' own open-pollinated varieties to improved varieties involving single-, double-, and three-way-cross hybrids. There is also something of a continuum of sources of those technologies, from local and foreign private, public, and jointly executed research.

The relationships between the type, source, and adoption of maize technologies, their

---

[1] Griliches (1958, p. 430) summed up his studies by stating that "To establish a case for public investment [in hybrid maize research] one must show that, in an area where social returns are high, private returns, because of the nature of the invention or of the relevant institutions, are not high enough relative to other private alternatives." See also Boyce and Evenson (1975, p. 65) for one of the earliest attempts to estimate the scope of private-sector agricultural research activities.

[2] During 1985-87, annual world maize production was 477 million tons and the area sown to maize was 129 million hectares. During that same period, wheat production was 520 million tons on 227 million hectares, and rice production was 467 million tons on 143 million hectares (*FAO Production Yearbook 1987*).

linkages with incentive structures facing the seed industry, and the efficiency of seed production, plus the transfer of technology through seed sales versus local adaptive research have been the subject of little systematic attention. These issues raise significant policy concerns in the context of public and private involvement, and their potential interaction, in the processes of generating and transferring agricultural technologies.[3]

The public-private-sector interaction is exceptional in the case of maize for two reasons. First, technological breakthroughs in more-developed countries were closely associated with the development of hybrids from which private seed companies could recover their investments. Second, unlike wheat or rice, maize is a major crop "... for which with some exceptions 'Green Revolutions' have bypassed small-holder agriculture in less-developed countries" (Heisey 1990, p. 245).

This chapter begins with an examination of the scope of maize seed production, trade, and research. Data on the 45 most important maize-producing countries in the world during 1960-85 are then used to assess the impact of public and private research, as well as seed trade, on maize productivity. Finally, policy implications are drawn from this analysis.

11.1 THE MAIZE SEED INDUSTRY

11.1.1 Typology of Maize Varieties

There are two main types of maize varieties, open-pollinated varieties (OPVs) and hybrids. OPVs include farmer-grown seed and improved certified varieties commercialized in the seed market. Hybrids can be classified into "conventional" and "nonconventional" (table 11.1). A conventional hybrid is derived from inbred lines. Single, double, and three-way crosses are examples of this category. If at least one of the parents is not an inbred line (or derived from an inbred line), the hybrid is called nonconventional. Varietal hybrids and top crosses are nonconventional hybrids. In less-developed countries, top-cross hybrids are the most common type used in the nonconventional group, and double crosses in the conventional category.

Open-pollinated, farmer-grown seed is sown on more than half the maize area in the less-developed countries (table 11.2). While improved OPVs cover only a small proportion (7%) of the maize area, they are understood to have an important quality-enhancing impact on the seed grown by farmers. When protected from cross-pollination, OPVs can maintain their genetic identity, with the important consequence that farmers do not have to buy seed every year.

---

3 For example, Norton and Davis (1981) in their review of methods to evaluate research, concluded that one of the major areas in need of methodological work is the public-private-sector interaction in agricultural research, including the transmission of research results to farmers. See Rausser et al. (1981) for a discussion of the public- and private-sector roles in agricultural research, and Ruttan (1982, p. 192) on the mix of public and private research in the development of plant varieties in the US.

Table 11.1: *Maize Varietal Types*

Open-pollinated varieties	Own seed	Local varieties saved and traded by farmers
	Marketed	Improved varieties sold usually as certified seed
Inbred line		Pure line created by self-pollination
Conventional hybrid	Single cross	A cross of two inbred lines
	Double cross	A cross of two single crosses
	Three-way cross	An inbred line and a single cross (female)
Nonconventional hybrids	Varietal	A cross of two varieties
	Top cross	A cross of a variety (female) and an inbred

Source: Echeverría (1988a).

Marketed (F1) hybrids are the result of first crosses among different varieties or lines. These specific crosses have to be remade each year to produce seed with the same genetic content. Experimental results under favorable temperate conditions suggest that yields decline by at least 20% when F2 or late-generation seed is used (Jugenheimer 1976). Given such potential losses under these growing conditions, farmers find it profitable to buy hybrid seed every year. However, in less-developed countries (especially under tropical conditions) the common practice of growing F2 and later generations may, in part, be explained by the low average yields and high year-to-year yield fluctuations experienced. Under such conditions, the yield decline from using seed of F2 or later generations is estimated to be rather less than 20%, and it is even lower when nonconventional hybrids are used (CIMMYT 1987a).[4]

Table 11.2 shows the percentage of total maize area planted to different varietal types: farmers' OPVs, marketed OPVs, and hybrids. Total improved area includes areas sown to marketed OPV and hybrid varieties. It is estimated that, during 1985-86, 88 million hectares (66% of the world total of 134 million hectares) were in this category. About one-third of the world maize area is under farmers' own seed, 62% with hybrids, and only 4% with marketed OPVs.

Table 11.2 also shows that of the more than 80 million hectares of maize in less-developed countries 45% are sown with improved seeds. This figure drops to only 27% of a revised less-developed country total when Argentina, Brazil, and China are excluded. The sub-Saharan Africa and Asia & Pacific regions have the lowest percentages of areas sown

---

4 Different varietal types have different yield potentials and production costs. Given that most of the hybrid maize breeding work has not been targeted toward the tropics and the more difficult growing conditions in these environments, the potential of hybrids in tropical regions is probably lower than in temperate regions. According to CIMMYT (1987a, p. 9), under US experimental conditions, single crosses yield about 30% more than improved open-pollinated varieties, while under tropical conditions they yield only 14% more. Seed production costs differ among varietal types primarily because of these differences in seed yield. For example, the seed yield of a single cross is only about half the seed yield of an improved variety.

Table 11.2:   *World Maize Area under Different Varietal Types, 1985-86*

| | | | Percentage area sown to | | |
| | | | | Marketed varieties | |
Region	Total area sown	Farmer's own OPVs	OPVs	Hybrids	Total[a]
	('000 ha)	%	%	%	%
Sub-Saharan Africa	15,274	76	8	16	24
China	18,050	28	0	72	72
Asia & Pacific, excl. China	17,750	78	14	8	22
Latin America & Caribbean	28,146	44	7	49	56
West Asia & North Africa	1,945	63	18	19	37
Less-Developed Countries (LDCs)	*81,165*	*55*	*7*	*38*	*45*
LDCs, excl. Argentina, Brazil, and China	*48,775*	*73*	*11*	*16*	*27*
MDC Non-Market Economies	12,125	5	0	95	95
MDC Market Economies	40,735	1	0	99	99
More-Developed Countries (MDCs)	*52,860*	*2*	*0*	*98*	*98*
World Total	134,025	34	4	62	66

Source: Constructed from CIMMYT (1987b).

Note: No precise count of the number of countries included in these (regional) totals is available because CIMMYT (1987b) only reported individual observations for the major maize producers and grouped the minor producers in a single category. Data may not add up exactly because of rounding.

[a] The total area sown to marketed varieties is the sum of the area sown to marketed open-pollinated varieties (OPVs) and hybrids. These values are underestimated in the table because some of the area under farmer's OPVs includes varieties derived from marketed OPVs.

with maize under improved varieties and hybrids. In less-developed countries, 38% of the total maize area is under hybrids, but only 16% when Argentina, Brazil, and China are excluded. Excluding these three important maize producers, 73% of the total maize area is still planted with farmers' own seed.[5]

Two conclusions can be derived from table 11.2. One is that much of the maize area in less-developed countries is not planted with improved seed. The second is that in the areas with improved seed, hybrids are more common than marketed OPVs.

Table 11.3 shows the relatively low value of the 2.1 million tons of maize seed sown annually in the less-developed countries. While this tonnage accounted for 69% of the total amount sown in the world, it represented only 25% of the total value of maize seed sown. This value-tonnage discrepancy stems from the substantial variation across varieties in the price of seed, due in large part to differences in their cost of production (see footnote 4). As

---

[5] See Timothy, Harvey, and Dowswell (1988) for a review of the development and spread of improved maize varieties and hybrids. See also Dalrymple (1985) on the development and spread of high-yielding varieties of wheat and rice in less-developed countries.

Table 11.3: *World Maize Seed Values of Different Varietal Types, 1985-86*

Region	Total seed sown	Total seed value	Percentage seed value sown to			
			Farmer's own OPVs [a]	Marketed varieties		
				OPVs	Hybrids	Total
	('000 tons)	*(millions US dollars)*	%	%	%	%
Sub-Saharan Africa	397	144	61	14	24	38
China	567	159	8	0	92	92
Asia & Pacific, excl. China	450	143	68	18	14	32
Latin America & Caribbean	647	316	19	6	75	81
West Asia & North Africa	58	14	50	19	30	49
Less-Developed Countries (LDCs)	*2,119*	*776*	*34*	*9*	*57*	*66*
LDCs, excl. Argentina, Brazil, and China	*1,239*	*525*	*22*	*6*	*72*	*78*
MDC Non-Market Economies	234	347	1	0	99	99
MDC Market Economies	720	1956	0	0	100	100
More-Developed Countries (MDCs)	*954*	*2,302*	*0*	*0*	*100*	*100*
World Total	3,073	3,078	9	2	89	91

Source: Constructed from CIMMYT (1987b).

Note: See table 11.2.

[a] OPV = Open-Pollinated Variety.

in table 11.2, there are also important variations among less-developed regions. For example, hybrid seed accounts for 92% of the total value of maize seed sown in China against only 14% in the rest of Asia & Pacific.

The use of seed of different varietal types is determined by economic as well as agronomic factors. In particular, the yield potential of each type in a given environment along with their relative prices play an important role in determining which seed is used. In addition, the ratio of seed to grain price is a useful indicator of the seed-type choices farmers make when deciding which type of seed to plant. CIMMYT (1987a) reports a fourfold ratio of seed price to grain price for improved varieties, fivefold for nonconventional hybrids, sevenfold for double and three-way crosses, and elevenfold for single crosses. Despite the variation across countries in the seed-to-grain price ratio, nonconventional hybrid seed prices were only about 20% above those of improved varieties.

11.1.2 Maize Seed Production

Since public and private involvement in the maize seed industries of less-developed countries has been little researched in the past, data are scarce. CIMMYT (1987a,b), however, recently surveyed the extent of public intervention in the maize seed industry of many less-developed countries (table 11.4).

Table 11.4: *Public- and Private- Sector Involvement in Maize Seed Industries, 1985-86*

			Countries with			
Region	Regulatory agency	Quality testing agency	Public seed company	Private seed company	Seed price control	Seed subsidy
	%	%	%	%	%	%
Sub-Saharan Africa (11)[a]	73	64	82	73	91	36
China	0	0	100	0	100	0
Asia & Pacific, excl. China (9)	100	100	89	78	44	67
Latin America & Caribbean (13)	69	85	85	92	54	8
West Asia & North Africa (3)	100	100	100	67	67	67
Less-Developed Countries (37)	*78*	*81*	*86*	*78*	*65*	*35*
MDC[b] Non-Market Economies (4)	100	100	100	0	100	25
MDC Market Economies (6)	83	100	0	100	0	0
More-Developed Countries (10)	*90*	*100*	*40*	*60*	*40*	*10*
Total Sample (47)	81	85	77	74	60	30

Source: Constructed from CIMMYT (1987b).

Note: Data represent percentage of countries reporting.

[a] Number of countries reporting is given in parentheses.
[b] MDC represents More-Developed Country.

In less-developed countries it is common to find public agencies regulating the seed industry, providing quality testing in order to certify cultivars, and also producing and marketing seeds. Most of the countries included in the CIMMYT survey have public seed-regulating and quality-testing agencies.[6] The more-developed market economies reported no public seed company, while 86% of less-developed countries did report one. The countries in the Latin American region show lower public involvement and higher private-sector presence than the rest of the less-developed countries. Maize seed price controls seem to be more common than seed subsidies in less-developed countries, although there is a wide variation by region.

Table 11.5 reveals the significance of private sales of maize seed in market economies during 1985-86, with 95% of all marketed maize seed being sold by private firms.[7] Overall these companies had a 64% share of improved OPV seed sales and a 98% share of hybrid seed sales. The percentage share of hybrid seed sold by private companies is altered when only less-developed countries are considered, with private shares of maize seed sales

Table 11.5: *Private-Sector Share of Maize Seed Sales in Market Economies, 1985-86*

Region	Private companies' share of sales		
	OPV[a]	Hybrid	Total marketed[b]
	%	%	%
Sub-Saharan Africa (10)[c]	48	95	84
Asia & Pacific, excl. China (9)	62	68	64
Latin America & Caribbean (13)	70	96	92
West Asia & North Africa (3)	62	56	59
Less-Developed Countries (35)	63	92	85
More-Developed Countries (6)	100	100	100
Total Sample (41)	64	98	95

Source: Constructed from CIMMYT (1987b).

Note: Non-market economies such as China and the Eastern European countries reported no private-sector sales in 1985-86. Including these countries in our regional averages would act to lower them substantially but add nothing to our understanding of the relative importance of the private sector in marketing maize seed. We therefore, opted to limit our sample to market economies only.

[a] OPV = Open-Pollinated Variety.

[b] Total marketed seed is the sum of marketed open-pollinated and hybrid seed sold by private and public companies.

[c] Number of countries reporting is shown in parentheses.

---

[6] Seed regulating and quality testing includes all quality control agencies such as seed certification, seed testing laboratories, and seed law enforcement and inspection organizations. Seed industry in this chapter refers to the entire complex of organizations, institutions, and individuals associated with plant breeding, production, processing, quality control, and supply of seeds.

[7] Based on a sample of 41 of the principal maize producing market-economies in the world.

accounting for 85% of all marketed maize seed and for 92% of hybrid seed.

Tables 11.4 and 11.5 suggest that, despite the existence of public seed enterprises in almost all less-developed countries with market economies, private companies produce and sell most of the hybrid seed and two-thirds of the improved OPVs. The role of the public sector in seed production and marketing has not been significant for a number of reasons, the lack of incentives to produce and market good-quality seed being one of the main ones.

Maize seed production and marketing is a complex process that involves seed growers and distributors at both the regional and local level. Key factors in this process are an appropriate reward to growers based on the quality of seed produced and seed market prices that cover processing and distribution costs.[8]

The retailing performance of many public seed companies is poor. In the absence of a profit motive, there is usually not much incentive to increase sales. On the other hand, the profits that foreign private firms can earn depend, in part, on the costs of entry and the size of the market. Given the high costs of entry and the small market size in many low-income agricultural regions, private firms may not be attracted. These constraints help to explain why the seed industries of many less-developed countries are still wedged between the low performance of their government enterprises and low levels of private activity.

11.1.3 Maize Seed Trade

Previous studies have stressed the importance of the transfer of knowledge and the development of local capacity to generate appropriate technologies for each region. They emphasized the environmental sensitivity of biological technology and the need to adapt technologies produced elsewhere to local conditions through local research. For a specific type of technology such as improved cultivars, they have largely discounted or overlooked the possibility of direct transfer through trade.

The seed trade makes possible a direct transfer of biological technology with or without the need for local adaptation.[9] Private seed companies, local and international, have not only played an important role in maize research, seed production, and distribution in a particular country or region, but they have also distributed the product of their research to a global market.

According to the USDA (1957 to 1987), total US seed exports have increased from less than 30,000 tons in the early 1950s to 250,000 tons by the early 1980s. In constant 1980 terms, the value of these exports has also increased from less than US\$ 15 million to more

---

[8] See Douglas (1980, p. 84) for alternative methods of developing seed enterprises in less-developed countries, focusing on public- and private-sector participation, and McMullen (1987) for a worldwide review of seed industries focusing on seed policies in the less-developed countries.

[9] See Hayami and Ruttan (1985, ch. 9) and Evenson and Kislev (1975b, chs. 3-4) for an analysis of international transfer of agricultural technology, and Englander (1981) for a detailed treatment of the extent of technological trade and adaptability of technology.

than US$ 100 million over the same period. But, the type of seed being exported and its destination have changed rather significantly over time.[10] Figure 11.1 shows US and EEC maize seed exports to the rest of the world. Total quantities exported grew from 18,000 tons in 1970 to more than 50,000 tons by 1985. US maize seed exports to Canada, Mexico, and Central America were more than 10,000 tons per year during the 1970s. This figure substantially decreased by 1985, while US exports to the EEC increased to more than 20,000 tons per year over the same period. Both Asia and the Middle East also became substantial importers of US maize seed during the 1970s and 1980s.

Figure 11.1: *USA and EEC maize seed exports to the rest of the world*

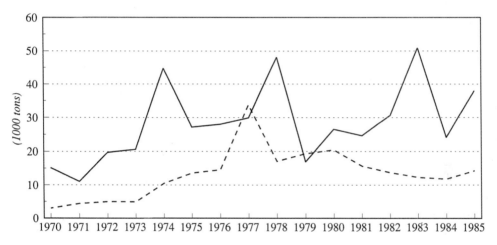

Legend: ———— USA; - - - - - EEC.

Source: USDA (1957 to 1987) and EUROSTAT (1966 to 1986).

Note: Membership of the EEC has increased over the period shown and may have biased the comparison over time.

The maize seed trade within the EEC has been more important than from the EEC to the rest of the world (figure 11.2). Table 11.6 shows maize seed production and trade flows of three leading maize producers in the EEC. France is the largest maize seed producer and exporter in Europe. Germany is the principal destination of French exports within the EEC, but an important share of German imports also comes from the US. Italy is the second largest

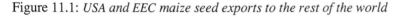

[10]Forage seed exports increased from 10,000 tons to 50,000 tons from 1950 to 1986. Grain seeds became the dominant type exported after the early 1970s, reaching 100,000 tons by the early 1980s. Exports of vegetable and other types of seeds have increased from less than 2,000 tons in the 1950s to more than 100,000 tons in the 1980s. North and Central America (mainly Canada and Mexico) and Europe have been the main US seed customers. Asia has become an important US customer since the mid 1970s, while South America and Africa have not imported significant quantities of US seed (USDA 1957 to 1987).

Figure 11.2: *Maize seed exports within the EEC and from the EEC to the rest of the world*

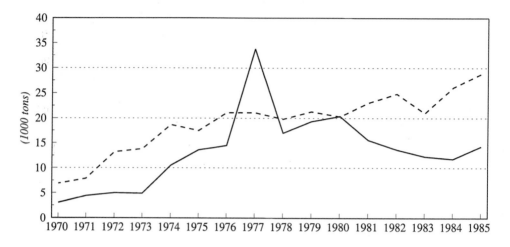

Legend: - - - - · within; ———— rest of the world.

Source: EUROSTAT (1966 to 1986).

Note: Membership of the EEC has increased over the period shown and may have biased the temporal comparison.

Table 11.6: *Seed Production and Trade of Main European Maize Producers, 1980-82 Average*

Country	Production	Exports		Imports	
		Intra-EEC	Rest of the world	Intra-EEC	Rest of the world
	(tons)	*(tons)*	*(tons)*	*(tons)*	*(tons)*
France	107,323	18,883	13,160	364	3,146
Italy	19,172	363	282	1,212	10,575
West Germany	3,766	641	328	18,423	11,892

Source: Adapted from McMullen (1987).

maize seed producer in the EEC, with most of its imports also coming from the US.

Maize seed exports from the US to the rest of the world were close to 36,000 tons in 1985-86. Using an average planting rate of 20 kg/ha, 1.8 million hectares could be planted directly with those exports. Table 11.7 was constructed in order to estimate the magnitude of US maize seed exports. Knowing the volume of seed imported and using planting rates specific to each country, the hybrid maize area that could be planted with imported US seed was calculated. The 14 selected countries are important US seed clients. The third column

Table 11.7: *Estimated Area Sown with Hybrid Maize Seed Imported from the US, 1985-86*

Country	Total hybrid seed sown	Seeds imported from the US[a]	Area sown with seed imported from the US[b]
	(tons)	*(tons)*	%
Italy	16,400	8,300	54
Greece	4,800	2,200	53
Turkey	6,000	1,400	24
South Korea	800	200	20
Chile	2,300	500	20
Mexico	20,000	3,700	18
Austria	3,800	400	11
Spain	9,400	1,000	11
Egypt	1,800	100	8
Hungary	32,900	2,300	7
Canada	24,000	1,600	6
Portugal	1,400	100	6
France	57,600	2,800	5
Japan	24,000	1,200	5

Source: CIMMYT (1987b) for hybrid seed sown, except for Mexico (Echeverría 1990a); USDA (1957 to 1987) for seed imported from the US.

[a] Hybrid seed imported from the US during 1984-86.

[b] Calculated from first two columns and CIMMYT (1987b) data on hybrid seed planting rates per country.

of table 11.7 shows that more than half of the maize area in Italy is planted with seed imported directly from the US. About one-fourth of Turkey's maize area under hybrids is also purchased from the US, while only 5% of the maize area in Japan and France is planted with such seed.

Direct transfer of technology embodied in maize seeds has been significant in many regions. The significance of the seed trade is, however, even more important than indicated above for two reasons. First, US companies (or US-owned companies) and multinational companies also conduct research and produce maize seed abroad. For example, seed sold in Western Europe is not only imported from the US but also from US-owned companies located in Eastern Europe.[11] Second, as shown in table 11.6, the US is not the only country producing and selling maize seed, as other countries such as France also produce and export maize seed within and outside the EEC.

---

[11] The volume of hybrid maize seed produced and marketed by Pioneer Hi-Bred International in Hungary, for example, is large enough to plant much more than the area shown for that country in data column 3 of table 11.7.

11.1.4 Maize Research[12]

In less-developed countries the public sector undertakes most of the biological, physiological, and genetic research on maize as well as the plant breeding research required to generate varieties and inbred lines. The private sector generally uses public inbred lines to develop their own hybrids, although a few large private firms develop their own inbred lines. Hybrids have some of the characteristics of a trade secret; hence, private companies with good control over their information have a clear incentive to invest in hybrid maize research.

The provision of improved germplasm and/or inbred lines by the public sector maintains competition, since most small seed companies rely on publicly developed lines to produce seed. Public research, in this sense, complements and stimulates private research. In the US the public sector has been the major force in the development of germplasm and inbred lines, while seed companies have specialized in the production and marketing of hybrid seed (Sprague 1980). Duvick (1984) reports that in 1970, 50% of the area planted with maize in the US used public inbred lines. By 1980 this area had decreased to 40%.[13]

Figure 11.3 indicates the principal interactions between public- and private-sector maize research and seed production. The actors in this relationship are national and international research centers, public seed companies, public regulatory agencies, national and international private companies, and farmers.

The maize programs of the international agricultural research centers do not intend to develop finished materials suited to a particular region but rather experimental varieties adapted to various *mega-environments*. These mega-environments are distinguished by overall ecology, length of growing season, and incidence of diseases and insect pests, among other factors (Cantrell 1986). The International Maize and Wheat Improvement Center (CIMMYT) has developed superior OPVs through recurrent selection in maize populations. Several special breeding projects (such as plant height reduction, early maturity, drought tolerance, and protein quality) have focused on improving particular traits. A hybrid program established in 1985 is also providing assistance to NARSs in hybrid breeding. International agricultural research centers release improved germplasm to NARSs as well as seed companies.

Many NARSs support their own maize research programs. The output of this research (a new variety, inbred line, or hybrid) is then often multiplied and marketed by a parastatal or local private company. The research programs of these local companies are rather

---

[12]The discussion of maize research throughout this chapter focuses on yield enhancing technologies embodied in seeds, although research is also directed to maintain yields; to improve nutritional content, palatability, and storability; and to reduce yield variability via drought, pest, and disease resistance.

[13]In the US the public sector undertakes more basic seed R&D than the private sector, and both sectors are active in applied R&D. The public sector conducts applied R&D on self-pollinating crops where returns to investments are difficult to appropriate. See Knudson (1990) for a discussion of the role of the US public sector in applied plant breeding research, with special reference to wheat.

Figure 11.3: *Public- and private-sector interaction in maize research and seed production*

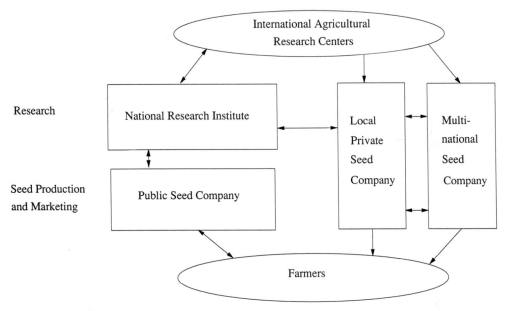

Source: Echeverría (1990a).

applied in nature and depend on genetic material provided by public research and/or multinational corporations. Multinational corporations conduct their own research and market seeds directly through joint ventures or by licensing.

At the local level, there are many institutional rules that shape the nature of public-private-sector interactions. In some cases this interaction can be restricted by law, as in Mexico where all materials arising from public research must pass to a public seed company for multiplication and marketing. In other cases, as in Guatemala, parastatals do not exist and local companies distribute publicly developed lines. In Argentina and Chile, however, almost all research and marketing is performed by local as well as multinational companies.

Figure 11.3 also characterizes the interactions that take place between local and multinational companies. Many local companies operate as licensees for multinational corporations or undertake joint ventures with them. This is an example of complementarity within the private sector, where multinational corporations provide most of the research capacity while local companies test and market promising materials.

Interactions between national research institutes and private seed companies occur in at least four different ways: (a) public organizations evaluate materials developed by private seed companies, (b) these private companies obtain or buy royalties for basic seed of public origin, (c) companies fund some public research, and (d) scientists move between the public and private sectors. There are also interactions at the policy level that link the public and the private sectors, involving seed laws, trade regulations, and input and output pricing policies, among other things.

Sehgal (1977) identified at least two areas where the relationship between international agricultural research centers and seed companies could be expanded. One suggestion was to exchange breeding materials on a more regular basis and the other was to include private varieties and hybrids in the worldwide evaluation network of the international agricultural research centers. An often neglected aspect of the relationship between public research institutes (international and national) and private seed companies concerns the objectives these institutions seek to attain. Maximizing expected profit is the main goal of a private company. Seed firms will therefore specialize in crops that justify R&D programs, i.e., where an economically significant part of the benefits generated from research are appropriable. This means that public research may be called for in those crops and cultivar types[14] where the private sector is less active.

Public maize research in less-developed countries was stimulated during the 1940s and 1950s by collaborative efforts among the Rockefeller and Ford Foundations, FAO, and USAID. The establishment of CIMMYT and the International Institute of Tropical Agriculture (IITA) in the mid-1960s institutionalized many of these earlier research and training efforts. CIMMYT and IITA develop and distribute improved germplasm, and variously support regional and national public breeding programs.

During the past two decades, CIMMYT, IITA, and many research systems in less-developed countries have expanded their investment in maize research (Timothy, Harvey, and Dowswell 1988). Quantitative evidence on the extent of these investments, however, is sketchy. Evenson (1987) estimated public research expenditures on maize by national institutes as a percentage of the value of the commodity during 1972-79 to be 0.44% in Africa, 0.21% in Asia, 0.18% in Latin America, and 0.23% for all less-developed countries combined. Compared with other commodities, this level of expenditure is relatively low. This is in contrast to research expenditures on maize at the international level. CGIAR center expenditures on maize were estimated by Evenson (1987) at 0.03% of total output value of maize in less-developed countries, which represented 11% of the total public expenditure on maize research.[15]

Private-sector activities concentrate on the development of hybrids and on the multiplication and marketing of seed. The efficiencies that arise through their profit orientation often gives private companies a comparative advantage in seed production, processing, and distribution. There are various types of private companies involved in maize R&D, ranging from multinational corporations such as Pioneer Hi-Bred, which develops its own inbred lines and markets seeds worldwide, to small local companies that specialize in evaluative trials and market materials developed elsewhere. In general, private companies give more emphasis to potentially high-yielding areas. The involvement of seed companies in maize

---

[14]Or particular regions and/or types of farmers.

[15]In 1981-85, total CGIAR expenditures averaged only 4.3% of total expenditures by NARSs in less-developed countries (chapter 9).

research will probably increase in the future. Their activities could complement those of NARSs working for the more marginal environments and supplying basic seed to local companies.

A few large-scale multinational corporations have been quite active in the past two decades in the direct transfer of technology through seed sales. They have also established important research programs in many maize-growing countries throughout the world. Most of these programs have developed during the past decade. Until the mid-1970s, the international R&D conducted by US maize companies was limited to the transfer and trial of varieties to temperate areas, such as Europe, and temperate less-developed countries, where US Corn-Belt hybrids perform well (Sehgal 1977). As improved germplasm for tropical and subtropical regions was developed, mainly by international agricultural research centers[16] and NARSs, companies have expanded both research and seed production in less-developed countries. For example, in Mexico and Guatemala, Echeverría (1990a) estimated that total expenditures on maize research by US companies and local private firms grew from negligible levels in the late 1970s to a 1987 figure of US$ 1.7 million and US$ 90,000, respectively.

11.1.5 Maize Productivity

World production increased almost threefold from 1955 to 1987 (figure 11.4).[17] In 1987, 458 million tons of maize were produced worldwide on 130 million hectares, yielding 3.6 tons per hectare. Much of this increment was achieved during the past 15 years, due primarily to growth in yields. Area increased by one-fourth while yields more than doubled during the period.

North America accounted for half of world maize production during the 1955-87 period. Output in this region was almost constant during the 1960s but increased quite rapidly during the 1970s and 1980s. The area planted to maize in North America remained fairly constant at around 40 million hectares, while yields increased substantially during the period.[18]

Maize production also increased substantially in Asia & Pacific and in Europe. In Asia, production has risen due to both growth in area cropped and also to an increase in yield per hectare, while dramatic yield increases largely explain the change in European production. Latin American & Caribbean and sub-Saharan Africa show relatively small

[16]CGIAR centers released a total of 238 maize varieties up to 1984: 61 in sub-Saharan Africa, 49 in Asia, 126 in Latin America, and 2 in North Africa and Middle East (CGIAR 1984, p. 24).

[17]Appendix table A11.1 shows maize production, area, and yield, as well as growth rates for the principal maize producing countries in the world, ranked by production during 1983-87.

[18]Cardwell (1982) attributes 58% of the increase in maize yield in the US over the past 50 years to the change from open-pollinated varieties to hybrids, combined with a genetic gain of 36.5 kg per hectare per year. The remaining yield increase of 40% was explained by, among other things, higher fertilizer applications, higher plant density, better soil drainage, fall plowing, and the increased use of herbicides.

Figure 11.4: *Maize production, area, and yield, 1955-87*

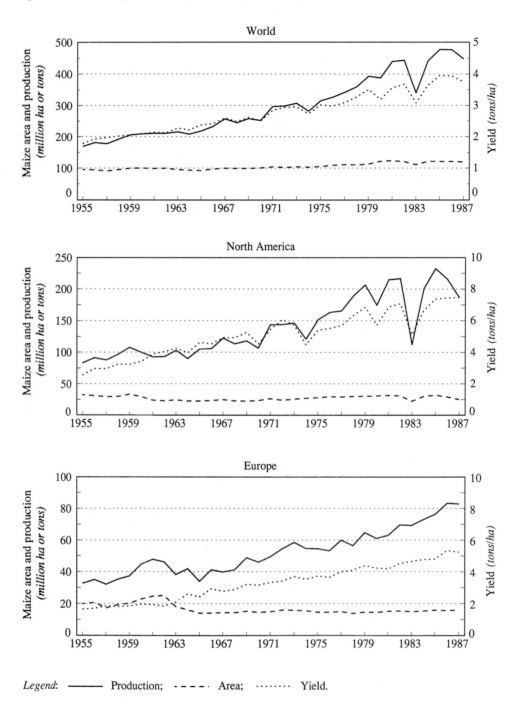

Figure 11.4: *Maize production, area, and yield, 1955-87 (Contd.)*

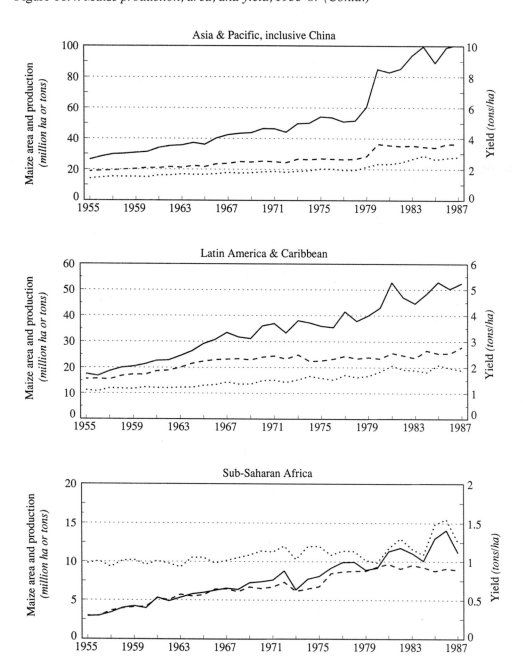

[a] The world total comprises 45 major maize producing countries (see table A11.1 for listing).
[b] Europe here comprises both market as well as nonmarket economies, including the USSR.

growth in production compared with the other regions, mainly because of the very modest growth in yields.

Almost 40% of the world's maize was produced in the US. Half of the remaining 60% was produced by only five countries: China, Brazil, Romania, USSR, and Mexico. Area and yields vary significantly across countries. Greece, with yields above 9 tons per hectare (in an area smaller than 200,000 hectares), and Zaire, with yields below 1 ton per hectare during 1983-87, represent the extreme countries. Most less-developed countries have maize yields below 3 tons per ha.

The annual rates of growth in production during 1960-85 for some of these countries are striking. For example Greece, Austria, and West Germany had more than 10% annual growth in maize production, and Canada and France experienced 9.2% and 6.8%, respectively. Most of this growth in production was achieved through improvements in yields. In China, yields grew at almost 5% annually, in Romania and Spain at 4.4%, and in Greece at an impressive 9.5%. On the other hand, many countries have had much slower growth rates.

The next section of this chapter is addressed to the question of why maize yields are so different across countries and over time.[19]

11.2   THE IMPACT OF PUBLIC AND PRIVATE RESEARCH ON MAIZE YIELDS

Maize output can be modeled as an aggregate production function of cultivated area, labor, and capital, as well as other variables such as weather. To estimate such a function requires periodic (usually annual) data on these variables on a regional or per country basis. The framework of this study is constrained by the lack of available data and their level of detail.[20]

Faced with the limitations of data, a yield response function is developed in order to examine the impact of public research, private research, and seed trade, among other things, on maize production. More specifically, this maize response function model relates output per hectare as a function of levels of seed imports, fertilizer use, and research investment, along with a variety of qualifying variables, such as land quality and weather.[21]

---

[19]The analysis focuses on absolute average yield differences across countries and time rather than on the variability of yields about their mean levels. See Anderson and Hazell (1989) for a review of evidence on the latter.

[20]See chapter 5 and the references therein for a discussion of data problems when measuring economic development and production activities, particularly in an international context.

[21]Although crop output per unit land as an indicator of agricultural productivity serves a useful purpose for comparisons over time and across regions, it is an incomplete measure. More complete total factor productivity indices measure output changes that are the net result of the contribution of all conventionally measured inputs. Yield-response functions have been utilized by Houck and Gallagher (1976), Otto and Havlicek (1981), and Menz and Pardey (1983) to estimate maize supply. These models express yields as a function of input and output prices, technical parameters, and other variables. The model developed here

11.2.1 A Response Function Model

In general, changes in maize yields reflect changes in the level of use and quality of variable inputs — particularly those quality changes associated with the development and adoption of new varieties — as well as improvements in crop management techniques. Yield changes also reflect the random effect of weather, disease, and other factors.

A maize yield production function of Cobb-Douglas form that expresses maize yields in country i at time t, y_{it}, as a function of $k = 1, \ldots, K$ inputs, x_k, can be written as follows:

$$y_{it} = \beta_{0it} + \sum_{k=1}^{K} \beta_{kit} x_{kit} + u_{it} \tag{11.1}$$

where β_{0it} is the intercept for the ith country at time t, a parameter to be estimated; β_{kit} is an unknown parameter to be estimated for the kth input for country i at time t, and represents the percentage change in yields in response to a percentage change in input use; x_{kit} is the value of the kth explanatory (independent) variable for country i at time $t;$ and u_{it} is a stochastic error term, assuming $E[u_{it}] = 0$ and $E[u_{it}^2] = \sigma_u^2$. The i units in (11.1) index the 45 countries listed in appendix table A11.1, the t time periods index the years from 1960 to 1985, the k units (for $k = 1, \ldots, 7$) index the explanatory variables of this function. All of the variables are defined logarithmically on a per country, per year, and per hectare basis.[22]

The explanatory variables utilized in this study to specify the yield response function 11.1 are seed imports, fertilizer use, quality of land, public-sector research, private research conducted by multinational companies, maize training conducted by CIMMYT, and climate.[23] Table 11.8 defines the units of measurement and sources of information used to construct these variables, as well as their statistical characteristics.

The quantity of seed imports attempts to capture the yield-enhancing effects of directly transferred technology embodied in maize seed. It includes imports only from the US, which is not a totally satisfactory measure of the seed trade impact since other countries also export maize seed, and US companies also produce seed within some other countries (section 11.1). The US, however, is the principal maize seed exporter and data are not readily available for the other exporting countries.

Maize output per unit of land is directly influenced by the level of fertilizer applica-

---

is, strictly, a relationship between inputs and output, and is not a function of prices.

[22]The function 11.1 is relatively easy to estimate by least squares because in logarithmic form it is linear in parameters. To achieve linearity, the dependent variable y_{it}, the intercept β_{0it}, and the vector of inputs x_{kit} in (11.1) are expressed in logs. Joint estimation of cross-section and time-series data improves the efficiency of estimation. See Judge et al. (1982) and Hsiao (1986) for a detailed analysis of "pooling" using time-series and cross-sectional data.

[23]By using a Cobb-Douglas function of the general form $y = \beta_0 \prod_k x_k^{\beta_k}$ we assume all inputs are technically complementary.

Table 11.8: *Descriptive Statistics for 45 Selected Countries, 1960-85*

Variable	Units	Mean	Standard deviation	Maximum	Minimum
Production	Thousand tons	6,806	22,081	225,575	20
Area	Thousand hectares	2,348	4,475	30,442	6
Yield	Tons per hectare	2.45	1.70	9.13	0.44
Seed imports	Tons[a]	648	2,278	27,420	0
Fertilizer use	Kg of total NPK application per hectare of arable land	68	90	471	0.1
Land quality	Index	94	27	180	59
Public research	No. of maize scientists	9	17	99	0
Public research	No. of maize publications[b]	74	187	1,393	0
Public research	No. of maize varieties[c]	25	38	183	0
Private research	Weighted no. of research stations[d]	0.7	1.5	9	0
Training	No. of trainees[e]	2.4	1.7	16	0
Climate	Dummy (0 or 1)[f]	0.5	0.5	1	0

Source: *FAO Production Yearbooks* for production, area, and yield; USDA (1957 to 1987) for seed imports; Martínez and Diamond (1982), *World Development Reports* and *FAO Fertilizer Yearbooks* for fertilizer application; Peterson (1987) for land quality index; Henderson (1960, 1968, 1976, 1984) for public research scientists; Boyce and Evenson (1975) for maize publications for 1948-1974, and CABI (1972 to 1986) for maize publications for 1972-1986; Henderson (1960, 1968, 1976, 1984) for public research varieties; Echeverría (1988b) for private research; CIMMYT (1987c) for training; FAO (1982) and Papadakis (1966) for climate.

Note: All variables are defined here per country (for the 45 selected countries listed in table A11.1) and per year (1960 to 1985). In the regressions they are expressed on the basis of hectares of maize sown.

[a] Data on seed imports are for all countries except the US, for the period 1967-1985.
[b] Abstracted in "Plant Breeding Abstracts" during 1948-1986. From 1948 to 1974 includes sorghum.
[c] Inbred lines, open-pollinated varieties, synthetics, and other breeding stocks (see table 11.1 for definitions).
[d] Information from eight multinational seed companies was included: Pioneer, Dekalb, Cargill, Northrup-King, Funk, Agrigenetics, Asgrow, and Pacific. Data are for all countries except the US.
[e] Of public research institutes in less-developed countries who received in-service maize training at CIMMYT headquarters in Mexico, from 1966 to 1985.
[f] 0 for "tropical" or "suitable and marginally suitable" maize growing conditions, and 1 for "temperate" or "very suitable conditions."

tion. Nitrogen and potassium are particularly important for maize. Fertilization increases ear weight and, for a given ear size, there is the possibility of increasing plant population with higher levels of fertilization. There is information on fertilizer application to maize for about one-third of the countries included in our sample (Martínez and Diamond 1982). For the rest of the countries the fertilizer variable includes nitrogen-phosphorus-potassium (NPK) use for all crops, rather than only for maize. This assumes that average fertilizer application rates approximate maize fertilization rates. Since we are working with some of the most important maize producing countries of the world, this may not be an unreasonable approximation.

A comparison of land productivity among countries should take into consideration the heterogeneity and intensity of use of agricultural land (chapter 5). In particular, countries in Asia irrigate their arable and permanently cropped land to a higher degree than most other countries, while sub-Saharan Africa has a quite small percentage of arable land under irrigation. While not entirely satisfactory for adjusting maize areas per se (see section 5.3.2) the land quality index developed by Peterson (1987) is a useful first approximation of cross-country land quality differentials and is included as a control variable.

Depending on the purpose at hand, the impact of research could be accounted for through due regard to either the inputs to or outputs from research. In the absence of a more comprehensive measure of research inputs, such as expenditures on maize research, a plausible alternative would be the number of maize breeders, adjusted by scientific expertise, in the public and private sectors in each country. Fortunately, there is some information available on the number of public-sector plant breeders for the principal maize producing countries in the world (Henderson 1960, 1968, 1976, 1984).

Two measures of research output were utilized to estimate the effect of the indigenous public research effort directed to maize within each country, namely, the number of research publications on maize and the number of maize varieties released in each country for which information is available. The number of scientific publications related to maize abstracted in "Plant Breeding Abstracts" is an indirect measure of research output.[24] This has the advantage that the publications are screened worldwide and meet international levels of research relevance. The major advantage is that the classification is done by commodity, by country, and by year. The proxy utilized in this study is the cumulative number of publications abstracted each year, measured as a stock since 1948, the first year in which information is available, per hectare of maize sown in each country for that particular year. Research publications as a measure of research output are probably biased downward in less- relative to more-developed countries as researchers in less-developed countries tend to publish less (and appear less frequently in abstracts) than do those in more-developed countries. The incentives to publish in many less-developed country NARSs are probably smaller than in more-developed research systems.

A second measure of research output utilized here is the number of maize inbred lines, OPVs, synthetics, and other breeding stocks released by public research organizations. This is a more direct measure than is the number of publications of the real output of maize research in each country. The variable included in the analysis is the cumulative number of varieties released, measured as a stock since 1948, the first year when information is available, per hectare of maize sown in each country for that particular year. To sum up,

---

[24]Boyce and Evenson (1975) used this proxy in their pioneering work on agricultural research programs. See Thorpe and Pardey (1990) for a recent discussion of the use of publication indicators in this context.

three measures of public-sector research are utilized in this study: personnel, publications, and varieties.

The private-research variable was the most problematic to proxy. The measure used is the number of research stations of the major multinational companies dealing with maize located in each country and multiplied by the number of years the stations had been in operation since 1960. This variable captures the local effort of multinational corporations involved in maize research and seed production but underestimates total private maize research in countries where local private research is more important, as it is in Europe, Brazil, and Argentina. A measure of the local private research effort is, however, difficult to obtain, particularly for as many as 45 countries. The analysis that follows does not, unfortunately, include a specific variable for local private research in each country.

Undoubtedly, the impact of research depends heavily on the quality of the research output. Both output measures of public-sector research utilized in this study, namely publications and varieties, account for the quantity of research undertaken, but the quality of this output depends on other factors, for example, the research capabilities of the maize researchers. A plausible measure for this is the number of NARS staff trained in maize production and breeding by the international agricultural research centers involved. For example, CIMMYT trained over 1000 maize scientists from more than 80 less-developed countries from 1966 to 1988 at its headquarters in Mexico.[25] The objective in these training activities is to transfer knowledge of how to improve maize research. This knowledge, embodied in the scientists, has had an output quality effect over and above what the number of publications and varieties can account for in our measures of public research discussed above.

A proxy for this training effect has been constructed based on the cumulative number of CIMMYT in-service trainees since 1966, the first year of CIMMYT's training activities, expressed per hectare of maize sown in each country for each year. This variable does not include in-country training, nor does it include other related activities such as visiting scientists, graduates, and postdoctoral fellows. In addition, it does not include the effect of similar training activities undertaken by IITA.

In order to control for innate differences in potential yields due to variations in growing conditions not captured by the land quality index, data from FAO (1982) and Papadakis (1966) were used to construct a dummy variable that partitioned the 45 countries in our sample into two climatic groups, namely, temperate and tropical.

A potentially important omitted variable relates to the human capital component of agricultural production. The important role of education in production, as well as the role of other investments in human capital in agriculture, is well documented (Welch 1978).

---

[25]These training activities focus in maize production and breeding and experiment station management. In-service trainees generally stay in Mexico a full crop cycle (five to seven months) and have been previously employed by their governments in agricultural research or extension for at least five years (Dowswell 1986).

Better-educated farmers are more likely to be able to search for and screen different technologies and to adapt the most beneficial ones to their particular agroeconomic conditions. Education, in this sense, affects crop productivity. The problem here is that the effect of education may already be captured somewhat in the other variables, such as use of improved cultivars and fertilizer. In addition, there is also a measurement problem since details on education are seldom available at the farm level.

Finally, two other important factors were not made explicit in the analysis. These involve complexities concerning the time lags between research activities and their potential impact, and the spillover effects related to cross-country flows of scientific knowledge and technologies. Certainly, an attempt to capture some aspects of the lagged effects of research on output was made by including varietal- and publication-based measures of research output in a cumulative fashion. However, this procedure clearly abstracts from many important issues concerning the precise shape and length of these lagged responses. In particular, the impact of research aimed at maintaining, rather than simply enhancing, past research-induced yield gains is not made explicit. With regard to spillover effects, only measures of the transfer of knowledge embodied in seeds exported from the US and embodied in maize researchers trained at CIMMYT were included. Regrettably, no other forms of transfer, such as exchange of research results and information sharing among countries could be addressed.

11.2.2 Analysis and Results

The model (11.1) is tested against a pooled cross-section and time series sample of 45 countries, listed in table A11.1, covering information for 1960-85. The regression results reported in table 11.9 are based on both a constant-coefficient and a fixed-effect specification of the model. The constant-coefficient specification treats all coefficients as constants with the disturbance term capturing differences among countries and through time. This means that there is a common intercept and a common set of slope coefficients for all countries and time periods.[26]

In the fixed-effect specification used here, the intercept term was free to vary across countries, while the slope coefficients were held constant. Since omitted variables may produce changes in the cross-section intercepts, the addition of dummy variables to the model corrects for those changes. This is a special case of covariance analysis. It provides a richer specification of the model by having a country-effect variable.[27]

---

[26] According to this, (11.1) becomes $y_{it} = \beta_0 + \Sigma_k \beta_k x_{kit} + u_{it}$. We assume $u_{it} \sim iid\,(0, \sigma_u^2)$ for all i,t, i.e., there is no serial correlation in the disturbances for any i, there is no dependence between the disturbances for different i's, and the disturbance has a constant variance for all i,t. All cross-section and time-series data are combined and ordinary least squares is applied to the entire data set.

[27] Then (11.1) becomes $y_{it} = \beta_{0i} + \Sigma_k \beta_k x_{kit} + u_{it}$. If we let $\beta_{0i} = \beta_0^* + v_i$, where β_{0i} is the intercept for the ith unit, β_0^* is the mean intercept, and v_i is the difference from the mean intercept for the ith unit, it could be written as $y_{it} = \beta_0^* + v_i + \Sigma_k \beta_k x_{kit} + u_{it}$. The estimation of this last specification depends on the assumptions

Table 11.9:   *Yield Response Function Results for 45 Countries, 1960-85*

| Explanatory variable | Constant coefficients | | | Fixed effects | |
| | All countries | | | Temperate | Tropical |
	Model 1	Model 2	Model 3	Model 4	Model 5
Constant	−0.309	−0.708	−0.998	−0.357	−0.394
Seed imports	0.052	0.046	0.023	0.042	0.009
	(0.016)	(0.011)	(0.019)	(0.009)	(0.008)
Fertilizer	0.183	0.200	0.298	0.215	0.324)
	(0.031)	(0.036)	(0.041)	(0.028)	(0.037)
Land quality	0.089	0.077	0.102	0.187	0.062
	(0.017)	(0.021)	(0.032)	(0.041)	(0.039)
Public research	0.009	0.018	0.031	0.026	0.012
	(0.017)	(0.013)	(0.015)	(0.014)	(0.008)
Private research	—	—	0.101	0.072	0.098
			(0.059)	(0.064)	(0.051)
Training	—	0.127	0.187	0.239	0.199
		(0.072)	(0.106)	(0.120)	(0.109)
Climate	0.383	0.291	0.402	—	—
	(0.072)	(0.060)	(0.081)		
Interaction					
Public research · private research	—	—	0.124	0.117	0.081
			(0.061)	(0.071)	(0.058)
Seed imports · private research	—	—	0.164	0.143	-0.099
			(0.089)	(0.097)	(0.065)
$\overline{R}^2$	0.74	0.76	0.81	0.86	0.73
Standard error of regression	0.32	0.38	0.40	0.22	0.27

Note: Dependent variable is yield. Figures in parentheses are standard errors. Models 1, 2, and 3 have 1118 degrees of freedom while model 4 and 5 have 546 degrees of freedom and 22 country-dummies.

The first three data columns of table 11.9 report pooled estimations using a constant-coefficient specification, and the last two columns are pooled estimations using a fixed-effect model. Model 1 includes the number of public-sector maize scientists as a measure of public research, model 2 includes the number of maize publications, while models 3, 4, and

---

made on v_i, assuming v_i are fixed leads to the "fixed effects," "dummy variable," or "covariance model." If we assume v_i are random then we are in the so called "random effects" or "error components model." In the fixed effects model utilized in this study, the u vector is assumed to be homoscedastic and non-autocorrelated so that ordinary least squares provides the best linear unbiased estimators.

5 use the number of maize varieties released as a measure of public research. Models 1 to 3 include the 44 selected countries listed in table A11.1.[28] In models 4 and 5, these 44 countries are subdivided according to climate into two groups of 22 countries each. Model 4 includes only temperate countries, and model 5 only tropical countries, so the climate variable is not included in these two models.

Constant-Coefficient Specification

The seed-imports coefficient was significant when the private-sector variable was not included (in models 1 and 2), but it had a low level of significance when the private-research variable was present in model 3. The estimated coefficients for the fertilizer, land-quality, training, and climate variables were significant in all regressions. As expected, higher fertilizer consumption rates and better quality land had a positive effect on yields. Climate is also an important determinant of yields, with maize yields higher in temperate as compared with tropical regions.

The public-research coefficient was neither significant in the first model where it was proxied by the number of scientists, nor in the second model where it was proxied by a measure of the number of publications. The coefficient became significant (although at a low level) when the alternative proxy measuring the number of local varieties released was used in model 3.

The coefficient of the private-research by multinational companies variable was significant when the analysis encompassed all countries (model 3). In this model, the public-private research and the interaction effects between seed imports and private research were positive.

The adjusted coefficients of determination, $\overline{R}^2$, were high (above 0.70), especially considering the functional form and the crudeness of the data utilized. The correlations among explanatory variables were not high, except for seed imports and private research, which had a correlation of 0.60, and public research (varieties released) and private research, which had a correlation of 0.49.

According to these results, the private transfer of technology in the form of imported maize seed was indeed important in explaining maize yields. The size of the contribution of seed imports to maize yields may be overstated because the measure of locally private research effort used in this analysis is not complete. It includes neither locally sponsored private research nor all of the multinationals doing maize research. The seed variable may be picking up some of the impact of genuinely local private research. When the number of local experiment stations of private multinational companies was included (model 3), the coefficient on the seed variable was reduced in size.

Although Italy, Greece, and a few other countries import substantial amounts of their

---

[28]These are the 45 countries listed except for the US because the seed imports variable accounts for trade from the US to the other 44 countries.

hybrid maize seed from the US, most countries import very little. The small amount of cropped area sown to imported US seed in most countries during the 1980s indicates that factors other than direct imports are contributing to increased yields. This supports the idea that seed imports could be a proxy for private research as well as for imports of other inputs such as herbicides and farm machinery.

Case studies from Asia and Latin America suggest two patterns in the relationship between seed trade and private research. In temperate regions, companies begin selling hybrids developed in the US after only a minimum amount of local testing. As their market expands, they invest in research to adapt the hybrids to a defined set of growing conditions. At first they import hybrid seed from the US because it is cheaper than establishing their own production and processing operation. As the market develops, private companies usually establish their own operations and then import seed from the US only to cover any shortfall of seed at the local level due to weather or other temporary factors. In the tropics, research either by the public or private sector is required first to develop suitable hybrids. Then companies begin selling seeds that are multiplied in subtropical parts of the US and eventually in the tropical country. In this way, then, trade and private research are closely related.

With respect to public research, when the number of lines released rather than the number of scientists or publications is used, the coefficient becomes significant (model 3). As noted above, previous research suggests that some local adaptive research is necessary, particularly for a site-sensitive plant such as maize. In temperate climates, public research still plays an important role in improving maize germplasm and crop management practices, although the private sector may now be investing more money than is the public. In most tropical countries, public investments in maize research are larger than private although there may be exceptions such as in the Philippines.

Public research has played a key role in producing new varieties and inbred lines. Although past studies, which have not had an explicit technology-transfer variable, have probably overestimated the importance of public research, the low significance of this variable is probably due to the way in which it was constructed in this study. Publication counts do not reflect real research output in many less-developed country NARSs. The number of maize varieties released by the public sector in each country, although probably a more representative measure than publication counts, is still a partial measure because technological spillover among similar agroclimatic regions could readily take place.

Fixed-Effect Specification

Models 4 and 5 show the results of using a fixed-effect specification. Specific country differences are now analyzed jointly with the effects of the other variables in the production function. Since the climate variable was highly significant in the previous models, the total of 44 countries is subdivided into two groups, temperate (model 4) and tropical (model 5), in order to sharpen an assessment of the effects of the trade and research variables on maize yields.

Imports of US maize seed had a significant effect on yields in temperate, but not tropical, countries. Fertilizer utilization and land quality were significant in both groups. The public-research coefficient was significant in the temperate group, but it was not so in the tropical. The public-research variable utilized in model 4 is the number of varieties released. The publications-based measure of research did not achieve significance in any of the regressions. Training was significant in both groups. The interaction effect between public and private research was significant in temperate but not in tropical countries.

These results confirm our earlier findings about the seed trade. The level of seed imports from the US plays an important role in accounting for yield levels in temperate, but not tropical, countries.[29] Private research by multinational companies is not significant in temperate countries, but it shows a low level of significance in tropical countries. The lack of significance for the temperate countries as a group may be explained by the fact that this variable varies much less in the temperate regions (where more-developed countries are located) than in the tropical ones since the level of private maize research is more homogeneous in more-developed countries than in less-developed countries. The impact of private research on maize productivity is underestimated in temperate countries, where local private research is important. Again, the poor measure of local private research is a problem. This may be one of the reasons why this variable shows significance only in tropical countries where, in general, local private activity is small.

Having separated countries into the two groups, the relationships between seed trade and private research can now be better understood. In temperate countries, to which most of the US seed is exported, the seed-import coefficient is significant. In the tropical countries as a group, where US seed exports do not appear to have a significant impact on local maize yields, private research is a significant explanator of maize yields, since more local research may be undertaken to adapt the temperate technology to tropical conditions.

In addition, the signs of the coefficients for the interaction terms between seed trade and private research seem to imply (although at low levels of significance) that trade substitutes for research in tropical countries but it complements private research in the temperate group.

The public-research coefficient had a low level of significance in our previous estimations for the total number of countries. Given the importance of public research, the results obtained when analyzing temperate and tropical countries separately are not surprising. Public research is significant in the temperate group where the effort has been greater and there is substantial cross-country variation. It is not significant, however, in the tropical countries analyzed in this study. According to the earlier discussion of the research complementarities among the sectors, it is difficult to justify any significant impact of

---

[29] Although tropical countries may benefit by importing and adopting maize seed from southern regions of the US such as Texas and Florida, temperate countries can more readily take advantage of the substantial germplasm developed for Corn Belt conditions.

private research when public research is not significant. Without public research, the impact of private research will probably be substantially reduced.[30]

There are at least three possible explanations for the low significance of the public-research variable in temperate countries and its apparent lack of significance in tropical countries. First, public maize research may be less important than is usually assumed or, alternatively, its importance may have decreased substantially during the past 25 years. Second, public maize research may not be accurately measured. A research output variable, such as number of publications or varieties released, measures real research effort more directly than a research input variable, such as expenditures or scientist years dedicated to maize research. But, as mentioned before, the research output variables utilized in this analysis have measurement problems too. Third, public research leads to private research. The increase in productivity due to private research is thus usually based on earlier public-sector results.

The findings of this study concerning the yield impact of public research create a real challenge for further research. If public maize research has not been significant in tropical countries, then more studies are needed to improve our understanding of similar cases in other crops or for other inputs where private research activities apparently have been important. The private research results in this study are not totally surprising, since private research is probably more important for maize than for any other crop. In this sense, these results are not easily extrapolated to other crops.

11.3 POLICY IMPLICATIONS

The frequently high payoffs to public investments in agricultural research have been extensively documented (e.g., Echeverría 1990b). Public research contributions to agricultural growth have been assessed, but usually without the role of private research being explicitly accounted for. Moreover, the interaction between public and private research has not been thoroughly examined, particularly in a quantitative fashion. The importance of trade in biological technology has also been largely overlooked. The main objectives in this chapter were to assess the relative importance of public and private investments in maize research and to analyze the impact of that research as well as the effect of seed trade on maize productivity.

Two factors appear decisive in determining private investments in maize research: potential market size and public-sector policies. Market size is determined by maize yield, type of farmer, total and improved maize area, the presence and market share of public seed companies, and the structure of the seed market. Public research and regulatory policies also influence the research investment strategies of private companies. A seed company will be

---

[30]In the US, for example, 54% of the maize crop planted in 1970 was based on six public inbred lines (Duvick 1984). By 1980 that figure had dropped to 40%. Although companies are relying less on public lines developed by US universities, the complementarity among the sectors is still important.

induced to invest when public research is strong and regulations are minimal. These regulations influence the availability of germplasm developed by the public sector. They also affect the approval process required to initiate research operations or certify new varieties. They include, as well, trade restrictions on germplasm and commercial seed. Through the seed trade, technology can readily be transferred directly from one region to another.

The efforts to develop agriculture by direct transfer of technology across various regions, however, have not been very successful. This is attributed, in general, to search and screening costs associated with discovering and evaluating new technologies, environmental specificity, and the absence of local research capacity at the applied level. Other factors such as government policies in the importing country also influence the transfer process. There are at least two major barriers that slow the direct transfer of technology in the form of seeds: import restrictions and insufficient as well as inefficient public-sector investment to borrow, adapt, and produce new seeds. Policies directed to improve the local seed industry may be more productive than regulations restricting seed imports.

World maize production has increased threefold over the past 30 years. This can be explained largely by an increase in yields rather than in area. Although average maize yields have increased worldwide, there is a wide variation among regions. Much of the maize area in less-developed countries is not planted with improved seed. In spite of this, hybrids are more common than open-pollinated varieties in the area planted with improved seeds. Many seed industries in less-developed countries seem to be characterized by the poor performance of the state seed companies and low levels of private activity. Fostering joint public- and private-sector efforts is one policy option for improving seed industries in most less-developed countries. Private involvement, as noted before, depends on potential market size and on regulations affecting seed research, production, and trade. A review of these considerations in two important maize producing countries indicates that the complementarities between the public and private sectors, in research and seed production, have a high payoff (Echeverría 1990a).

This chapter has analyzed, among other topics, some of the relationships between public and private maize research and seed production. As private research activities expand, public research directed towards more basic areas (such as improving germplasm for the production of varieties or inbred lines and developing improved agronomic practices) will foster the complementarity between the sectors. To achieve this, however, public-research results should be freely available to all private companies, especially local ones. Public systems may also enhance competition by directing their research towards more applied areas, i.e., producing varieties or hybrids for specific regions. This means that an assessment of the different sources of maize technologies available is needed in each particular case.

The complementarity of those efforts is an essential input in agricultural development. The results presented here suggest that farmers can benefit by the private sector taking a larger role in developing, transferring, and marketing improved maize seed. Since public research and regulatory policies have an important effect on private-sector investments in R&D, those policies should be directed to strengthening the public research programs, to training scientists, and to keeping research, seed production, and marketing regulations to a minimum in order to assure competition yet maintain quality.

11.4   CONCLUSIONS

When reviewing the significance of public and private investments in maize research, it was found that the public sector does most of the basic research in areas such as biology and genetics, as well as the plant breeding research to create improved varieties and inbred lines. The private sector does mostly applied research, producing hybrids from their own lines (in the case of large companies) or from publicly developed lines (in the case of smaller companies). Since hybrids have the characteristic of a trade secret, private firms have incentives to invest in hybrid maize research. The provision of improved germplasm and/or inbred lines by the public sector maintains competition. Public research complements and stimulates private research.

Differences in maize yield between countries and over time were modeled in this chapter using a Cobb-Douglas production function of seed imports, fertilizer consumption, land quality, public and private research, training, and climate. Data from 45 countries were employed in the analysis, selected according to a ranking by maize production in 1983-87. The time period investigated was from 1960 to 1985.

The regression estimates show positive and significant coefficients for most variables, including public-private research and interaction effects between seed imports and private research. The results of the analysis suggest that seed imports, public and private research, land quality, and training provided by CIMMYT, as well as fertilizer and climate, significantly explained the variation in maize yields in the 45 principal maize producing countries of the world during 1960 to 1985. When the countries were divided into two groups, temperate and tropical, it was found that seed imports from the US and public research were significant only in temperate countries. Private research was significant only in tropical countries. Temperate countries, therefore, benefitted by importing maize seed. In tropical countries, where seed imports were not significant in explaining maize productivity, private research carried out by multinational companies is significant. Presumably, this is because more research is required to adapt technologies to tropical conditions. Given the potential complementarities between public and private research, the significance of private research (and the lack of significance of public research) in tropical countries is somewhat puzzling.

It is important to note that the analysis presented in this chapter is based on a simple, highly stylized model. The reality of R&D, particularly in the case of maize, is undoubtedly more complex. However, lack of data as well as the analytical difficulty of specifying a more complete model leave, for the time being, few alternatives.

The results presented in this chapter document the impact of private research activities for a crop that received early attention by the private sector as witnessed by the investments in hybrid breeding that initally took place in the US during the first half of this century. More recently, privately executed and funded research has expanded to include other crops and regions (chapter 10).

In the short term for the more-developed countries and in the medium term for the less-developed countries there is likely to be an accelerating trend toward privatizing agricultural research. Moreover, the nature of the technologies being developed in the public and private domain is also likely to undergo substantial change (chapter 12). Taken together, these changes will reshape the conduct of agricultural R&D in less-developed countries, the relationship between less- and more-developed country (public and private) research activities, and the policy agenda facing public agricultural research institutions.

Annex table A11.1 (next page)

Source: *FAO Production Yearbooks*.

[a] Countries are ranked by average maize production during 1983-87.

[b] Average annual percent growth rates for 1961-65 to 1983-87 calculated as $x_t = x_0 [1 + (g/100)]^t$; where x_t = average of data for ending period; x_0 = average of data for base period; t = number of years from the midpoint of one period to that of the other; g = average annual percent growth rate.

Table A11.1: *Maize Production, Area, Yield, and Growth Rate of 45 Selected Maize Producing Countries*

Country[a]	1983-87			1961-65 / 1983-85 Growth rates[b]		
	Production	Area	Yield	Production	Area	Yield
	(000 ton)	*(000 ha)*	*(tons per ha)*	%	%	%
USA	182,960	26,449	6.83	2.9	0.7	2.2
China	70,562	18,900	3.73	5.8	1.0	4.8
Brazil	21,832	12,009	1.80	3.5	1.9	1.5
Romania	16,522	3,050	5.41	3.9	-0.5	4.4
USSR	13,535	4,224	3.21	0.2	-1.7	19.0
Mexico	12,668	7,798	1.64	3.0	0.8	2.2
France	11,425	1,787	6.39	6.8	3.2	3.5
Yugoslavia	10,669	2,318	4.60	3.1	-0.3	3.4
Argentina	10,408	3,117	3.32	3.5	0.4	3.1
India	7,381	6,726	1.29	2.5	1.2	1.4
Hungary	6,876	1,132	6.07	3.3	-0.7	4.0
Canada	6,568	1,076	6.12	9.2	8.1	1.0
Italy	6,388	896	7.16	2.9	-0.7	3.6
South Africa	6,325	3,993	1.59	0.1	-0.1	0.2
Indonesia	4,878	2,780	1.75	3.0	-0.4	3.5
Thailand	3,952	1,663	2.36	3.1	6.9	1.1
Philippines	3,829	3,526	1.08	4.6	2.6	2.0
Egypt	3,599	796	4.53	3.0	0.8	2.2
Spain	2,940	477	6.08	4.1	-0.3	4.4
North Korea	2,681	432	6.20	4.1	2.6	1.5
Bulgaria	2,431	541	4.42	2.9	-0.07	3.6
Kenya	2,248	1,537	1.45	2.9	1.6	1.2
Greece	1,942	210	9.26	0.3	0.7	9.5
Turkey	1,916	559	3.42	2.6	-0.9	3.5
Tanzania	1,883	1,634	1.14	4.6	2.4	2.2
Zimbabwe	1,743	1,341	1.32	3.4	3.0	0.4
Austria	1,697	209	7.64	10.4	6.9	3.3
Ethiopia	1,486	864	1.72	2.2	0.4	1.8
Nigeria	1,421	1,145	1.59	3.5	2.8	0.7
Malawi	1,328	1,178	1.13	2.5	1.5	0.9
West Germany	1,135	182	6.22	15.1	12.2	2.6
Guatemala	1,053	701	1.51	2.8	0.7	2.1
Zambia	1,038	558	1.86	0.8	-2.5	3.4
Pakistan	1,032	806	1.28	3.3	2.3	1.0
Czechoslovakia	985	196	5.01	3.2	0.2	3.0
Venezuela	855	480	1.76	1.4	-0.9	2.3
Colombia	834	587	1.42	0.2	-1.1	1.4
Nepal	813	586	1.40	-0.4	0.8	-1.2
Afghanistan	805	474	1.70	0.6	-0.3	0.9
Peru	772	397	1.94	1.6	-0.5	2.1
Zaire	726	844	0.86	3.6	2.4	1.2
Chile	670	116	5.89	5.8	2.0	3.8
Paraguay	620	433	1.40	5.5	5.7	-0.2
Portugal	577	300	2.01	-0.4	-1.5	1.2
Vietnam	529	392	1.35	2.4	1.9	0.5

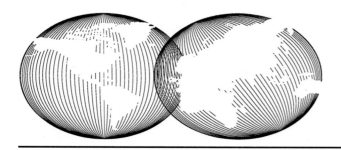

PART V

EMERGING ISSUES

Chapter 12

Challenges to Agricultural Research in the 21st Century

Vernon W. Ruttan[1]

In this final chapter, I discuss some of the challenges facing the global agricultural research system as we move into the first decades of the next century. Before doing so, however, I would first like to place my remarks within the intellectual climate that has conditioned our thinking about the relationships among environmental, technological, and institutional change during the second half of the 20th century. I will then report on some of the findings for research that have emerged from several recent "consultations" that I have organized around the issues of (a) biological and technical constraints on crop and animal productivity and (b) resource and environmental constraints on sustainable growth in agricultural production.

12.1 TECHNOLOGY, INSTITUTIONS, AND THE ENVIRONMENT

The research that is conducted in our universities, our research institutes, and our agricultural experiment stations is valued primarily for its contributions to technical and institutional change. The demand for advances in knowledge in the natural sciences is derived primarily from society's demand for technical change. The demand for advances in knowledge in the social sciences and humanities, and in related professional fields, is derived primarily from the demand for institutional change and more effective institutional performance. There are several ways of characterizing the significance of technical change (chapter 5). For purposes of this chapter, however, it is sufficient to note that technical

1 Earlier drafts of this chapter were presented at the Symposium on "Technology and Economics," National Academy of Engineering, Washington, DC, April 4, 1990, and at the Dewhirst Symposium, University of Arizona, Tucson, January 26, 1990. The author is indebted to (a) participants in the CURA-HHH Institute Seminar on Global Warming and Its Effects on the Upper Midwest (January 16, 1990), (b) participants in the History of Science and Technology Seminar (February 9, 1990) at the University of Minnesota; and (c) Earl D. Kellogg, Robert D. Munson, and Pierre Crosson for comments on earlier drafts.

change permits the substitution of knowledge for resources; it permits the substitution of more abundant for less abundant resources; and it releases the constraints on growth imposed by inelastic resource supplies.

But technical change is itself the product of institutional innovation. Whitehead (1925, p. 96) insisted that the greatest invention of the 19th century was the institutionalization of the process of invention — the invention of the research university, the industrial research laboratory, and the agricultural experiment station. There is a lag in the development of institutional innovations needed to achieve an incentive-compatible institutional infrastructure, i.e., institutions capable of achieving compatibility between individual, organizational, and social objectives. One of the effects of this lag is that the by-products of technical change (what the resource economists refer to as residuals) are now filling the landscape with garbage and the earth, water, and atmosphere with chemicals. I am prepared to insist that the contribution of advances in natural and social science knowledge to technical and institutional change has enabled modern society to achieve a more productive and better balanced relationship to the natural world than in the ancient civilizations or in earlier stages of Western industrial civilization (Ruttan 1971). But the relationship between advances in knowledge, resource utilization, and human well-being continues to be uneasy. We are, for example, in the midst of the third wave of social concern about the relationship between natural resources and the sustainability of improvements in human well-being since World War II, and the fifth since Malthus.

The first post-war wave of concern, in the late 1940s and early 1950s, focused primarily on the quantitative relations between resource availability and growth — the adequacy of land, water, energy, and other natural resources to sustain growth. The reports of the President's Water Resources Policy Commission (1950) and the President's Materials Policy Commission (1952) were the landmarks of the early post-war resource assessment studies generated by this wave of concern in the US. One response to this first wave of concern was technical change. A stretch of high prices has not yet failed to induce the new knowledge and new technologies needed to locate new deposits, promote substitution, and enhance productivity. If the Materials Policy Commission were writing today, it would have to conclude that there has been abundant evidence "of the nonevident becoming evident; the expensive, cheap; and the inaccessible accessible."

The second wave of concern occurred in the late 1960s and early 1970s. In this second wave, the earlier concern with the potential "limits to growth" imposed by natural resource scarcity was supplemented by concern about the capacity of the environment to assimilate the multiple forms of pollution generated by growth. An intense conflict was emerging between the two major sources of demand for environmental services. One was the rising demand for environmental assimilation of residuals derived from growth in commodity production and consumption , i.e., asbestos in our insulation, pesticides in our food, smog in the air, and radioactive wastes in the biosphere. The second was the rapid growth in consumer demand for environmental amenities — for direct consumption of environmental services — arising out of rapid growth in per capita income and high income elasticity of demand for such environmental services as access to natural environments and freedom from pollution and congestion. The response to these concerns, still incomplete, was the design of local incentive-compatible institutions designed to force individual firms and other organizations to bear the costs arising from the externalities generated by commodity production.

Since the mid-1980s, these two earlier concerns have been supplemented by a third. These more recent concerns center around the implications for environmental quality, food production, and human health of a series of environmental changes that are occurring on a transnational scale, issues such as global warming, ozone depletion, acid rain, and others. The institutional innovations needed to respond to these will be more difficult to design. They will, like the sources of change, need to be transnational or, at the very least, international. Experience with attempts to design incentive-compatible transnational regimes, such as the Law of the Sea Convention, or even the somewhat more successful Montreal Protocol on reduction of CFC emissions, suggests that the difficulty of resolving free-rider and distributional equity issues imposes a severe constraint on how rapidly effective transnational regimes to resolve these new environmental concerns can be put in place.

It is of interest that, with each new wave of concern, the issues that dominated the earlier waves have been recycled. The result is that, while the intensity of earlier concerns has receded, in part due to the technical and institutional changes that have evolved, the set of concerns about the relationships between natural resources, environment, and sustainable growth in agricultural production has broadened.

By the end of the 1980s, concerns about the impacts of agricultural intensification widened (chapter 3). In the 1970s these concerns had initially focused on the effects of pesticides and non-point sources of pollution on natural environments and on the safety of farm workers and consumers. During the 1980s, there were growing concerns about the effects of more intensive agricultural production on (a) resource degradation through erosion, salinization, and depletion of groundwater; and (b) the quality of surface and groundwater through runoff and leaching of plant nutrients and pesticides. Terms that had initially been introduced by the populist critics of agricultural research — such as alternative, low-input, regenerative, and sustainable agriculture — began to enter the vocabulary of those responsible for allocating resources to agricultural research. After a period of initial resistance, some leaders of the agricultural research community moved to embrace this new set of concerns. The recently issued report by the Committee on the Role of Alternative Farming Methods on Modern Production Agriculture (National Research Council 1989) has been viewed as a landmark in this conversion. In my judgment, it is more appropriately viewed as a political document designed to capture the initiative from the populist critics of institutionalized agricultural research.

It seems quite clear to me that, by the end of the first decade of the next century, the agricultural research landscape will look much different than it does today. Nor will pressures for the revision of research priorities arising from scientific, societal, and environmental change recede.

12.2   BIOLOGICAL AND TECHNICAL CONSTRAINTS

During the past year and a half, I have had the opportunity, with support from the Rockefeller Foundation, to organize a series of small "consultations" on the agricultural research priorities that might be expected to emerge as we move into the early decades of the next century. The first of these consultations was organized around the topic, "Biological and Technical Constraints on Crop and Animal Productivity" (Ruttan 1989a). The second consultation was organized around the issues of "Resource and Environmental Constraints on Sustainable Growth in Agricultural Production" (Ruttan forthcoming). The issues discussed in the consultations were not confined to domestic US priorities.

Those familiar with the evidence on long-term declines in agricultural commodity prices or with media attention that has been devoted to the "new biotechnology" may find it difficult to comprehend why anyone should be concerned about the possibilities of a lag in either agricultural production or productivity over the next several decades. Let me justify my concern with just four observations: (a) the yields obtained on maximum yield trials at the International Rice Research Institute are today no higher than the mid-1960s; (b) maize yields in the United States continue to increase at about one bushel per year, but this is a much smaller rate of increase than 30 years ago; (c) the projected timings of biotechnology impacts on agricultural production continue to recede — impacts expected in this decade have now receded into the next; and (d) support for national agricultural research has weakened in a significant number of debt-plagued less-developed countries (chapter 7) and in eastern Europe and the USSR.

12.2.1 Advances in Conventional Technology as the Primary Source of Growth

Advances in conventional technology will remain the primary source of growth in crop and animal production over the next quarter century. Almost all increases in agricultural production in the future must come from further intensification of agricultural production on land that is presently devoted to crop and livestock production. Until well into the second decade of the next century, the necessary gains in crop and animal productivity will continue to be generated by improvements resulting from conventional plant and animal breeding and from more intensive and efficient use of technical inputs, including chemical fertilizers, pest-control chemicals, and higher-quality animal feeds. The productivity gains from conventional sources are likely to come in smaller increments than in the past. If they are to be realized, higher plant populations per unit area, new tillage practices, improved pest and disease control, more precise application of plant nutrients, and advances in soil and water management will be required. Gains from these sources will be crop-, animal- and location-specific. They will require closer articulation between the suppliers and users of new knowledge and new technology. These sources of productivity gains will be extremely knowledge- and information-intensive. If they are to be realized, research and technology transfer efforts in the areas of information and management technology must become increasingly important sources of growth in crop and animal productivity. In the short run, taken here to mean the next several decades, no other sources of growth in production will become available that will be adequate to meet the demands arising from growth in population and income that will be placed on agricultural production in both the more- and less-developed countries. This conclusion is that both national and international agricultural research systems will find it productive to increase the proportion of research resources devoted to improving agronomic practices relative to plant breeding.

12.2.2 Advances in Conventional Technology Will Be Inadequate

Advances in conventional technology will be inadequate to sustain the demands that will be placed on agriculture as we move into the second decade of the next century and beyond. Advances in crop yields have come about primarily by increasing the ratio of grain to straw rather than by increasing total dry-matter production. Advances in animal feed efficiency have come by decreasing the proportion of feed consumed that is devoted to animal maintenance and increasing the proportion used to produce usable animal products. There are severe physiological constraints to continued improvement along these conventional paths. These constraints are most severe in those areas that have already achieved the highest levels of productivity, as in Western Europe, North America, and parts of East Asia.

The impact of these constraints can be measured in terms of declining incremental response to energy inputs, in the form of a reduction in both the incremental yield increases from higher levels of fertilizer application and the incremental savings in labor inputs from the use of larger and more powerful mechanical equipment. One consequence is that in countries that have achieved the highest levels of output per hectare or output per animal unit, an increasing share of both public- and private-sector research budgets is being devoted to maintenance research — the research needed to sustain existing productivity levels. If the incremental returns to agricultural research should decline, it will impose a higher priority on efficiency in the organization of research and on the allocation of research resources.

12.2.3 Issues to Be Met over the Next Two Decades

Reorient the Organization of Agricultural Research

A reorientation of the way we organize agricultural research will be necessary in order to realize the opportunities for technical change being opened up by advances in microbiology and biochemistry. Advances in basic science, particularly in molecular biology and bio-chemistry, have created and are continuing to open up new possibilities for supplementing traditional sources of growth in plant and animal productivity. Many possibilities were discussed at the consultation, ranging from the transfer of growth hormones into fish to conversion of lignocellulose into edible plant and animal products.

The realization of these possibilities will require reorganizing the performance of agricultural research. An increasing share of the new knowledge generated by research will reach producers in the form of proprietary products or services. This means that the incentives must exist to draw substantially more private-sector resources into agricultural research. Public-sector research organization will have to increasingly move from a "little science" to a "big science" mode of organization. Examples include the Rockefeller Foundation-sponsored collaborative research program on the biotechnology of rice and the University of Minnesota program on the biotechnology of maize. In the absence of more focused research efforts, it seems likely that the promised gains in agricultural productivity from biotechnology will continue to recede.

Expansion of Research Capacity

Efforts to institutionalize agricultural research capacity in less-developed countries must be intensified. Levels of crop and animal productivity in most less-developed countries remain well below the levels that are potentially feasible. Access to conventional sources of productivity growth, such as from advances in plant breeding, agronomy, and soil and water management, will require the institutionalization of a substantial agricultural research capacity. In a large number of less-developed countries, this capacity is just beginning to be put in place. A number of countries that experienced substantial growth in capacity during the 1960s and 1970s have experienced an erosion of capacity in the 1980s (chapter 7). Even a relatively small country, producing a limited range of commodities under a limited range of agroclimatic conditions, will require a cadre of 250 to 300 agricultural scientists. Countries that do not acquire an adequate agricultural research capacity will not be able to meet the demands placed on their farmers as a result of population and income growth. Research systems that do not generate a resource- and productivity-enhancing capacity will fail to sustain public support.

Fragile Resource Areas

There are substantial possibilities for developing sustainable agricultural production systems in a number of fragile resource areas. Research underway in the tropical rain forests of Latin America and in the semi-arid tropics of Africa suggest the possibility of developing sustainable agricultural systems with substantially enhanced productivity. It is unlikely, however, and perhaps undesirable, that these areas become important components of the global food supply system. But enhanced productivity is important to those who reside in these areas, now and in the future. It is important that the research investment in the areas of soil and water management and in farming systems be intensified in these areas.

Depletion of Energy and Mineral Resources

Over the very long run, energy and mineral nutrition can be expected to emerge as increasingly serious constraints on agricultural production. During the past century technical change has been directed along alternative paths by relative resource endowments. Countries where land was relatively scarce or expensive, such as Japan, placed a major emphasis on biological technology — in effect, inventing around the land resource constraint. Countries where labor was relatively scarce or expensive, such as the United States, placed greater emphasis on advancing mechanical technology — in effect, inventing around the labor constraint. Over the next half century, energy derived from liquid fuels is likely to become a serious constraint. It is also possible that the reserves of phosphate will decline to levels that will result in much higher relative prices for phosphatic fertilizer. It is likely that it will be necessary to allocate substantial research resources to invent around these two constraints.

Regulatory Regimes

The rationalization of regulatory regimes will become an increasingly important factor in determining the profitability of research investments and international competitiveness in agricultural production. Incentives for private-sector agricultural research appear to be quite sensitive to uncertainty about changes in regulatory regimes and the administration of regulations. Incentives for research and the potential gains from research investment are dampened when use of technology is restricted for reasons other than the assurance of health and safety. Consumers may press for regulation in the interests of aesthetic concerns. Producers may press for regulation to protect themselves from domestic or international competition. Pressure to achieve greater consistency among national regulatory regimes is likely to become an increasingly important factor in international trade negotiations. It will be necessary to devote substantial research efforts to identifying and quantifying the scientific, technical, economic, and psychological information needed to rationalize regulatory regimes in the future.

Preservation of Genetic Resources

A major effort to assemble and characterize available plant and animal genetic resources is essential in order to make the transition from the now-conventional biological technology of the 20th century to a biotechnology-based agriculture for the 21st century. A major constraint in the development of a cost-effective strategy for collection and preservation of genetic resources is an adequate characterization of the materials in *in situ* locations and in *ex situ* collections. A crop plant genome-mapping program is essential if we are going to make effective use of the genetic engineering techniques that are available now and that will become available in the future.

New Crops and Animals

Research on alternative crops and animals that can be introduced into production systems can become a useful source of growth in some areas. On a local or regional basis, the development and incorporation of minor cultivars and species could make important nutritional and economic contributions. It is unlikely that alternative crops or animals will emerge to substantially replace existing crop cultivars or animal species in production systems. It would be wishful thinking to expect any new developments as significant as the expansion of soybean production during the past half century.

Basic Research in the Tropics

There is a need to establish a substantial capacity for basic biological research and training in the tropical less-developed countries. There are a series of basic biological research agendas that are important for applied research and technology development in health and agriculture in the tropics that receive, and are likely to continue to receive, inadequate attention in the more-developed countries located in the temperate regions. There is also a need for closer articulation between training in applied science and technology and training in basic biology. When such institutes are established, they should be more closely linked with existing universities than the series of agricultural research institutes established by the Consultative Group on International Agricultural Research (CGIAR).

12.3   RESOURCE AND ENVIRONMENTAL CONSTRAINTS

12.3.1 An Overview

As we look even further into the next century, there is a growing concern, as noted earlier, about the impact of a series of resource and environmental constraints that may seriously impinge on our capacity to sustain growth in agricultural production. One set of concerns centers on the environmental impacts of agricultural intensification. These include ground-water contamination from plant nutrients and pesticides, soil erosion and salinization, the growing resistance of insect pests, pathogens, and weeds to present methods of control, and the contribution of agricultural production and land use changes to global climate change. The second set of concerns stems from the effects of industrial intensification on global climate change. It will be useful, before presenting some of the findings of the second consultation, to briefly characterize our state of knowledge about global climate change.

There can no longer be any question that the accumulation of carbon dioxide (CO_2) and other greenhouse gasses — principally methane (CH_4), nitrous oxide (N_2O), and chlorofluorocarbons (CFCs) — has set in motion a process that will result in some rise in average global surface temperatures over the next 30 to 60 years. There is substantial disagreement about whether warming due to greenhouse gasses has already been detected. And there continues to be great uncertainty about the increases in temperature that can be expected to occur at any particular date or location in the future.

The bulk of carbon dioxide emissions come from fossil fuel consumption. Carbon dioxide accounts for roughly one-half of radiative forcing (figure 12.1). Biomass burning, cultivated soils, natural soils, and fertilizers account for close to one-half of nitrous oxide emissions. Most of the known sources of methane are a by-product of agricultural activities — principally enteric fermentation in ruminant animals, release of methane from rice production and other cultivated wetlands, and biomass burning. Estimates of nitrous oxide and methane sources have a very fragile empirical base. Nevertheless, it appears that agriculture and related land use could account for somewhere in the neighborhood of 25% of radiative forcing. On a regional basis, the United States contributes about 20%, and Western and Eastern Europe and the USSR about 30%, of radiative forcing by all greenhouse gasses. In the near future contributions to radiative forcing from the Third World will exceed that of the OECD and what used to be called the centrally planned economies.

Figure 12.1: *Contributions to increases in radiative forcing in the 1990s*

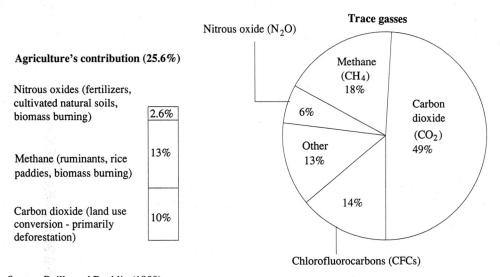

Source: Reilly and Bucklin (1989).
Note: Radiative forcing measures the contribution to surface temperature change by the several greenhouse gasses on the basis of their radiative properties.

During the consultation, Rayner (1990, pp. 110-18), as well as several others, characterized the alternative policy approaches to the threat of global warming as *preventionist* and *adaptionist*. It seems clear that a preventionist approach could involve about five policy options. They include reduction in fossil fuel use or capture of CO_2 emissions at the point of fossil fuel combustion, reduction in the intensity of agricultural production, reduction of biomass burning, expansion of biomass production, and energy conservation.

The simple enumeration of these policy options should be enough to introduce considerable caution about assuming that radiative forcing will be limited to anywhere near present levels. Let me be more specific. Fossil fuel use will be driven, on the demand side, largely by the rate of economic growth in the Third World and by improvements in energy efficiency in the more-developed and the centrally planned economies. On the supply side, it will be constrained by the rate at which alternative energy sources will be substituted for fossil fuels. Of these only energy efficiency and conservation are likely to make any significant contribution over the next generation. And the speed with which it will occur will be limited by the pace of capital replacement. Any hope of significant reversal of agricultural intensification, reduction in biomass burning, or increase in biomass absorption is unlikely to be realized within the next generation. The institutional infrastructure or institutional resources that would be required do not exist and will not be put in place rapidly enough to make a significant difference.

The possibilities for energy conservation make it fairly easy for me to be cautiously optimistic about endorsing a preventionist approach in dealing with the industrial sources of climate forcing, at least in the presently industrialized countries. I see little alternative, however, to an adaptionist approach in attempting to assess how agricultural research portfolios should respond to the implications of global climate change. It also forces me to agree, as Abrahamson (1989) insisted during the consultation, that we will not be able to rely on a technological fix to the global warming problem. The fixes, whether driven by preventionist or adaptionist strategies, must be both technological and institutional. An adaptionist strategy for agriculture implies moving as rapidly as possible to design and put in place the institutions needed to remove the constraints that intensification of agricultural production are currently imposing on sustainable increases in agricultural production. I am referring, for example, to the policies and institutions needed to rationalize water use in the western United States or to deal with groundwater management (including contamination) in both more- and less-developed countries. If we are successful in putting such policies and institutions in place, we will then be in a better position to respond to the more uncertain changes that will emerge as a result of future global climate change.

Let me now turn to some of the research implications that emerged from the consultation.

12.3.2　Emerging Research Implications

Initiate Research on the Design of Incentive-Compatible Institutions

A major research program on incentive-compatible institutional design should be initiated. The first research priority is to begin a large-scale program of research on the design of institutions capable of implementing incentive-compatible resource management policies and programs. By incentive-compatible institutions, I mean institutions capable of achieving compatibility between individual, organizational, and social objectives. A major source of the global warming and environmental pollution problem is the direct result of the operation of institutions that induce behavior by individuals and public agencies that is not compatible with societal development (some might say survival) goals. In the absence of more efficient incentive-compatible institutional design, the transaction costs involved in ad hoc approaches are likely to be enormous. Substantial basic research will be required to support a successful program of applied research and institutional design.

Increase Research on Alternative Scenarios

A serious effort to develop alternative scenarios for land use, farming systems, and food systems in the 21st century should be initiated. A clearer picture of the demands that are likely to be placed on agriculture over the next century and of the ways in which agricultural systems might be able to meet such demands has yet to be produced. World population could rise from the present five billion level to the range of 10 to 20 billion. The demands that will be placed on agriculture will also depend on the rate of growth of income, particularly in the poor countries where consumers spend a relatively large share of income on subsistence — food, clothing, energy, and housing. The resources and technology that will be used to increase agricultural production by a multiple of three to six will depend on both the constraints on resource availability that are likely to emerge and the rate of advance in knowledge. Advances in knowledge can permit the substitution of more abundant for increasingly scarce resources and reduce the resource constraints on commodity production. Past studies of the potential effects of climate change on agriculture have given insufficient attention to adaptive change in nonclimate parameters. But application of advances in biological and chemical technology, which substitute knowledge for land, and advances in mechanical and engineering technology, which substitute knowledge for labor, have in the past been driven by increasingly favorable access to energy resources by declining prices of energy. It is not unreasonable to anticipate that there will be strong incentives by the early decades of the next century to improve energy efficiency in agricultural production and utilization.

Particular attention should be given to alternative and competing uses of land. Land-use transformation, from forest to agriculture, is presently contributing to radiative forcing through release of CO_2 and methane into the atmosphere. Conversion of low-intensity agricultural systems to forest has been proposed as a method of absorbing CO_2. There will also be increasing demands on land use for protecting watersheds and producing biomass energy.

Alternative farming systems will also be needed. Therefore, research on environmentally compatible farming systems should be intensified. In agriculture, as in the energy field, there are a number of technical and institutional innovations that could have both economic and environmental benefits. Among the technical possibilities is the design of new third- or fourth-generation chemical, biorational, and biological pest management technologies. Another is the design of land-use technologies and institutions that will contribute to reduction of erosion, salinization, and groundwater pollution.

In addition to alternative land use and farming systems scenarios, considerable attention should be given to alternative food systems. A food-system perspective should become an organizing principle for improvements in the performance of existing systems and for the design of new systems. The agricultural science community should be prepared, by the second quarter of the next century, to contribute to the design of alternative food systems. Many of these alternatives will include the use of plants other than the grain crops that now account for a major share of world feed and food production. Some of these alternatives will involve radical changes in food sources. Ragoff and Rawlins (1987) have described one such system based on lignocellulose, both for animal production and human consumption.

Strengthen Environmental Monitoring Capacity

The capacity to monitor the agricultural sources and impacts of environmental change should be strengthened. It is a matter of serious concern that only in the past decade and a half it has been possible to estimate the productivity effects and magnitude of soil loss in the United States. Even rudimentary data on soil loss are almost completely unavailable in most less-developed countries. The same point holds, with even greater force, for ground-water pollution, salinization, species loss, and others. It is time to design the elements of a comprehensive system to monitor agriculturally related resources and to establish priorities for implementation. Data on the effects of environmental change on the health of individuals and communities is even less adequate. The monitoring effort should include a major focus on the effects of environmental change on human populations.

Lack of firm knowledge about the contribution of agricultural practices to the methane and nitrous oxide sources of greenhouse forcing was mentioned at numerous times during the consultation. Much closer collaboration between production-oriented agricultural scientists, ecologically trained biological scientists, and physical scientists who have been traditionally concerned with global climate change is essential. This effort should be explicitly linked with the monitoring efforts currently being pursued under the auspices of the International Geosphere-Biosphere Programs (IGBP).

In addition, the modeling of the sources and impacts of climate change must become more sophisticated. One of the problems with both the physical and economic modeling efforts is that they have tended to be excessively resistant to advances in micro-level knowledge, including failure to take into consideration possibilities of responses to climate change from agricultural research and the response behavior of decision-making units such as governments, agricultural producers, and consumers.

Increasing Importance of Water Resources

The design of technologies and institutions to achieve more efficient management of surface and groundwater resources will become increasingly important. During the 21st century, water resources will become an increasingly serious constraint on agricultural production. Agricultural production is a major source of decline in the quality of both surface and groundwater. Limited access to clean and uncontaminated water supplies is a major source of disease and poor health in many parts of the less-developed world and in the centrally planned economies. Global climate change can be expected to have a major differential impact on water availability, water demand, erosion, salinization, and flooding. The development and introduction of technologies and management systems that enhance the efficiency of water use represents a high priority, both because of short- and intermediate-run constraints on water availability and the longer-run possibility of seasonal and geographical shifts in water availability. The identification, breeding, and introduction of water-efficient crops for dryland and saline environments is a potentially important aspect of achieving greater efficiency of water use.

Environmental Consequences of Agricultural Policies

Immediate efforts should be made to reform agricultural commodity and income support policies. In both more- and less-developed countries, producers' decisions on land management, farming systems, and the use of technical inputs (such as fertilizers and pesticides) are influenced by government interventions such as price supports and subsidies, programs to promote or limit production, and tax incentives and penalties. It is increasingly important that such interventions be designed to take into account the environmental consequences of intervention-induced decisions by land owners and producers.

12.4   CONCLUDING PERSPECTIVE

In this concluding section, I return to the problem of whether our public and private agricultural research systems will respond to the new challenges and opportunities of (a) releasing the biological and technical constraints on crop and animal productivity, (b) ameliorating the contribution of the agricultural sector to environmental degradation, and (c) enabling the agricultural sector to adapt to those environmental changes that emerge in response to the intensification of industrial production. Issues of both scientific and political capacity are involved. A failure, particularly over the past decade, of NARSs in both more- and less-developed countries to keep pace with the growing demands placed upon them have left the research systems of the 1990s in a weakened position to respond to either (let alone both) sets of concerns.

Appendix

Appendix:　*National Agricultural Research Expenditure and Personnel Estimates, 1961-85*

Countries	Total agricultural research expenditures (millions 1980 PPP dollars per year)					Total number of researchers (full-time equivalents)				
	1961-65	1966-70	1971-75	1976-80	1981-85	1961-65	1966-70	1971-75	1976-80	1981-85
Nigeria	20.6	36.7	62.8	104.5	80.1	172	306	348	903	1003
Benin	3.3	5.3	5.1	4.0	2.3	9	14	22	35	47
Burkina Faso	1.0	1.5	2.3	13.0	17.4	6	9	17	97	120
Cameroon	2.0	2.4	3.0	5.4	15.4	43	82	97	105	176
Cape Verde	0.2	0.3	0.3	0.2	0.2	3	4	4	9	16
Chad	2.4	2.4	3.3	2.7	1.6	13	20	31	19	28
Côte d'Ivoire	10.2	18.3	25.5	27.2	28.8	68	113	180	190	201
Gambia	0.4	0.5	0.6	2.0	2.8	4	6	7	32	62
Ghana	6.4	6.8	4.5	4.9	2.9	70	98	111	126	147
Guinea	2.3	1.7	3.1	6.0	8.8	23	17	28	101	177
Guinea-Bissau	0.5	0.5	0.5	0.5	0.8	4	4	4	4	8
Liberia	0.5	0.4	0.6	2.4	5.2	7	8	17	25	33
Mali	1.4	1.3	1.6	4.3	13.8	11	11	24	86	275
Mauritania	0.6	0.9	1.5	1.4	0.6	4	6	5	9	12
Niger	1.2	1.6	1.8	2.2	1.9	6	9	12	16	57
Senegal	8.0	12.0	13.6	12.3	14.7	60	56	63	103	174
Sierra Leone	1.8	2.0	2.3	1.3	1.4	20	23	27	29	46
Togo	0.8	1.3	1.9	7.1	5.9	6	9	18	39	58
Western Africa, excl. Nigeria	42.9	59.1	71.4	96.9	124.6	356	487	667	1023	1636
Burundi	0.7	1.5	1.5	2.7	4.4	11	18	20	33	56
Central African Republic	2.4	3.7	3.9	2.1	2.1	14	24	34	22	22
Congo	0.9	1.1	1.1	2.2	2.6	21	27	24	39	73
Gabon	1.2	1.2	1.1	2.0	2.6	6	7	7	21	24
Rwanda	1.4	2.0	2.0	1.3	2.1	9	8	16	23	34
Sao Tome & Principe	0.1	0.2	0.2	0.1	0.2	2	3	3	2	3
Zaire	6.8	4.2	6.7	5.7	4.0	45	29	64	37	43
Central Africa	13.5	14.0	16.6	16.1	17.9	108	115	169	177	255

Appendix: *National Agricultural Research Expenditure and Personnel Estimates, 1961-85 (Contd.)*

	Total agricultural research expenditures					Total number of researchers				
	1961-65	1966-70	1971-75	1976-80	1981-85	1961-65	1966-70	1971-75	1976-80	1981-85
	(millions 1980 PPP dollars per year)					*(full-time equivalents)*				
Angola	5.8	7.8	4.3	6.4	4.3	29	39	37	36	28
Botswana	0.6	1.0	2.1	3.1	5.8	4	14	24	33	56
Lesotho	0.2	0.3	0.8	2.3	6.0	2	3	9	13	18
Madagascar	5.8	13.5	11.3	11.2	6.6	58	64	71	49	82
Malawi	2.5	3.3	3.7	3.5	4.9	22	27	38	65	82
Mauritius	2.0	3.8	3.7	3.8	5.4	28	40	55	81	100
Mozambique	6.5	6.5	7.0	5.3	7.9	32	32	25	46	77
Swaziland	0.3	0.7	1.8	0.7	3.1	6	10	12	8	14
Zambia	2.3	3.0	3.9	4.2	4.0	25	47	76	68	110
Zimbabwe	8.3	10.6	14.6	17.7	16.6	107	116	133	119	166
Southern Africa	*34.2*	*50.4*	*53.2*	*58.1*	*64.8*	*312*	*391*	*478*	*518*	*732*
Comoros	0.3	0.5	0.6	0.8	1.0	4	7	8	12	14
Ethiopia	2.5	3.8	5.2	9.6	11.8	10	35	53	64	136
Kenya	12.9	19.2	26.3	32.1	27.1	129	209	332	320	462
Seychelles	0.1	0.1	0.1	0.1	0.3	1	1	1	4	7
Somalia	2.0	3.1	0.4	0.4	0.4	8	11	9	19	31
Sudan	7.1	14.2	12.3	14.7	12.1	45	67	76	164	206
Tanzania	7.1	16.9	20.5	18.7	19.7	107	99	127	210	276
Uganda	6.4	9.2	7.5	7.0	12.5	71	114	148	113	185
Eastern Africa	*38.3*	*67.1*	*72.9*	*83.5*	*84.9*	*375*	*543*	*755*	*906*	*1316*
Sub-Saharan Africa	*149.5*	*227.2*	*276.9*	*359.1*	*372.3*	*1323*	*1841*	*2416*	*3526*	*4941*
China	271.4	296.2	485.4	689.3	933.7	6966	9900	11563	20048	32224

Appendix: *National Agricultural Research Expenditure and Personnel Estimates, 1961-85 (Contd.)*

	Total agricultural research expenditures (millions 1980 PPP dollars per year)					Total number of researchers (full-time equivalents)				
	1961-65	1966-70	1971-75	1976-80	1981-85	1961-65	1966-70	1971-75	1976-80	1981-85
Afghanistan	1.7	2.3	2.9	4.7	4.0	24	30	44	84	80
Bangladesh	21.8	18.0	28.2	46.8	68.4	296	409	635	1141	927
India	116.1	171.3	253.8	392.4	450.0	2939	4245	5666	6910	8389
Laos	1.5	1.5	1.5	1.5	1.5	29	29	29	29	29
Myanmar	1.5	1.2	3.4	13.9	12.0	21	25	55	192	267
Nepal	2.6	3.2	9.0	11.9	10.7	61	96	202	304	446
Pakistan	13.5	12.8	18.3	42.2	74.3	893	1403	1551	2834	2972
Sri Lanka	6.7	12.7	12.8	17.8	21.4	74	105	148	244	391
South Asia	*165.4*	*223.0*	*329.8*	*531.0*	*642.3*	*4337*	*6342*	*8329*	*11738*	*13502*
Brunei	0.1	0.2	0.7	0.9	1.4	2	3	10	13	20
Hong Kong	2.1	1.5	2.4	2.3	2.2	21	15	24	23	22
Indonesia	40.7	57.3	73.9	114.9	141.1	415	433	450	1005	1349
Korea, Republic of	8.9	24.6	25.7	28.0	50.0	521	776	887	1052	1356
Malaysia	14.0	29.4	71.2	74.1	110.8	151	171	295	663	811
Philippines	17.6	27.0	41.9	35.4	28.6	375	519	973	1390	1965
Singapore	1.1	4.1	1.5	2.1	2.8	7	8	10	*14*	*19*
Taiwan	20.3	30.0	33.8	57.1	72.0	405	600	700	1142	1607
Thailand	39.0	67.8	55.5	56.3	77.8	308	488	585	1343	*1676*
Southeast Asia	*143.8*	*241.9*	*306.7*	*371.2*	*486.8*	*2205*	*3013*	*3932*	*6645*	*8824*
Cook Islands	0.1	0.1	0.2	0.3	0.2	2	2	3	4	4
Fiji	1.4	3.5	3.7	3.5	5.0	12	29	31	30	38
French Polynesia	0.4	0.4	0.4	0.4	0.4	3	3	3	3	3
Guam	0.1	0.2	0.2	0.9	0.8	1	2	2	7	10
New Caledonia	1.0	1.4	1.4	1.4	1.8	8	11	11	12	14
Papua New Guinea	3.7	4.0	7.7	18.0	20.3	59	62	107	90	140

Appendix: *National Agricultural Research Expenditure and Personnel Estimates, 1961-85 (Contd.)*

	Total agricultural research expenditures					Total number of researchers				
	1961-65	1966-70	1971-75	1976-80	1981-85	1961-65	1966-70	1971-75	1976-80	1981-85
	(millions 1980 PPP dollars per year)					*(full-time equivalents)*				
Solomon Islands	0.5	0.4	0.6	0.6	0.6	7	7	8	11	11
Tonga	0.0	0.0	0.1	0.4	0.4	1	1	4	6	13
Tuvalu	0.0	0.0	0.2	0.2	0.2	0	0	1	1	1
Vanuatu	0.1	0.2	0.2	0.2	0.3	2	3	3	6	8
Western Samoa	0.2	0.3	0.3	0.2	0.5	3	5	5	8	9
Pacific	*7.6*	*10.4*	*15.0*	*26.1*	*30.5*	*98*	*125*	*177*	*177*	*251*
Asia & Pacific, excl. China	*316.7*	*475.4*	*651.5*	*928.3*	*1159.6*	*6641*	*9480*	*12439*	*18559*	*22576*
Antigua	0.1	0.1	0.1	0.1	0.2	2	5	5	2	6
Bahamas	0.3	0.5	0.8	1.1	1.7	4	7	11	14	22
Barbados	0.5	1.6	1.4	2.1	1.8	3	20	29	37	50
Bermuda	0.3	0.2	0.4	0.4	0.3	4	3	5	6	4
Dominica	0.1	0.2	0.3	0.1	0.2	2	3	7	5	6
Dominican Republic	0.6	0.7	1.6	4.3	4.0	8	10	12	74	136
Grenada	0.0	0.1	0.0	0.1	0.1	1	2	1	2	3
Guadeloupe	0.5	0.8	1.0	1.4	1.7	4	7	8	12	14
Haiti	0.9	0.9	0.8	0.8	1.7	39	40	42	25	32
Jamaica	2.5	4.0	4.9	3.3	2.4	44	65	79	55	48
Martinique	0.2	0.4	0.6	0.6	0.8	2	3	5	5	7
Montserrat	0.1	0.2	0.3	0.3	0.3	1	2	3	3	3
Puerto Rico	9.0	11.8	12.5	11.0	8.4	132	129	105	104	77
St. Kitts-Nevis	0.3	0.3	0.0	0.1	0.1	4	3	1	2	5
St. Lucia	0.1	0.5	0.9	1.0	1.8	2	7	13	14	21
St. Vincent	0.1	0.2	0.1	0.2	0.2	3	4	3	5	5
Trinidad & Tobago	1.6	1.4	2.7	3.8	4.0	27	21	34	47	58
Virgin Islands (US)	0.0	0.0	0.2	0.2	0.5	0	0	2	4	6
Caribbean	*17.3*	*23.9*	*28.7*	*30.8*	*29.9*	*282*	*331*	*363*	*415*	*502*

Appendix: *National Agricultural Research Expenditure and Personnel Estimates, 1961-85 (Contd.)*

	Total agricultural research expenditures					Total number of researchers				
	1961-65	1966-70	1971-75	1976-80	1981-85	1961-65	1966-70	1971-75	1976-80	1981-85
	(millions 1980 PPP dollars per year)					*(full-time equivalents)*				
Belize	0.2	0.4	0.5	0.6	0.7	5	8	11	13	16
Costa Rica	2.2	2.6	3.4	4.3	2.8	51	43	60	90	114
El Salvador	2.9	3.5	3.5	4.5	5.4	49	58	77	107	131
Guatemala	2.0	2.2	6.0	8.0	7.3	22	22	63	112	160
Honduras	1.4	1.7	2.3	1.4	2.6	21	34	56	50	65
Mexico	11.5	11.5	36.4	80.1	129.0	192	239	444	797	1058
Nicaragua	3.0	3.0	3.0	4.4	5.1	24	24	29	51	65
Panama	0.9	1.2	1.0	3.0	6.1	7	21	23	43	115
Central America	*24.0*	*26.1*	*56.2*	*106.2*	*159.0*	*370*	*449*	*763*	*1262*	*1723*
Argentina	47.6	56.5	67.6	72.4	61.7	520	711	867	920	1062
Bolivia	1.7	4.7	4.0	5.0	2.3	29	61	51	75	104
Brazil	60.3	114.8	173.4	317.5	292.3	657	1469	2181	2648	3794
Chile	11.1	18.9	27.2	26.2	26.9	134	187	228	263	271
Colombia	33.3	46.7	40.5	35.7	47.8	326	398	559	400	454
Ecuador	3.4	6.2	16.2	17.2	13.3	34	54	138	180	211
Guyana	0.4	1.2	3.9	3.3	4.3	7	18	38	49	50
Paraguay	0.6	0.8	2.1	2.1	10.2	10	13	26	41	86
Peru	6.1	20.7	19.4	11.6	20.3	121	174	217	256	262
Suriname	0.9	0.9	0.9	0.9	0.8	13	13	19	17	22
Uruguay	2.0	3.2	4.0	4.5	4.1	33	56	75	72	77
Venezuela	20.4	30.5	42.5	46.1	35.9	130	189	316	392	383
South America	*187.7*	*305.2*	*401.6*	*542.3*	*519.9*	*2014*	*3342*	*4713*	*5314*	*6774*
Latin America & Caribbean	*229.1*	*355.1*	*486.6*	*679.3*	*708.8*	*2666*	*4122*	*5840*	*6991*	*9000*

Appendix: *National Agricultural Research Expenditure and Personnel Estimates, 1961-85 (Contd.)*

	Total agricultural research expenditures (millions 1980 PPP dollars per year)					Total number of researchers (full-time equivalents)				
	1961-65	1966-70	1971-75	1976-80	1981-85	1961-65	1966-70	1971-75	1976-80	1981-85
Algeria	7.8	5.4	13.5	19.8	21.3	49	62	117	229	305
Egypt	16.8	27.5	23.3	31.8	44.7	569	1431	2070	2748	4246
Libya	8.7	12.9	17.1	14.7	20.1	97	100	103	112	127
Morocco	10.3	13.2	17.5	18.5	25.2	117	108	111	184	217
Tunisia	3.3	7.7	9.9	8.7	14.7	39	101	113	91	121
North Africa	*46.9*	*66.7*	*81.3*	*93.5*	*126.0*	*870*	*1803*	*2514*	*3364*	*5016*
Cyprus	1.4	1.7	2.3	2.7	3.5	19	29	45	51	57
Iran	23.3	86.7	94.0	71.1	82.3	322	405	563	518	493
Iraq	6.4	7.5	9.3	24.6	37.9	101	117	162	352	542
Israel	16.2	16.8	22.4	41.1	45.5	300	376	417	458	550
Jordan	1.2	1.5	1.5	2.0	1.5	20	23	27	58	57
Kuwait	0.3	0.5	0.6	0.8	1.2	2	3	4	5	8
Lebanon	3.0	3.8	4.8	2.9	2.9	41	71	120	67	67
Oman	0.3	0.5	0.6	0.8	3.9	3	5	6	10	42
Qatar	0.4	0.8	1.0	1.4	1.5	2	4	5	7	7
Saudi Arabia	4.2	6.7	9.1	11.5	23.4	42	67	91	115	171
Syria	3.0	6.0	11.9	5.1	6.6	15	69	112	153	217
Turkey	18.4	47.7	57.3	76.0	107.4	397	479	630	783	1612
United Arab Emirates	0.3	0.4	0.7	1.0	1.3	3	4	6	7	12
Yemen, PDR	1.0	1.7	2.4	4.6	8.1	10	17	19	25	69
Yemen, Arab Republic	0.6	0.9	1.6	2.0	2.3	10	15	24	46	77
West Asia	*80.0*	*183.0*	*219.4*	*247.7*	*329.4*	*1287*	*1683*	*2232*	*2655*	*3980*
West Asia & North Africa	*126.9*	*249.7*	*300.7*	*341.2*	*455.4*	*2157*	*3485*	*4746*	*6019*	*8995*
Less-Developed Countries	*1093.6*	*1603.7*	*2201.0*	*2997.3*	*3629.8*	*19753*	*28829*	*37004*	*55143*	*77737*

Appendix: *National Agricultural Research Expenditure and Personnel Estimates, 1961-85 (Contd.)*

	Total agricultural research expenditures (millions 1980 PPP dollars per year)					Total number of researchers (full-time equivalents)				
	1961-65	1966-70	1971-75	1976-80	1981-85	1961-65	1966-70	1971-75	1976-80	1981-85
Japan	404.4	573.1	780.6	891.2	1021.6	12535	13123	13798	13747	14779
Australia	131.6	165.1	229.4	189.8	236.1	2118	2662	3519	4254	4579
New Zealand	29.5	43.6	60.6	68.8	76.6	509	616	776	1138	1324
Australia & New Zealand	161.1	208.7	290.1	258.6	313.7	2627	3278	4294	5392	5902
Denmark	28.9	30.6	27.5	25.0	33.7	413	444	411	405	457
Finland	15.5	21.2	20.4	28.4	32.4	242	284	341	388	405
Iceland	1.5	2.4	3.1	4.9	3.5	39	43	42	49	77
Norway	21.1	31.3	37.5	52.0	57.8	410	449	551	665	759
Sweden	22.5	36.5	46.4	45.6	54.3	415	534	650	810	1013
Northern Europe	89.6	122.0	134.8	155.8	181.7	1519	1753	1996	2317	2711
Austria	6.3	12.3	12.9	16.5	18.3	145	222	222	274	285
Belgium	20.1	33.7	30.2	44.3	41.7	610	647	568	546	496
France	56.8	109.0	175.2	208.8	241.4	1143	1321	1558	1871	2361
Germany, Fed. Rep. of	147.7	195.3	284.4	256.2	253.4	2283	2592	2643	2151	2125
Ireland	18.1	21.4	25.0	27.5	24.7	276	321	345	393	400
Netherlands	56.7	118.3	154.3	190.0	189.7	832	908	1249	1366	1630
Switzerland	11.6	19.5	21.3	15.0	18.9	130	155	189	216	286
United Kingdom	136.4	204.3	276.3	300.7	346.8	2220	2569	3113	3567	3814
Western Europe	453.8	713.8	979.8	1058.9	1134.8	7639	8733	9887	10384	11396
Greece	8.9	8.9	17.0	21.3	25.0	253	281	371	425	460
Italy	46.5	54.3	67.7	84.1	181.3	995	1016	1200	1793	2327
Portugal	11.2	14.8	18.9	23.3	22.1	298	333	376	351	449
Spain	21.9	19.0	38.2	61.3	88.4	589	502	637	973	1249
Southern Europe	88.4	97.1	141.9	190.0	316.8	2135	2132	2584	3542	4485

Appendix: *National Agricultural Research Expenditure and Personnel Estimates, 1961-85 (Contd.)*

	Total agricultural research expenditures					Total number of researchers				
	1961-65	1966-70	1971-75	1976-80	1981-85	1961-65	1966-70	1971-75	1976-80	1980-85
	(millions 1980 PPP dollars per year)					*(full-time equivalents)*				
Canada	148.8	278.4	258.3	315.9	421.4	1879	2198	2252	2317	2737
United States	844.7	1064.0	1140.8	1301.0	1423.9	12061	12822	13313	13903	14366
North America	993.5	1342.4	1399.1	1616.9	1845.2	13940	15020	15565	16220	17103
More-Developed Countries	2190.7	3057.2	3726.3	4171.4	4812.9	40395	44039	48123	51602	56376
Total	3284.3	4660.9	5927.3	7168.7	8442.7	60148	72868	85126	106745	134113

Source: Pardey and Roseboom (1989a); Fan (1991b) — China; and Pardey, Eveleens, and Hallaway (1991) — US.

Note: The nonitalicized five-year averages are based on directly observed estimates as reported in the sources described above. Italicized country-level figures are not based on direct estimates but were derived using various shortcut procedures as described in section 5.4.4, chapter 5. We caution against over interpreting individual country-level observations without first consulting our data sources and the references and documentation contained therein.

References

Abdullah, S. "The Missing Link in the Infrastructure of Science: A Case Study in Agricultural Research in Malaysia." Paper presented at the International Conference on Science Indicators for Developing Countries, organized by STD-ORSTOM. Paris, 15-19 October 1990.

Abrahamson, D. E. *The Challenge of Global Warming*. Washington, D.C.: Island Press for National Resources Defense Council, 1989.

Adelman, I., J. M. Bournieux, and J. Waelbroeck. *Agricultural Development-Led Industrialization in a Global Perspective*. Agricultural and Research Economics Working Paper No. 435. Berkeley: University of California, 1986.

Adusei, E. O., and G. W. Norton. "The Magnitude of Agricultural Maintenance Research in the USA." *Journal of Production Agriculture* Vol. 3, No. 1 (January-March 1990): 1-6.

Ahmad, S. *Approaches to Purchasing Power Parity and Real Product Comparisons Using Shortcuts and Reduced Information*. World Bank Staff Working Paper No. 418. Washington, D.C.: World Bank, September 1980.

Akiyama, T., and R. C. Duncan. *Analysis of the World Cocoa Market*. World Bank Staff Commodity Working Paper No. 8. Washington, D.C.: World Bank, 1982.

Alston, J. M., G. W. Edwards, and J. W. Freebairn. "Market Distortions and Benefits from Research." *American Journal of Agricultural Economics* Vol. 70, No. 2 (1988): 281-288.

Alston, J. M., and J. Mullen. "Incentives for Research into Traded Goods: The Case of Australian Wool." University of California, Davis, and New South Wales Department of Agriculture, Orange, 1989. Mimeo.

Anais, G. "INRA — The French National Institute for Agronomic Research." In *Proceedings of the Seminar on Strengthening Regional Agricultural Research in the Caribbean*. Ede, The Netherlands: CTA, 1985.

Anderson, J. R. "Allocation of Resources in Agricultural Research." *Journal of the Australian Institute of Agricultural Science* Vol. 38, No. 1 (1972): 7-13.

Anderson, J. R. "Risk Efficiency in the Interpretation of Agricultural Production Research." *Review of Marketing and Agricultural Economics* Vol. 42, No. 3 (1974): 31-84.

Anderson, J. R. "Essential Probabilistics in Modelling." *Agricultural Systems* Vol. 1, No. 3 (1976): 219-231.

Anderson, J. R. "Impacts of Climatic Variability in Australian Agriculture: A Review." *Review of Marketing and Agricultural Economics* Vol. 47, No. 3 (1979): 147-177.

Anderson, J. R. "On Risk Deductions in Public Project Appraisal." *Australian Journal of Agricultural Economics* Vol. 27, No. 3 (1983): 45-52.

Anderson, J. R. "Economic Impacts." In *Handbook of Applied Meteorology*, edited by D. D. Houghton. New York: Wiley, 1985a.

Anderson, J. R. *International Agricultural Research Centres: Achievements and Potential*. Washington, D.C.: CGIAR Secretariat, 1985b.

Anderson, J. R. Review of "Assessment of Agricultural Research Priorities: An International Perspective", by J. S. Davis, P. A. Oram, and J. G. Ryan. *Agricultural Science* Vol. 2, No. 2 (1989a): 45-46.

Anderson, J. R. "Reconsiderations on Risk Deductions in Public Project Appraisal." *Australian Journal of Agricultural Economics* Vol. 33, No. 3

(1989b): 136-140.

Anderson, J. R. *Forecasting, Uncertainty, and Public Project Appraisal.* Policy, Planning and Research Working Paper WPS 154. Washington, D.C.: World Bank, International Commodity Markets Division, 1989c.

Anderson, J. R. "A Framework for Understanding the Mixed Impact of 'Improved' Agricultural Technologies in Africa." In *Research Issues in Agricultural Technology in sub-Saharan Africa: A Workshop,* edited by S. Gnagey and J. R. Anderson. World Bank Discussion Paper. Washington, D.C.: World Bank, 1991.

Anderson, J. R., and P. B. R. Hazell, eds. *Variability in Grain Yields: Implications for Agricultural Research and Policy in Developing Countries.* Baltimore: Johns Hopkins University Press, 1989.

Anderson, J. R., and K. A. Parton. "Techniques for Guiding the Allocation of Resources among Rural Research Projects: State of the Art." *Prometheus* Vol. 1, No. 1 (1983): 180-201.

Anderson, J. R., J. L. Dillon, and J. B. Hardaker. *Agricultural Decision Analysis.* Ames: Iowa State University Press, 1977.

Anderson, J. R., R. W. Herdt, and G. M. Scobie. *Science and Food: The CGIAR and its Partners.* Washington, D.C.: World Bank, 1988.

Anderson, K. "Economic Growth, Structural Change and the Political Economy of Protection." In *The Political Economy of Agricultural Protection,* edited by K. Anderson and Y. Hayami. Sydney: Allen and Unwin, 1986.

Anderson, K. "China's Economic Growth, Changing Comparative Advantages and Agricultural Trade." University of Adelaide, Adelaide, Australia, July 1990. Mimeo.

Anderson, K., and Y. Hayami, eds. *The Political Economy of Agricultural Protection.* Sydney: Allen and Unwin, 1986.

Anon. "The Agricultural Sector of the Netherlands Antilles." Country paper presented at the Workshop on Agricultural Research Policy and Management, Port of Spain, Trinidad, 26-30 September 1983.

Antle, J. M. "Infrastructure and Aggregate Agricultural Productivity: International Evidence." *Economic Development and Cultural Change* Vol. 31, No. 3 (April 1983): 609-619.

Antony, G., and J. R. Anderson. "A Framework for the Ex Ante Analysis of Priorities for Papua New Guinea Agricultural Research." UNE-ACIAR draft interim report. Armidale: Department of Agricultural Economics and Business Management, University of New England, February 1988. Mimeo.

APO. *Productivity Measurement and Analysis: Asian Agriculture.* Tokyo: Asian Productivity Organization, 1987.

Appelbaum, E., and E. Katz. "Transfer Seeking and Avoidance: On the Full Social Costs of Rent Seeking." *Public Choice* Vol. 48 (1986): 175-181.

Appelbaum, E., and E. Katz. "Seeking Rents by Setting Rents: The Political Economy of Rent Seeking." *Economic Journal* Vol. 97 (1987): 685-699.

Arnon, I. *The Planning and Programming of Agricultural Research.* Rome: FAO, 1975.

Arnon, I. *Agricultural Research and Technology Transfer.* London: Elsevier Science, 1989.

Arrow, K. J., and R. C. Lind. "Uncertainty and the Evaluation of Public Investment Decisions." *American Economic Review* Vol. 60, No. 3 (1970): 364-378.

Avery, D. "United States Farm Dilemma: The Global Bad News Is Wrong." *Science* Vol. 230, No. 4724 (October 25 1985): 408-412.

Bailey, E. E., and A. F. Friedlaender. "Market Structure and Multiproduct Industries." *Journal of Economic Literature* Vol. 20, No. 3 (September 1982: 1024-1048.

Balassa, B. *Adjusting to External Shocks: The Newly Industrialized Developing Countries in 1974-76 and 1979-81.* Discussion Paper No. DRD89. Washington, D.C.: IBRD, May 1984.

Balassa, B. "Policy Responses to Exogenous Shocks in Developing Countries." *American Economic Review* Vol. 76, No. 2 (1986): 244-248.

Ball, V. E. "Output, Input, and Productivity Measurement in U.S. Agriculture, 1948-79." *American Journal of Agricultural Economics* Vol. 67 (1985): 475-486.

Ballenger, N., J. Dunmore, and T. Lederer. *Trade Liberalization in World Markets.* Agricultural Information Bulletin No. 516. Washington, D.C.: USDA, ERS, May 1987.

Bates, R. "Governments and Agricultural Markets in Africa." In *The Role of Markets in the World Food Economy,* edited by D. G. Johnson and G. E. Schuh. Boulder: Westview Press, 1983.

Batie, S. "Sustainable Development: Challenges to the Profession of Agricultural Economics."

American Journal of Agricultural Economics Vol. 71, No. 5 (December 1989): 1083-1101.

Baum, W. C. *Partners against Hunger: The Consultative Group on International Agricultural Research*. Washington, D.C.: World Bank, 1986.

Baumol, W. "Macroeconomics of Unbalanced Growth: The Anatomy of Urban Crisis." *American Economic Review* Vol. 57, No. 3 (June 1967): 415-426.

Baumol, W. J. "Productivity Growth, Convergence, and Welfare: What the Long-Run Data Show." *American Economic Review* Vol. 76, No. 5 (December 1986): 1072-1085.

Baumol, W. J., and K. S. Lee. "Contestable Markets, Trade, and Development." *World Bank Research Observer* Vol. 6, No. 1 (January 1991): 1-17.

Baumol, W. J., and E. N. Wolff. "Productivity Growth, Convergence, and Welfare: Reply." *American Economic Review* Vol. 78, No. 5 (December 1988): 1155-1159.

Baumol, W. J. , J. C. Panzar, and R. D. Willig. *Contestable Markets and the Theory of Industry Structure*. Revised edition. San Diego: Harcourt Brace Jovanovich, 1988.

Beattie, B. R., and C. R. Taylor. *The Economics of Production*. New York: John Wiley & Sons, 1985.

Beck, H. *A Description of Research and Development in the UK Agricultural Sector*. Reading, UK: Department of Agricultural Economics and Management, University of Reading, 1987.

Becker, G. "The Theory of Competition among Pressure Groups for Political Influence." *Quarterly Journal of Economics* Vol. 93 (August 1983): 372-400.

Behrman, J. M., and W. A. Fischer. *Overseas R & D Activities of Transnational Companies*. Boston: Oelgeschlager, Gunn, and Hain, 1980.

Bengston, D. N. "A Price Index for Deflating State Agricultural Experiment Station Research Expenditures." *Journal of Agricultural Economics Research* Vol. 41, No. 4 (Fall 1989): 12-20.

Bengston, D. N., and H. Gregersen. *Forestry Research Capacity in Developing Countries: A Review of Issues and Findings*. Center for Natural Resource Policy and Management Studies Working Paper No. 4. St. Paul: Department of Forest Resources, University of Minnesota, March 1988.

Bennell, P. *Agricultural Researchers in Sub-Saharan Africa: A Quantitative Overview*. Working Paper No. 5. The Hague: ISNAR, September 1986.

Bhagwati, N. J. "Directly Unproductive Profit Seeking Activities." *Journal of Political Economy* Vol. 90 (1982): 988-1002.

Bhagwati, N. J., and T. N. Srinivasan. "Revenue Seeking: A Generalization of the Theory of Tariffs." *Journal of Political Economy* Vol. 88 (1980): 1069-1087.

Bhagwati, N. J., R. A. Brecher, and T. N. Srinivasan. "DUP Activities and Economic Theory." In *Neoclassical Political Economy*, edited by D. Colander. Cambridge, Mass.: Ballinger Publishing Company, 1984.

Billing, K. J. *Zimbabwe and the CGIAR Centers: A Study of Their Collaboration in Agricultural Research*. CGIAR Study Paper No. 6. Washington, D.C.: World Bank, 1985.

Binswanger, H. P. "A Microeconomic Approach to Induced Innovation." *Economic Journal* Vol. 84. (December 1974): 940-958.

Binswanger, H. P., and P. Pingali. "Technological Priorities for Farming in Sub-Saharan Africa." *World Bank Research Observer* Vol. 3, No. 1 (January 1988): 81-98.

Binswanger, H. P. "The Policy Response of Agriculture." In *Proceedings of the World Bank Annual Conference on Development Economics, 1989*. Washington, D.C.: World Bank, 1989.

Binswanger, H. P., and J. G. Ryan. "Efficiency and Equity Issues in Ex Ante Allocation of Research Resources." *Indian Journal of Agricultural Economics* Vol. 32, No. 3 (1977): 217-231.

Binswanger, H. P., V. W. Ruttan, et al. *Induced Innovation: Technology, Institutions, and Development*. Baltimore: Johns Hopkins University Press, 1978.

Binswanger, H. P., Maw-Ching Yang, A. Bowers, and Y. Mundlak. "On the Determinants of Cross-Country Supply." *Journal of Econometrics* Vol. 36 (1987): 111-131.

Blandford, D. "Instability in World Grain Markets." *Journal of Agricultural Economics* Vol. 34, No. 3 (1983): 379-395.

Bollard, A., D. Harper, and M. Theron. *Research and Development in New Zealand: A Public Policy Framework*. Research Monograph 39. Wellington: New Zealand Institute of Economic Research, 1987.

Boserup, E. *The Conditions of Agricultural Growth*. London: Allen and Unwin, 1965.

Boserup, E. *Population and Technological Change — A Study of Long-Term Trends*. Chicago: University of Chicago Press, 1981.

Boyce, J. K. "Agricultural Research in Indonesia, The Philippines, Bangladesh, South Korea and India: A Documentary History." University of Minnesota, St. Paul, July 1980. Mimeo.

Boyce, J. K., and R. E. Evenson. *National and International Agricultural Research and Extension Programs*. New York: Agricultural Development Council, 1975.

Brandao, A. S. P., and J. L. Carvalho. "A Comparative Study of Agricultural Pricing Policies: The Case of Brazil." In *The Political Economy of Agricultural Price Policy in Selected Latin American Countries*, edited by A. Krueger et al. Baltimore: Johns Hopkins University Press, 1989.

Braun, J. von, and H. de Haan. *The Effects of Food Price and Subsidy Policies on Egyptian Agriculture*. IFPRI Research Report No. 42. Washington, D.C.: IFPRI, November 1983.

Braverman, A., and R. Kanbur. "Urban Bias and the Political Economy of Agricultural Reform." *World Development* Vol. 15, No. 9 (1987): 1179-1187.

Bremer, J., T. Babb, J. Dickinson, P. Gore, E. Hyman, and A. Madaline. *Fragile Lands: A Theme Paper on Problems, Issues, and Approaches for Development of Humid Tropical Lowlands and Steep Slopes in the Latin American Region*. Washington, D.C.: Development Alternatives, Inc., June 1984.

Brennan, J. P. *An Economic Investigation of Wheat Breeding Programs*. Agricultural Economics Bulletin No. 35. Armidale: University of New England, 1988.

Brockway, L. H. *Science and Colonial Expansion — The Role of the British Botanic Gardens*. New York: Academic Press, 1979.

Brogan, B., and J. Remenyi, eds. *Commodity Price Stabilisation in Papua New Guinea — A Work in Progress Seminar*. Discussion Paper No. 27. Port Moresby: Institute for National Affairs, January 1987.

Buchanan, J. M. "Rent Seeking and Profit Seeking." In *Toward a Theory of Rent Seeking Society*, edited by J. M. Buchanan, R. D. Tollinson, and G. Tullock. College Station: Texas A&M University Press, 1980.

Bumb, B. L. *Global Fertilizer Perspective, 1960-1995: The Dynamics of Growth and Structural Change*. Muscle Shoals: International Fertilizer Development Center, 1989.

Butler, L. J., and B. W. Marion. *The Impacts of Patent Protection on the U.S. Seed Industry and Public Plant Breeding*. Madison: University of Wisconsin Press, 1985.

Byerlee, D. "Food for Thought: Technological Challenges in Asian Agriculture in the 1990's." Background paper prepared for the Conference of Asian and Near East Bureaus' Agricultural and Rural Development Officers, USAID, Rabat, Morocco, 19-24 February 1989.

Byerlee, D. "Technical Change, Productivity, and Sustainability in Irrigated Wheat Systems of Asia: Emerging Issues." Paper presented at the 1990 Annual Meetings of the American Agricultural Economics Association, Vancouver, Canada, 1990.

CAAS. *Thirty Years of the Chinese Academy of Agricultural Sciences, 1957-87*. Beijing: Chinese Academy of Agricultural Sciences, 1987.

CAB. *List of Research Workers in the Agricultural Sciences in the Commonwealth and in the Republic of Ireland 1969*. Slough, U.K.: Commonwealth Agricultural Bureau, 1969.

CAB. *List of Research Workers in Agricultural Sciences in the Commonwealth 1981*. Slough, U.K.: Commonwealth Agricultural Bureau, 1981.

CABI. *CAB Abstracts*. Wallingford, UK: Commonwealth Agricultural Bureau International, multiple years (1972 to 1986).

Cantrell, R. P. "The Past Experience and Future Course of the CIMMYT Maize Program." In *Proceedings of the 41st Annual Corn and Sorghum Industry Research Conference*. Washington, D.C.: American Seed Trade Association, 1986.

Capalbo, S. M., and J. M. Antle, eds. *Agricultural Productivity: Measurement and Explanation*. Washington, D.C.: Resources for the Future, 1988.

Cardwell, V. B. "Fifty Years of Minnesota Corn Production: Sources of Yield Increase." *Agronomy Journal* Vol. 74 (1982): 984-990.

CARIS. *Agricultural Research in Developing Countries*. Volume 1: Research Institutions. Rome: FAO, 1978.

Carr, S. J. *Technology for Small-Scale Farmers in Sub-Saharan Africa: Experience with Food Crop*

Production in Five Major Ecological Zones. World Bank Technical Paper No. 109. Washington, D.C.: World Bank, 1989.

Carrasquillo, C. Y. "The Role of the Subject-Matter Specialist in the Puerto Rico Agricultural Extension Service." Ph.D. diss., North Carolina State University, Raleigh, 1984.

Carter, H. "The Agricultural Sustainability Issue: An Overview and Research Assessment." In *The Changing Dynamics of Global Agriculture*, edited by E. Javier and U. Renborg. The Hague: ISNAR, 1988.

Casas, J., ed. *Agricultural Research in Countries of the Mediterranean Region.* Paris: International Centre for Advanced Mediterranean Agronomic Studies, September 1988.

Cavallo, D. "Agriculture and Economic Growth: The Experience of Argentina 1913-1984." Paper presented at the 20th Conference of Agricultural Economists, Buenos Aires, 1988.

Caves, D. W., L. R. Christensen, and W. E. Diewert. "Multilateral Comparisons of Output, Input, and Productivity Using Superlative Index Numbers." *Economic Journal* Vol. 92 (1982): 73-86.

Caves, R. E. *Multinational Enterprise and Economic Analysis.* New York: Cambridge University Press, 1982.

Central Office of Information. *Britain and the Developing Countries — Research Institutions.* Central Office of Information Reference Pamphlet 103. London: Her Majesty's Stationery Office, 1972.

CGIAR. *Report of the Review Committee.* Washington, D.C.: CGIAR Secretariat, 1977.

CGIAR. *Second Review of the CGIAR.* Washington, D.C.: CGIAR Secretariat, 1981.

CGIAR. *Annual Report 1983-1984.* Washington, D.C.: CGIAR Secretariat, 1984.

CGIAR Secretariat. "Program Budget Analyses 1974-1984 CGIAR Centers." CGIAR Secretariat, Washington, D.C., October 1982. Mimeo.

CGIAR Secretariat. "Statistics on Expenditure by International Agricultural Research Centers 1960-1987, CGIAR Contributions 1972-1982." CGIAR Secretariat, Washington, D.C., August 1983a. Mimeo.

CGIAR Secretariat. "Program Budget Analyses 1980-1985 CGIAR Centres." CGIAR Secretariat, Washington, D.C., October 1983b. Mimeo.

CGIAR Secretariat. "1987 Funding Requirements of CGIAR Centers." CGIAR Secretariat, Washing-ton, D.C., September 1986. Mimeo.

CGIAR Secretariat. "Printout Expenditure Report Menu." CGIAR Secretariat, Washington, D.C., August 1988a. Mimeo.

CGIAR Secretariat. *Relationships between Non-Associated Centers and the CGIAR.* Washington, D.C.: CGIAR Secretariat, 1988b.

CGIAR Secretariat. "Funding of the CGIAR — Retrospective 1983-88." CGIAR Secretariat, Washington, D.C., June 1988c. Mimeo.

CGIAR Secretariat. "Trends in CGIAR Operating Expenditures: 1983-1988." CGIAR Secretariat, Washington, D.C., 1989a. Mimeo.

CGIAR Secretariat. "CGIAR 1988 Financial Report." CGIAR Secretariat, Washington, D.C., May 1989b. Mimeo.

CGIAR Secretariat. "1990 Funding Requirements of CGIAR Centers." CGIAR Secretariat, Washington, D.C., 1989c. Mimeo.

CGIAR Secretariat. "Analysis of CG Expenditures by Commodity/Region." CGIAR Secretariat, Washington, D.C., August 1989d. Mimeo.

CGIAR Secretariat. *1986-1990 Expenditures, Staffing and Funding of the Non-Associated Centres.* Background paper for TAC. Washington, D.C.: CGIAR Secretariat, 1990a.

CGIAR Secretariat. "1991 Funding Requirements of CGIAR Centers." CGIAR Secretariat, Washington, D.C., May 1990b. Mimeo.

CGIAR Secretariat. "Appendix to Unknown Document." CGIAR Secretariat, Washington, D.C., 1990c.

Christensen, P. "Historical Roots for Ecological Economics — Biophysical versus Allocative Approaches." *Ecological Economics* Vol. 1, No. 1 (1989): 17-36.

Christiansen, R. E. *The Impact of Economic Development on Agricultural Trade Patterns.* Washington, D.C.: USDA, ERS, IED, January 1987.

CIMMYT. *1986 CIMMYT World Maize Facts and Trends: The Economics of Commercial Maize Seed Production in Developing Countries.* Mexico, D.F.: CIMMYT, 1987a.

CIMMYT. *Economics Program Database.* 1985-1986. Mexico, D.F.: CIMMYT, 1987b.

CIMMYT. *Training Program Database.* 1966-1985. Mexico, D.F.: CIMMYT, 1987c.

CIMMYT. *CIMMYT 1988 Annual Report.* Mexico, D.F.: CIMMYT, 1989.

CIRAD. *From GERDAT to CIRAD*. Paris: CIRAD, 1987.

Clague, C. "Short-Cut Estimates of Real Income." *Review of Income and Wealth* Vol. 32, No. 3 (September 1986): 313-331.

Clark, C. *Mathematical Bioeconomics*. New York: Wiley, 1976.

Cochran, M. J., L. J. Robison, and W. Lodwick. "Improving the Efficiency of Stochastic Dominance Techniques Using Convex Set Stochastic Dominance." *American Journal of Agricultural Economics* Vol. 67, No. 2 (1985): 289-295.

Colander, D. C., ed. *Neoclassical Political Economy: The Analysis of Rent Seeking and DUP Activities*. Cambridge, Mass.: Ballinger Publishing Company, 1984.

Conesa, A. P., and J. Casas. *Overview of the French System of Agricultural and Agro-Food Research*. Staff Note 88-37e. The Hague: ISNAR, December 1986.

Conway, G. "Agroecosystems Analysis." *Agricultural Administration* Vol. 20 (1985): 31-55.

Cooper, St. G. C. *Agricultural Research in Tropical Africa*. Nairobi: East African Literature Bureau, 1970.

Corbo, V., J. de Melo, and J. Tybout. "What Went Wrong with the Recent Reforms in the Southern Cone." *Economic Development and Cultural Change* Vol. 34 (1986): 607-640.

Cornelius, J. C., J. E. Ikerd, and A. G. Nelson. "A Preliminary Evaluation of Price Forecasting Performance by Agricultural Economists." *American Journal of Agricultural Economics* Vol. 63, No. 4 (1981): 712-714.

Craig, B. J., and P. G. Pardey. *Multidimensional Output Indices*. Staff Paper Series P90-63. St. Paul: University of Minnesota, October 1990a.

Craig, B. J., and P. G. Pardey. *Patterns of Agricultural Development in the United States*. Staff Paper Series P90-72. St. Paul: University of Minnesota, December 1990b.

Crawford, J. G. "Development of the International Agricultural Research System." In *Resource Allocation and Productivity in National and International Agricultural Research*, edited by T. M. Arndt, D. Dalrymple, and V. W. Ruttan. Minneapolis: University of Minnesota Press, 1977.

Crosby, E. "A Survey of U.S. Agricultural Research by Private Industry III." In *Policy for Agricultural Research*, edited by V. W. Ruttan and C. E. Pray.

Boulder: Westview, 1987.

Dahab, S. "The Agricultural Machinery and Implement Industry in Brazil: Its Historical Development and Inventive Activity." Ph.D. diss., Economics Department, Yale University, New Haven, 1985.

Dalrymple, D. G. "The Development and Adoption of High-Yielding Varieties of Wheat and Rice in Developing Countries." *American Journal of Agricultural Economics* Vol. 67, No. 5 (1985): 1067-1073.

Dasgupta, P., and G. Heal. *Economic Theory and Exhaustible Resources*. Cambridge: Cambridge University Press, 1979.

Davidson, W. H. "The Location of Foreign Direct Investment Activity: Country Characteristics and Experience Effects." *Journal of International Business Studies* Vol. 11, No. 2 (Fall 1980): 9-22.

Davis, C. G. "Agricultural Research and Agricultural Development in Small Plantation Economies: The Case of the West Indies." *Social and Economic Studies* Vol. 24, No. 1 (1975): 117-152.

Davis, J. S. "A Comparison of Procedures for Estimating Returns to Research Using Production Functions." *Australian Journal of Agricultural Economics* Vol. 25, No. 1 (1981): 60-72.

Davis, J. S., P. A. Oram, and J. G. Ryan. *Assessment of Agricultural Research Priorities: An International Perspective*. Canberra: ACIAR and IFPRI, 1987.

Delgado, C., and J. Mellor. "A Structural View of Policy Issues in African Agricultural Development: Reply." *American Journal of Agricultural Economics* Vol. 69, No. 2 (May 1987): 389-391.

Deolalikar, A., and R. Evenson. "Technology Production and Technology Purchase in Indian Industry: An Econometric Analysis." Economic Growth Center, Yale University, New Haven, 1988. Mimeo.

Diewert, W. E. "Superlative Index Numbers and Consistency in Aggregation." *Econometrica* Vol. 46, No. 4 (July 1978): 883-900.

Divisia, F. *Economie Rationnelle*. Paris: Gaston Doin, 1928.

Dornbusch, R., and A. Reynoso. "Financial Factors in Economic Development." *American Economic Review* Vol. 79, No. 2 (May 1989): 204-209.

Douglas, J. E., ed. *Successful Seed Programs: A Planning and Management Guide*. Boulder: Westview Press, 1980.

Dowswell, C. R. *Strengthening National Research Programs through Training: A Twenty Year Progress Report.* Mexico, D.F.: CIMMYT, 1986.

Drachoussof, M. V. "Historique des Recherches en Agronomie Tropicale Africaine." In *Amélioration et Protection des Plantes Vivrières Tropicales,* edited by C. A. Saint Pierre. Paris: John Libbey Eurotext, 1989.

Drechsler, L. "Weighting of Index Numbers in Multilateral International Comparisons." *Review of Income and Wealth* Vol. 19, No. 1 (March 1973): 17-35.

Duncan, R. C. "Evaluating Returns to Research in Pasture Improvement." *Australian Journal of Agricultural Economics* Vol. 16 (1972): 153-168.

Duncan, R. C., and C. Tisdell. "Research and Technical Progress: The Returns to Producers." *Economic Record* Vol. 47, No. 117 (1971): 124-129.

Duvick, D. N. "Genetic Diversity in Major Farm Crops on the Farm and in Reserve." *Economic Botany* Vol. 38, No. 2 (1984): 161-178.

Dyer, P. T., and G. M. Scobie. *The Payoff to Investment in Agroforestry Research: A Preliminary Report.* Discussion Paper 4/84. Hamilton, New Zealand: MAF Economics Division, 1984.

Dyer, P. T., G. M. Scobie, and S. R. Davis. *The Payoff to Investment in a Recombinant DNA Research Facility at Ruakura: A Monte Carlo Simulation Study.* Discussion Paper 1/84. Hamilton, New Zealand: MAF Economics Division, 1984.

ECE and FAO. *Expenses and Income of Agriculture in European Countries and North America 1966-1975.* New York: United Nations, 1981.

ECE and FAO. *Output and Inputs in Agriculture of Countries in the ECE Region, 1980-1987.* New York: United Nations, 1989.

Echeverría, R. G. "Public and Private Sector Investments in Agricultural Research: The Case of Maize." Ph.D. diss., University of Minnesota, St. Paul, 1988a.

Echeverría, R. G. Unpublished interviews with US Private Companies. 1988b.

Echeverría, R. G. *Public and Private Investments in Maize Research in Mexico and Guatemala.* CIMMYT Economics Working Paper 90/03. Mexico, D.F.: CIMMYT, 1990a.

Echeverría, R. G. "Assessing the Impact of Agricultural Research." In *Methods for Diagnosing Research System Constraints and Assessing the Impact of Agricultural Research – Volume II, Assessing the Impact of Agricultural Research,* edited by R. G. Echeverría. The Hague: ISNAR, 1990b.

Edwards, G. W., and J. W. Freebairn. "The Social Benefit from an Increase in Productivity in a Part of an Industry." *Review of Marketing and Agricultural Economics* Vol. 50, No. 2 (1982): 193-210.

Edwards, G. W., and J. W. Freebairn. "The Gains from Research into Tradeable Commodities." *American Journal of Agricultural Economics* Vol. 66, No. 1 (1984): 41-49.

Eicher, C. K. "Building African Scientific Capacity for Agricultural Development." *Agricultural Economics* Vol. 4, No. 2 (June 1990): 117-143.

Eicher, C. K., and D. C. Baker. *Research on Agricultural Development in Sub-Saharan Africa: A Critical Survey.* MSU International Development Paper No. 1. East Lansing: Michigan State University, Department of Agricultural Economics, 1982.

Eisemon, T. O., C. H. Davis, and E. M. Rathgeber. "Transplantation of Science to Anglophone and Francophone Africa." *Science and Public Policy* Vol. 12, No. 4 (August 1985): 191-202.

Elias, V. *Government Expenditures on Agriculture and Agricultural Growth in Latin America.* IFPRI Research Report 50. Washington, D.C.: IFPRI, 1985.

Englander, A. S. "Technology Transfer and Development in Agricultural Research Programs." Ph.D. diss., Yale University, New Haven, 1981.

Erlich, P. "The Limits to Substitution: Meta-Resource Depletion and a New Economic-Ecological Paradigm." *Ecological Economics* Vol. 1, No. 1 (1989): 9-16.

Europa Publications. *Africa South of the Sahara 1990.* London: Europa Publications, 1990.

EUROSTAT. *EUROSTAT External Trade Analytical Tables.* Luxembourg: Office for Official Publications of the European Communities, multiple years (1966 to 1986).

EUROSTAT. *Multilateral Measurements of Purchasing Power and Real GDP.* The Hill Report. Luxembourg: Office for Offical Publications of the European Communities, 1982.

Evenson, D. D., and R. E. Evenson. "Legal Systems and Private Sector Incentives for the Invention of Agricultural Technology in Latin America." In *Technical Change and Social Conflict in Agricul-*

ture: Latin American Perspectives, edited by M. Piñeiro and E. Trigo. Boulder: Westview Press, 1983.

Evenson, R. E. "The Contribution of Agricultural Research to Production." *Journal of Farm Economics* Vol. 49, No. 5 (1967): 1415-1425.

Evenson, R. E. "Intellectual Property Rights, Agribusiness Research and Development: Implications for the Public Agricultural Research System." *American Journal of Agricultural Economics* Vol. 65, No. 4 (1983): 967-975.

Evenson, R. E. "Observations on Brazilian Agricultural Research and Productivity." In *Brazilian Agriculture and Agricultural Research*, edited by L. Yeganiantz. Brasília: EMBRAPA, 1984.

Evenson, R. E. *The International Agricultural Research Centers: Their Impact on Spending for National Agricultural Research and Extension.* CGIAR Study Paper No. 22. Washington, D.C.: World Bank, 1987.

Evenson, R. E. "Spillover Benefits of Agricultural Research: Evidence from U.S. Experience." *American Journal of Agricultural Economics* Vol. 71, No. 2 (1989): 447-454.

Evenson, R. E., and Y. Kislev. *Investment in Agricultural Research and Extension: A Survey of International Data.* Center Discussion Paper No. 124. Connecticut: Yale University, Economic Growth Center, August 1971.

Evenson, R. E., and Y. Kislev. "Investment in Agricultural Research and Extension: A Survey of International Data." *Economic Development and Cultural Change* Vol. 23 (April 1975a): 507-521.

Evenson, R. E., and Y. Kislev. *Agricultural Research and Productivity.* New Haven: Yale University Press, 1975b.

Evenson, R. E., C. E. Pray, and G. M. Scobie. "The Influence of International Research on the Size of National Research Systems." *American Journal of Agricultural Economics* Vol. 67, No. 5 (December 1985): 1074-1079.

Evenson, R. E., and J. Putnam. "Intellectual Property Management." In *Agricultural Biotechnology: Opportunities for International Development*, edited by G. J. Persley. Wallingford: CAB International, 1990.

Fan, S. *Regional Productivity Growth in China's Agriculture.* Boulder: Westview Press, 1990.

Fan, S. "Effects of Technological Change and Institutional Reform on Production Growth in Chi-

nese Agriculture." *American Journal of Agricultural Economics* Vol. 73, No. 2 (May 1991a).

Fan, S. "Institutional and Quantitative Development of the Chinese Agriculture Research System." ISNAR, The Hague, 1991b. Mimeo.

FAO. *FAO Production Yearbook.* Volumes 11-43. Rome: FAO, multiple years (1958 to 1990).

FAO. *FAO Fertilizer Yearbook.* Volumes 15-38. Rome: FAO, multiple years (1965 to 1989).

FAO. *Handbook of Economic Accounts for Agriculture.* Rome: FAO, 1974.

FAO. *FAO Trade Yearbook 1976.* Rome: FAO, 1977.

FAO. *FAO Yearbook of Forest Products.* Rome: FAO, 1980.

FAO. *Report on the Agro-Ecological Zones Project.* Volume 15. Rome: FAO, 1982.

FAO. *Preliminary Results of a Survey of Forestry Research Capabilities in the Asia/Pacific Region with Particular Emphasis on Fuelwood and Wood Energy.* Rome: FAO, 1984a.

FAO. *Preliminary Results of a Survey of Forestry Research Capabilities in the Latin America Region with Particular Emphasis on Fuelwood and Wood Energy.* Rome: FAO, 1984b.

FAO. "Survey of Wood Energy Research and Development Capabilities in Africa." Paper presented at the Technical Consultation on Wood Energy Research and Development in Africa, Addis Ababa, Ethiopia, 27-30 November, 1984c.

FAO. *Agricultural Research Systems in the Asia-Pacific Region.* Bangkok: FAO Regional Office for Asia and the Pacific, 1986a.

FAO. *Intercountry Comparisons of Agricultural Production Aggregates.* FAO Economic and Social Development Paper 61. Rome: FAO, 1986b.

FAO. *Agriculture toward 2000.* Revised version. Rome: FAO, 1987a.

FAO. *FAO Annual Demographic Estimates: 1961-88.* AGROSTAT Diskettes. Rome: FAO, 1987b.

FAO. *FAO Fertilizer Consumption Data: 1961-88.* AGROSTAT Diskettes. Rome: FAO, 1990a.

FAO. *FAO Agricultural Tractors in Use Data: 1961-88.* AGROSTAT Diskettes. Rome: FAO, 1990b.

FAO. *FAO Trade Yearbook 1988.* Rome: FAO, 1990c.

FAO. *Agricultural Research Systems in the Near East and North Africa.* Rome: FAO Research and Technology Development Division, 1990d.

Feder, G., and G. T. O'Mara. "On Information and Innovation Diffusion: A Bayesian Approach." *American Journal of Agricultural Economics* Vol.

64, No. 1 (1982): 145-147.

Feder, G., and R. Slade. "The Acquisition of Information and the Adoption of New Technology." *American Journal of Agricultural Economics* Vol. 66, No. 3 (1984): 312-320.

Feder, G., R. E. Just, and D. Zilberman. "Adoption of Agricultural Innovations in Developing Countries: A Survey." *Economic Development and Cultural Change* Vol. 35, No. 1 (1986): 255-298.

Fei, J. C. H., and G. Ranis. *Development of the Labor Surplus Economy: Theory and Policy.* Homewood, Ill.: Irwin, 1964.

Fellner, W. "Two Propositions of Induced Innovations." *Economic Journal* Vol. 71 (1961): 305-308.

Fields, G. S. "Changes in Poverty and Inequality in Developing Countries." *World Bank Research Observer* Vol. 4, No. 2 (July 1989): 167-185.

Fishel, W. L., ed. *Resource Allocation in Agricultural Research.* Minneapolis: University of Minnesota Press, 1971.

Fisher, B. S. "The Impact of Changing Marketing Margins on Farm Prices." *American Journal of Agricultural Economics* Vol. 63, No. 2 (May 1981): 261-263.

Fox, G. "Is the United States Really Underinvesting in Agricultural Research." *American Journal of Agricultural Economics* Vol. 67, No. 4 (November 1985): 807-812.

Fox, G. "Models of Resource Allocation in Public Agricultural Research: A Survey." *Journal of Agricultural Economics* Vol. 38, No. 3 (1987): 449-462.

Freebairn, J. W. "An Evaluation of Outlook Information for Australian Agricultural Commodities." *Review of Marketing and Agricultural Economics* Vol. 46, No. 3 (1978): 294-314.

Freebairn, J. W. "Drought Assistance Policy." *Australian Journal of Agricultural Economics* Vol. 27, No. 3 (1983): 185-199.

Freebairn, J. W., J. S. Davis, and G. W. Edwards. "Distribution of Research Gains in Multistage Production Systems." *American Journal of Agricultural Economics* Vol. 64, No. 1 (1982): 39-46.

Freeman, A. M. *The Benefits of Environmental Improvement.* Baltimore: Johns Hopkins University Press, 1979.

Gardner, B. L. "The Farm-Retail Price Spread." *American Journal of Agricultural Economics* Vol. 57, No. 3 (August 1975): 399-409.

Gardner, B. L. *Price Supports and Optimal Spending on Agricultural Research.* Seminar Paper 90-01. Adelaide: Centre for International Economic Studies, University of Adelaide, May 1989.

Gerschenkron, A. "Economic Backwardness in Historical Perspective." In *The Progress of Underdeveloped Areas,* edited by B. F. Hoselitz. Chicago: University of Chicago Press, 1952.

Goode, R. *Government Finance in Developing Countries.* Washington, D.C.: Brookings Institution, 1984.

Gorter, H. de, and D. Zilberman. "Public Good Inputs." *American Journal of Agricultural Economics* Vol. 72, No. 1 (1990): 131-137.

Graham-Tomasi, T. *Uncertainty, Information, and Irreversible Investment.* Staff Paper P85-26. St. Paul: Department of Agricultural and Applied Economics, University of Minnesota, 1985.

Graham-Tomasi, T. "Valuation of Nontimber Services of Forests: Economic Concepts in a Policy Context." In *Valuing the Contributions of Forests to Human Welfare: Theory and Practice,* edited by H. M. Gregersen and A. L. Lundgren. St. Paul: Forestry for Sustainable Development Program, University of Minnesota, 1990a.

Graham-Tomasi, T. "Techniques for the Valuation of the Services of Natural Forests." In *Valuing the Contributions of Forests to Human Welfare: Theory and Practice,* edited by H. M. Gregersen and A. L. Lundgren. St. Paul: Forestry for Sustainable Development Program, University of Minnesota, 1990b.

Grantham, G. "The Shifting Locus of Agricultural Innovation in Nineteenth-Century Europe: The Case of the Agricultural Experiment Stations." *Research in Economic History* Vol. 3 (1984): 191-214.

Greene, D., and T. Roe. "Political Economy of Agricultural Pricing Policy in the Dominican Republic." In *The Political Economy of Agricultural Price Policy in Selected Latin American Countries,* edited by A. Krueger et al. Baltimore: Johns Hopkins University Press, 1989.

Griliches, Z. "Hybrid Corn: An Exploration in the Economics of Technological Change." *Econometrica* Vol. 25, No. 4 (1957a): 501-522.

Griliches, Z. "Specification Bias in Estimates of Production Functions." *Journal of Farm Economics* Vol. 39 (1957b): 8-20.

Griliches, Z. "Research Costs and Social Returns: Hybrid Corn and Related Innovations." *Journal of Political Economy* Vol. 66 (1958): 419-431.

Griliches, Z. "Issues in Assessing the Contribution of Research and Development to Productivity Growth." *Bell Journal of Economics* Vol. 10 (1979): 92-116.

Griliches, Z. "Returns to Research and Development Expenditures in the Private Sector." In *New Developments in Productivity and Analysis Measurement*, edited by J. W. Kendrick and B. N. Vaccara. Chicago: University of Chicago Press, 1980.

Griliches, Z., ed. *R & D Patents and Productivity.* Chicago: University of Chicago Press, 1984.

Gunasena, H. P. M. "Agricultural Education in Sri Lanka." In *Proceedings of Scientific Agricultural Manpower in Asia*, edited by H. P. M. Gunasena and H. M. G. Herath. Sri Lanka: National Agricultural Society, 1985.

Guttman, J. "Interest Groups and the Demand for Agricultural Research." *Journal of Political Economy* Vol. 86, No. 3 (1978): 467-484.

Hanemann, M. "Information and the Concept of Option Value." *Journal of Environmental Economics and Management* Vol. 16 (1989): 23-37.

Hardaker, J. B., and E. M. Fleming. "Agricultural Research Problems in Small Developing Countries: Case Studies from the South Pacific Island Nations." *Agricultural Economics* Vol. 3, No. 4 (1989): 279-292.

Harriri, G. *Organization and Structure of Arab National Agricultural Research Systems.* ISNAR Staff Notes No. 88-9. The Hague: ISNAR, September 1988.

Hayami, Y., and R. W. Herdt. "Market Price Effects of Technological Change on Income Distribution in Semisubsistence Agriculture." *American Journal of Agricultural Economics* Vol. 59, No. 2 (1977): 245-256.

Hayami, Y., and K. Inagi. "International Comparisons of Agricultural Productivities." *Farm Economist* Vol. 11, No. 10 (1969): 407-419.

Hayami, Y., and V. W. Ruttan. *Sources of Agricultural Productivity Differences among Countries: Resource Accumulation, Technical Inputs and Human Capital.* Staff Paper P69-24. St. Paul: Department of Agricultural and Applied Economics, University of Minnesota, November 1969.

Hayami, Y., and V. W. Ruttan. *Agricultural Development: An International Perspective.* Baltimore: Johns Hopkins University Press, 1971.

Hayami, Y., and V. W. Ruttan. *Agricultural Development: An International Perspective.* Revised Edition. Baltimore: Johns Hopkins University Press, 1985.

Hayami, Y., and S. Yamada. "Agricultural Research Organization in Economic Development: A Review of the Japanese Experience." In *Agriculture in Development Theory*, edited by L. G. Reynolds. New Haven: Yale University Press, 1975.

Hayami, Y., B. B. Miller, W. W. Wade, and S. Yamashita. *An International Comparison of Agricultural Production and Productivities.* Technical Bulletin 277. St. Paul: Agricultural Experiment Station, University of Minnesota, 1971.

Hazell, P. B. R., and C. Ramasamy. *Green Revolution Reconsidered: The Impact of the High-Yielding Rice Varieties in South India.* Baltimore: Johns Hopkins University Press, 1991.

Hazell, P. B. R., C. Pomareda, and A. Valdés, eds. *Crop Insurance for Agricultural Development: Issues and Experience.* Baltimore: Johns Hopkins University Press, 1986.

Headrick, D. R. *The Tentacles of Progress: Technology Transfer in the Age of Imperialism, 1850-1940.* Oxford: Oxford University Press, 1988.

Heady, E. O. *Economics of Agricultural Production and Resource Use.* Englewood Cliffs: Prentice-Hall, 1952.

Heisey, P. W. "Comment: Maize Research in Malawi." *Journal of International Development* Vol. 2, No. 2 (1990): 243-253.

Henderson, C. B. *Inbred Lines of Corn Released to Private Growers from State and Federal Agencies and Lists of Double Crosses, by States.* Third revision. Champaign: Illinois Seed Producers Association, April 1960.

Henderson, C. B. *Inbreds, Breeding Stocks, Maize Investigations and Academic Research Personnel.* Maize Research and Breeders Manual No. 6. Champaign: Illinois Foundation Seeds, Inc, April 1968.

Henderson, C. B. *Inbreds, Breeding Stocks, Maize Chromosomes and Genes, Maize Investigations and Academic Research Personnel.* Maize Research and Breeders Manual No. 8. Champaign: Illinois Foundation Seeds, Inc, December 1976.

Henderson, C. B. *Inbreds, Breeding Stocks, Maize Investigations and Academic Research Personnel.* Maize Research and Breeders Manual No. 10. Champaign: Illinois Foundation Seeds, Inc., 1984.

Hertford, R., and A. Schmitz. "Measuring Economic Returns to Agricultural Research." In *Resource Allocation and Productivity in National and International Research*, edited by T. M. Arndt, D. G. Dalrymple, and V. W. Ruttan. Minneapolis: University of Minnesota Press, 1977.

Heston, A., and R. Summers. "What We Have Learned about Prices and Quantities from International Comparisons: 1987." *American Economic Review* Vol. 78, No. 2 (May 1988): 467-473.

Hicks, J. R. *Theory of Wages.* London: MacMillan, 1932.

Hirshleifer, J. "The Private and Social Value of Information and the Reward to Inventive Activity." *American Economic Review* Vol. 61 (1971): 561-574.

Hobbs, J., J. R. Anderson, J. L. Dillon, and H. Harris. "The Effects of Climatic Variations on Agriculture in the Australian Wheatbelt." In *The Impact of Climatic Variations on Agriculture, Vol. 2 Assessments in Semi-Arid Regions*, edited by M. L. Parry, T. R. Carter, and N. T. Konijn. Dordrecht: Kluwer, 1988.

Houck, J. P. "Foreign Agricultural Assistance: It's Mostly a Good Thing for U.S. Farmers." *Choices* (First Quarter 1987): 19.

Houck, J. P., and P. W. Gallagher. "The Price Responsiveness of US Corn Yields." *American Journal of Agricultural Economics* Vol. 58, No. 4 (1976): 731-734.

House of Commons. *The Overseas Development Administration's Scientific and Special Units.* Fourth Report from the Foreign Affairs Committee, Session 1982-83. London: Her Majesty's Stationery Office, 1983.

Hsiao, C. *Analysis of Panel Data.* New York: Cambridge University Press, 1986.

Huffman, W. E., and R. E. Evenson. *The Development of U.S. Agricultural Research and Education: An Economic Perspective – Part II.* Staff Paper No. 169. Ames: Iowa State University, 1987.

Huici, N. *La Industria de la Maquinaria Agrícola en Argentina.* PROAGRO Document No. 9. Buenos Aires: CISEA, 1984.

IBPGR. *IBPGR Annual Report 1987.* Rome: IBPGR, 1988.

IBPGR. *IBPGR Annual Report 1988.* Rome: IBPGR, 1989.

Idachaba, F. S. *Agricultural Research Policy in Nigeria.* Research Report No. 15. Washington, D.C.: IFPRI, 1980.

Idachaba, F. S. "Agricultural Research in Nigeria: Organization and Policy." In *Policy for Agricultural Research*, edited by V. W. Ruttan and C. E. Pray. Boulder: Westview Press, 1987.

IDRC. *Multilateral Research Institutions in the Third World.* IDRC Manuscript Report 129e. Ottawa: IDRC, 1986.

ILO. *International Standard Classification of Occupation.* Revised edition. Geneva: ILO, 1986.

IMF. *Trade Policy Issues and Developments.* Occasional Paper 38. Washington, D.C.: IMF, 1985.

IMF. *World Economic Outlook.* Washington, D.C.: IMF, April 1989a.

IMF. *International Financial Statistics Yearbook 1989.* Washington, D.C.: IMF, 1989b.

INIA. *Qué es el INIA?* Mexico, D.F.: Instituto Nacional de Investigaciones Agrícolas, n.d.

Intal, P. S., and J. H. Power. "Government Interventions and Philippine Agriculture." In *The Political Economy of Agricultural Price Policy in Selected Latin American Countries*, edited by A. Krueger et al. Baltimore: Johns Hopkins University Press, 1989.

IRRI. *Rice Research and Production in China: An IRRI Team's View.* Los Baños: IRRI, 1978.

ISNAR. "Inventory of CGIAR Activities in Sub-Saharan Africa." ISNAR, The Hague, 1986. Mimeo.

ISNAR. "ISNAR Accounts December 31, 1987." ISNAR, The Hague, 1987. Mimeo.

ISNAR. "ISNAR Accounts December 31, 1988." ISNAR, The Hague, 1988. Mimeo.

ISRA/ISNAR. *Ressources Humaines de l'ISRA: Situation Actuelle et Implications Financières de Politiques Salariales Alternatives.* The Hague: ISNAR, 1989.

Jacobs, E., and M. Gutierrez. *La Industria de Semillas en Paises Semi-Industrializados: Los Casos de Argentina y Brasil.* The Hague: ISNAR, 1986.

Jaffe, S. A. *A Price Index for Deflation of Academic R&D Expenditures.* NSF 72-310. Washington, D.C.: National Science Foundation, May 1972.

Jaffe, A. B. "Technological Opportunity and Spillovers of R&D: Evidence from Firms' Patents, Profits and Market Value." *American Economic Review* Vol. 76 (1986): 984-1001.

Jaffe, A. B. "Characterizing the 'Technological Position' of Firms, with Application to Quantifying Technological Opportunity and Research Spillovers." *Research Policy* Vol. 18, No. 2 (April 1989): 87-97.

Jamieson, B. M. "Resource Allocation to Agricultural research in Kenya from 1963 to 1978." Ph.D. diss., University of Toronto, 1981.

Jain, H. K. *Organization and Structure in National Agricultural Research Systems.* ISNAR Working Paper No. 21. The Hague: ISNAR, 1989.

Jalan, B. M., ed. *Problems and Politics in Small Countries.* New York: St. Martin's Press, 1982.

Janvry, A. de, and E. Sadoulet. "Growth and Equity in Agriculture-Led Growth." Paper presented at the Meeting of the International Economics Association, New Delhi, India, December 1986a.

Janvry, A. de, and E. Sadoulet. "The Conditions for Harmony between Third World Agricultural Development and U.S. Farm Exports." *American Journal of Agricultural Economics* Vol. 68, No. 5 (December 1986b): 1340-1346.

Janvry, A. de, and E. Sadoulet. "The Conditions for Compatibility between Aid and Trade in Agriculture." *Economic Development and Cultural Change* Vol. 37, No. 1 (October 1988): 1-32.

Janvry, A. de, and E. Sadoulet. *The Political Feasibility of Rural Poverty Reduction.* Giannini Foundation Paper. Berkeley: University of California, 1989.

Janvry, A. de, E. Sadoulet, and M. Fafchamps. *Agrarian Structure, Technical Innovations and the State.* Giannini Foundation Paper. Berkeley: University of California, 1987.

Jeffries, C. *A Review of Colonial Research 1940-1960.* London: Her Majesty's Stationery Office, 1964.

Johnson, D. G. "Agricultural Research Policy in Small Developing Countries." In *Managing Renewable Natural Resources in Developing Countries,* edited by C. W. Howe. Boulder: Westview Press, 1982.

Jorgenson, D. W. "The Development of a Dual Economy." *Economic Journal* Vol. 71 (June 1961): 309-334.

Jorgenson, D. W., and L. Lau. "An Economic Theory of Agricultural Household Behaviour." Paper presented to the 4th Far Eastern Meeting of the Econometric Society, Tokyo, Japan, 1969.

Judd, M. A., J. K. Boyce, and R. E. Evenson. *Investing in Agricultural Supply.* Discussion Paper No. 442. New Haven: Economic Growth Center, Yale University, June 1983.

Judd, M. A., J. K. Boyce, and R. E. Evenson. "Investing in Agricultural Supply: The Determinants of Agricultural Research and Extension Investment." *Economic Development and Cultural Change* Vol. 35, No. 1 (October 1986): 77-113.

Judge, G. G., R. C. Hill, W. E. Griffiths, H. Lütkepohl, and T. C. Lee. *Introduction to the Theory and Practice of Econometrics.* New York: John Wiley & Sons, 1982.

Jugenheimer, R. W. *Corn: Improvement, Seed Production and Uses.* New York: John Wiley & Sons, 1976.

Just, R. E., and R. D. Pope. "Stochastic Specification of Production Functions and Economic Implications." *Journal of Econometrics* Vol. 7, No. 1 (1978): 67-86.

Kamien, M. I., and N. L. Schwartz. *Market Structure and Innovation.* Cambridge, UK: Cambridge University Press, 1982.

Kellogg, E., R. Kodl, and P. Garcia. "The Effects of Agricultural Growth on Agricultural Imports." *American Journal of Agricultural Economics* Vol. 68, No. 5 (December 1986): 1347-1352.

Kerin, J., and P. Cook. *Research, Innovation & Competitiveness: Policies for Reshaping Australia's Primary Industries and Energy Portfolio Research and Development.* Canberra: Australian Government Printing Office, 1989.

Khaldi, N. *Evolving Food Gaps in the Middle East/North Africa: Prospects and Policy Implications.* IFPRI Research Paper No. 47. Washington, D.C.: IFPRI, December 1984.

Khamis, S. H. "Suggested Methods for Consistent Temporal-Spatial Comparisons." In *World Comparison of Incomes, Prices and Products,* edited by J. Salazar-Carillo and D. S. Prasada Rao. Amsterdam: North-Holland, 1988.

Kim, I. H. "Agricultural Research in Korea." In *National Agricultural Research Systems in Asia,* edited by A. H. Moseman. New York: Agricultural Development Council, 1971.

King, R. P., and L. J. Robison. "An Interval Approach to Measuring Decision Maker Preferences." *American Journal of Agricultural Economics* Vol. 63, No. 3 (1981): 510-520.

Kislev, Y., and W. Peterson. "Prices, Technology, and Farm Size." *Journal of Political Economy* Vol. 90, No. 3 (1982): 578-595.

Knoblauch, H. C., E. M. Law, W. P. Meyer, B. F. Beacher, R. B. Nestler, and B. S. White, Jr. *State Agricultural Experiment Stations — A History of Research Policy and Procedure.* USDA Miscellaneous Publication No. 904. Washington, D.C.: US Government Printing Office, 1962.

Knudson, M. K. "The Role of the Public Sector in Applied Breeding R & D: The Case of Wheat in the USA." *Food Policy* Vol. 15, No. 3 (1990): 209-217.

Kravis, I. B. "The Three Faces of the International Comparison Project." *World Bank Research Observer* Vol. 1, No. 1 (January 1986): 3-26.

Kravis, I. B., A. Heston, and R. Summers. *International Comparisons of Real Product and Purchasing Power.* Baltimore: Johns Hopkins University Press, 1978.

Kravis, I. B., A. Heston, and R. Summers. *World Product and Income: International Comparisons of Real Gross Product.* Baltimore: Johns Hopkins University Press, 1982.

Kravis, I. B., Z. Kenessey, A. Heston, and R. Summers. *A System of International Comparisons of Gross Product and Purchasing Power.* Baltimore: Johns Hopkins University Press, 1975.

Krueger, A. O. "The Political Economy of the Rent-Seeking Society." *American Economic Review* Vol. 64, No. 3 (1974): 291-303.

Krueger, A. O. *Foreign Trade Regimes and Economic Development: Liberalization Attempts and Consequences.* Cambridge, Mass.: Ballinger Press, 1978.

Krueger, A. O. *Exchange Rate Determination.* New York: Cambridge University Press, 1985.

Krueger, A. O. "Aid in the Development Process." *World Bank Research Observer* Vol. 1, No. 1 (1986): 57-78.

Krueger, A. O. "Government Failures in Development." *Journal of Economic Perspectives* Vol. 4, No. 3 (Summer 1990): 9-23.

Krueger, A. O., M. Schiff, and A. Valdés. "Agricultural Incentives in Developing Countries: Measuring the Effects of Sectoral and Economywide Policies." *World Bank Economic Review* Vol. 2, No. 3 (1988): 255-271.

Kumar, N. "Technology Imports and Local Research and Development in Indian Manufacturing." *The Developing Economies* Vol. 25, No. 3 (September 1987): 220-233.

Kuznets, S. "Economic Growth and the Contribution of Agriculture: Notes on Measurement." *International Journal of Agrarian Affairs* Vol. 3 (April 1961): 55-75.

Kwagoe, T., and J. Hayami. "An Intercountry Comparison of Agricultural Production Efficiency." *American Journal of Agricultural Economics* Vol. 67, No. 1 (February 1985): 87-92.

Lal, R. "Managing the Soils of Sub-Saharan Africa." *Science* Vol. 236 (May 1987): 1069-1076.

Lee, B. M. S., and A. Bui-lan. "Use of Errors of Prediction in Improving Forecast Accuracy: An Application to Wool in Australia." *Australian Journal of Agricultural Economics* Vol. 26, No. 1 (1982): 49-62.

Lee, J., and M. Shane. *U.S. Agricultural Interests and Growth in Developing Countries: The Critical Linkage.* Washington, D.C.: USDA, ERS, June 1985.

Lele, U. J., B. H. Kinsey, and A. O. Obeya. "Building Agricultural Research Capacity in Africa: Policy Lessons for the MADIA Countries." Paper prepared for the joint TAC/CGIAR Center Directors Meeting, Rome, Italy, June 1989.

Lesser, W. H., and R. T. Masson. *An Economic Analysis of the Plant Variety Protection Act.* Washington, D.C.: American Seed Trade Association, 1983.

Levich, R. M. "Empirical Studies of Exchange Rates: Price Behaviour, Rate Determination and Market Efficiency." In *Handbook of International Economics*, edited by R. W. Jones and P. B. Kenan. Amsterdam: North Holland, 1985.

Levin, R. C., and P. C. Reiss. "Tests of a Schumpeterian Model of R&D and Market Structure." In *R&D, Patents, and Productivity*, edited by Z. Griliches. Chicago: University of Chicago Press, 1984.

Levin, R. C., W. M. Cohen, and D. C. Mowery. "Appropriability, Opportunity, and Market Structure: New Evidence on Some Schumpeterian Hypotheses." *American Economic Review* Vol. 75, No. 2 (1985): 204.

Lewis, J. P. *External Funding of Development-Related Research: A Survey of Some Major Donors.* IDRC Manuscript Report 160e. Ottawa: IDRC, September 1987.

Lewis, W. A. "Economic Development with Unlimited Supplies of Labor." *Manchester School of Economic and Social Studies* Vol. 22 (May 1954): 139-191.

Lin, J. Y. "Public Research Resource Allocation in Chinese Agriculture: A Test of Induced Technological Innovation Hypotheses." *Economic Development and Cultural Change* (forthcoming).

Lindner, R. K. "Adoption as a Decision Theoretic Process." Ph.D. diss., University of Minnesota, St. Paul, 1981.

Lindner, R. K., and A. J. Fischer. *Risk Aversion, Information Quality and the Innovation Adoption Time Lag.* Economics Department Working Paper 81-17. Adelaide: University of Adelaide, 1980.

Lindner, R. K., and F. G. Jarrett. "Supply Shifts and the Size of Research Benefits." *American Journal of Agricultural Economics* Vol. 60, No. 1 (1978): 48-58.

Lindner, R. K., and F. G. Jarrett. "Supply Shifts and the Size of Research Benefits: Reply." *American Journal of Agricultural Economics* Vol. 62, No. 4 (1980): 841-844.

Lindner, R. K., P. G. Pardey, and F. G. Jarrett. "Distance to Information Source and the Time Lag to Early Adoption of Trace Element Fertilizers." *Australian Journal of Agricultural Economics* Vol. 26, No. 2 (1982): 98-113.

Lipton, M. L. *Why Poor People Stay Poor.* London: Temple Smith, 1977.

Lipton, M. L. "The Place of Agricultural Research in the Development of Sub-Saharan Africa." *World Development* Vol. 16, No. 10 (October 1988): 1231-1257.

Lipton, M. L., with R. Longhurst. *New Seeds and Poor People.* Baltimore: Johns Hopkins University Press, 1989.

Long, J. B. de. "Productivity Growth, Convergence, and Welfare: Comment." *American Economic Review* Vol. 78, No. 5 (December 1988): 1138-1154.

Lu, L. *Current Status and Policy of Agricultural Services and Technology Development in the People's Republic of China.* A Consultancy Report prepared for FAO. Bangkok: RAPA, 1985.

Lucas, R. E. "Adjustment Cost and the Theory of Supply." *Journal of Political Economy* Vol. 75, Part 1 (1967): 321-334.

Lundgren, A. L., L. S. Hamilton, and N. Vergara. *Strategies for Improving the Effectiveness of Asia-Pacific Forestry Research for Sustainable Development.* Honolulu: Environment and Policy Institute, East-West Center, 1986

Lynam, J., and R. Herdt. "Sense and Sustainability: Sustainability as an Objective in International Agricultural Research." *Agricultural Economics* Vol. 3, No. 4 (1989): 381-398.

MacDonald, A. S. "Exchange Rates for National Expenditure on Research and Development." *Economic Journal* Vol. 83 (June 1973): 477-494.

MacLaren, D. "Agricultural Policy Uncertainty and the Risk Averse Firm." *European Review of Agricultural Economics* Vol. 74, No. 4 (1980): 395-411.

MacLaren, D. "The Output Response of the Risk Averse Firm: Some Comparative Statics for Agricultural Policy." *Journal of Agricultural Economics* Vol. 34, No. 1 (1983): 45-56.

Maddock, N. "Privatizing Agriculture: Policy Options in Developing Countries." *Food Policy* Vol. 12 (1987): 295-298.

Magrath, W. *The Challenge of the Commons: The Allocation of Nonexclusive Resources.* Environment Department Working Paper No. 14. Washington, D.C.: World Bank, 1989.

Mangundojo, S. "Agricultural Research in Indonesia." In *National Agricultural Research Systems in Asia*, edited by A. H. Moseman. New York: Agricultural Development Council, 1971.

Mansfield, E., J. Rapoport, J. Schnee, S. Wagner, and M. Hamburger. *Research and Innovation in the Modern Corporation.* New York: Norton, 1971.

Mansfield, E. "Social and Private Rates of Return from Industrial Innovations." *Quarterly Journal of Economics* Vol. 91 (1977): 221-240.

Mansfield, E. "R&D and Innovation: Some Empirical Findings." In *R&D Patents and Productivity*, edited by Z. Griliches. Chicago: The University of Chicago Press, 1984.

Mansfield, E. "Price Indexes for R&D Inputs, 1969-1983." *Management Science* Vol. 33, No. 1 (January 1987): 124-129.

Mansfield, E., A. Romeo, and L. Switzer. "R&D Price Indexes and Real R&D Expenditures in the United States." *Research Policy* Vol. 12 (1983):

105-112.

Marcano, L. "Latin America & Caribbean." In *The Role of International Associations in Strengthening National Agricultural Research*. The Hague: ISNAR, 1982.

Markandya, A., and D. Pearce. *Environmental Considerations and the Choice of Discount Rate in Developing Countries*. Environment Department Working Paper No. 3. Washington, D.C.: World Bank, 1988.

Marsden, J. S., G. E. Martin, D. J. Parham, T. J. Ridsdill Smith, and B. G. Johnson. *Returns on Australian Agricultural Research*. Canberra: CSIRO, 1980.

Martínez, A., and R. B. Diamond. *Fertilizer Use Statistics in Crop Production*. Technical Bulletin No. 24. Muscle Shoals: International Fertilizer Development Center, 1982.

Masefield, G. B. *A History of the Colonial Agricultural Service*. Oxford: Clarendon Press, 1972.

McCalla, A. F. *A Possible Expansion of the CGIAR: A Draft Outline of Possible Approaches for TAC and the CGIAR*. Rome: TAC Secretariat, FAO, 1988.

McKelvey, J. J. "Agricultural Research." In *The African World: A Survey of Social Research*, edited by R. A. Lystad. London: Pall Mall Press, 1965.

McLean, I. W. "Economic Wellbeing." In *The Australian Economy in the Long Run*, edited by R. Maddock and I. W. McLean. New York: Cambridge University Press, 1987.

McMullen, N. *Seeds and World Agricultural Progress*. Washington, D.C.: National Planning Association, 1987.

Meadows, D., et al. *The Limits to Growth*. New York: Universe Books, 1972.

Mellor, J. W. "Agricultural Development: Opportunities for the 1990's." Paper presented at International Centres Week, IFPRI, Washington, D.C., 1988.

Menon, K. P. A. "Building Agricultural Research Organisations — The Indian Experience." In *National Agricultural Research Systems in Asia*, edited by A. H. Moseman. New York: Agricultural Development Council, 1971.

Menz, K. M., and P. G. Pardey. "Technology and US Corn Yields: Plateaus and Price Responsiveness." *American Journal of Agricultural Economics Vol. 62, No. 3 (1983): 558-562.*

Mergen, F., R. E. Evenson, M. A. Judd, and J. Putnam. *Forestry Research: A Provisional Global Inventory*. Discussion Paper No. 503. New Haven: Economic Growth Center, Yale University, May 1986.

Mergen, F., R. E. Evenson, M. A. Judd, and J. Putnam. "Forestry Research: A Provisional Global Inventory." *Economic Development and Cultural Change* Vol. 37, No. 1 (October 1988): 149-171.

Meyer, J. "Second Degree Stochastic Dominance with Respect to a Function." *International Economic Review* Vol. 18, No. 2 (1977): 477-487.

Mikkelsen, K. W. "Inventive Activity in Philippine Industry." Ph.D. diss., Yale University, New Haven, 1984.

Miranowski, J. A., and G. A. Carlson. "Economic Issues in Public and Private Approaches to Preserving Pest Susceptibility." In *Pesticide Resistance: Strategies and Tactics for Management*. Washington, D.C.: National Academy of Sciences, National Academy Press, 1986.

Mitra, P. "A Description of Adjustment to External Shocks: Country Groups." In *Stagflation, Savings, and the State of Perspectives on the Global Economy*, edited by D. Lal and M. Wolf. New York: Oxford University Press, 1986.

Mundlak, Y. *The Aggregate Agricultural Supply*. Working Paper No. 85-11. Rehovot, Israel: Center for Agricultural Economic Research, 1985.

Nankani, G. T. *The Intercountry Distribution of Direct Foreign Investment in Manufacturing*. New York: Garland, 1979.

National Research Council. *Alternative Agriculture*. Report by the Committee on the Role of Alternative Farming Methods on Modern Production Agriculture, Board on Agriculture. Washington, D.C.: National Academy Press, 1989.

Negishi, T. "Welfare Economics and the Existence of an Equilibrium for a Competitive Economy." *Metroeconomica* Vol. 12 (1960): 92-97.

Nelson, R. R. "Uncertainty, Learning and the Economics of Parallel Research and Development Efforts." *Review of Economics and Statistics* Vol. 43, No. 4 (1961): 351-364.

Newbery, D. M. G., and J. E. Stiglitz. *The Theory of Commodity Price Stabilization — A Study in the Economics of Risk*. Oxford: Clarendon, 1981.

Nguyen, D. "On Agricultural Productivity Differences among Countries." *American Journal of Agricultural Economics* Vol. 61, No. 3 (August

1979): 565-570.

Norton, G. W., and J. S. Davis. "Evaluating Returns to Agricultural Research: A Review." *American Journal of Agricultural Economics* Vol. 63, No. 4 (1981): 685-699.

Norton, G. W., J. Ortiz, and P. G. Pardey. "The Impact of Foreign Assistance on Agricultural Growth." *Economic Development and Cultural Change* (forthcoming).

Norton, G. W., and P. G. Pardey. *Priority-Setting Mechanisms for National Agricultural Research Systems: Present Experience and Future Needs.* ISNAR Working Paper No. 7. The Hague: ISNAR, 1987.

NSF. *Experimental Input Price Indexes for Research and Development, Fiscal Years 1961-65.* NSF Report 70-7. Washington, D.C.: National Science Foundation, 1970.

NSF. *Research and Development in Industry: 1987.* NSF Report 89-323. Washington, D.C.: National Science Foundation, 1989.

Nugent, J. B. "Applications of the Theory of Transactions Costs and Collective Action to Development Problems and Policy." Paper presented at the Conference on the Role of Institutions in Economic Development, Cornell University, Ithaca, New York, 1986.

ODA. *British Overseas Aid — Agricultural Research (Crop and Soil Sciences) 1974-1978.* Overseas Research Publication No. 26. London: Her Majesty's Stationery Office, 1979.

ODA. *Report on Research and Development.* London: Overseas Development Administration, multiple years (1980 to 1987).

OECD. *The Instability of Agricultural Commodity Markets.* Paris: OECD, 1980.

OECD. *The Measurement of Scientific and Technical Activities — "Frascati Manual" 1980.* Paris: OECD, 1981.

OECD. *Development Cooperation in the 1990s — Efforts and Policies of the Members of the Development Assistance Committee.* Paris: OECD, December 1989.

OECD. *OECD Development Cooperation.* Paris: OECD, multiple years (1977, 80, 83, 85, and 87).

Oehmke, J. F. "Persistent Underinvestment in Public Agricultural Research." *Agricultural Economics* Vol. 1, No. 1 (1986): 53-65.

Olson, M. *The Rise and Decline of Nations.* New Haven: Yale University Press, 1982.

Oram, P. A., and V. Bindlish. *Resource Allocation to National Agricultural Research: Trends in the 1970s — A Review of Third World Systems.* The Hague: ISNAR, 1981.

ORSTOM. *Rapport d'Activité.* Paris: ORSTOM, 1986.

Ossa, C. "The Diversity in Growth Rates among Developing Countries." *Development Policy Review* Vol. 8 (1990): 115-129.

Otto, D., and J. Havlicek. "Some Preliminary Results of Estimating the Impacts of Research Investment on Corn, Wheat and Sorghum." In *Evaluation of Agricultural Research,* Miscellaneous Publication No. 8, edited by G. W. Norton, W. L. Fishel, A. A. Paulsen, and W. B. Sundquist. Minneapolis: Minnesota Agricultural Experiment Station, April 1981.

Panzar, J. C., and R. D. Willig. "Economies of Scope." *American Economic Review* Vol. 72, No. 2 (May 1981): 268-272.

Papadakis, J. *Climates of the World and Their Agricultural Potentialities.* Buenos Aires: J. Papadakis, 1966.

Parasram, S. *Centre of Excellence Based in Trinidad, West Indies — A Historical View.* ISNAR Staff Notes No. 90-85. The Hague: ISNAR, April 1990.

Pardey, P. G. "Public Sector Production of Agricultural Knowledge." Ph.D. diss., University of Minnesota, St. Paul, June 1986.

Pardey, P. G. "The Agricultural Knowledge Production Function: An Empirical Look." *Review of Economics and Statistics* Vol. 81, No. 3 (1989): 453-461.

Pardey, P. G., and B. J. Craig. "Causal Relationships between Public Sector Agricultural Research Expenditures and Output." *American Journal of Agricultural Economics* Vol. 71, No. 1 (February 1989): 9-19.

Pardey, P. G., and J. Roseboom. *ISNAR Agricultural Research Indicator Series: A Global Data Base on National Agricultural Research Systems.* Cambridge, U.K.: Cambridge University Press, 1989a.

Pardey, P. G., and J. Roseboom. "A Global Evaluation of National Agricultural Research Investments: 1960-1985." In *The Changing Dynamics of Global Agriculture*, edited by E. Javier and U. Renborg. The Hague: ISNAR, 1989b.

Pardey, P. G., B. J. Craig, and M. L. Hallaway. "US Agricultural Research Deflators: 1890-1985." *Research Policy* Vol. 18, No. 5 (October 1989):

289-296.

Pardey, P. G., W. M. Eveleens, and M. L. Hallaway. *A Statistical History of US Agricultural Research: 1889 to 1986.* St. Paul: Center for International Food and Agricultural Policy, University of Minnesota, forthcoming 1991.

Pardey, P. G., M. S. Kang, and H. Elliott. "The Structure of Public Support for National Agricultural Research: A Political Economy Perspective." *Agricultural Economics* Vol. 3, No. 4 (December 1989): 261-278.

Pardey, P. G., J. Roseboom, and B. J. Craig. "A Yardstick for International Comparisons: An Application to National Agricultural Research Expenditures." *Economic Development and Cultural Change* (forthcoming).

Parton, K. A., J. R. Anderson, and J. P. Makeham. *Evaluation Methods of Australian Wool Production Research.* Agricultural Economics Bulletin No. 29. Armidale: University of New England, 1984.

Pasour, E. C., Jr., and M. A. Johnson. "Bureaucratic Productivity: The Case of Agricultural Research Revisited." *Public Choice* Vol. 39, No. 2 (1982): 301-317.

Perrin, R. K., K. A. Hunnings, and L. A. Ihnen. *Some Effects of the U.S. Plant Variety Protection Act of 1970.* Economic Research Report No. 46. Raleigh: Department of Economics and Business, North Carolina State University, 1983.

Perrings, C. "Environmental Bonds and Environmental Research in Innovative Activities." *Ecological Economics* Vol. 1, No. 1 (1989): 95-110.

Persley, G. J., ed. *Agricultural Biotechnology: Opportunities for International Development.* Oxon, UK: CAB International, 1990.

Peskin, H. *Accounting for Natural Resource Degradation in Developing Countries.* Environmental Department Working Paper No. 13. Washington, D.C.: World Bank, 1989.

Peterson, W. L. "A Note on the Social Returns to Private Research and Development." *American Journal of Agricultural Economics* Vol. 58 (1976): 324-326.

Peterson, W. L. "International Farm Prices and the Social Cost of Cheap Food Policies." *American Journal of Agricultural Economics* Vol. 61, No. 1 (February 1979): 12-21.

Peterson, W. L. *Land Quality and Prices.* Staff Paper Series P84-29. St. Paul: Department of Agricul-tural and Applied Economics, University of Minnesota, December 1984.

Peterson, W. L. *International Land Quality Indexes.* Staff Paper Series P87-19. St. Paul: Department of Agricultural and Applied Economics, University of Minnesota, 1987.

Peterson, W. L. "International Supply Response." *Agricultural Economics* Vol. 2 (1988): 365-374.

Peterson, W. L. "Rates of Return on Capital: An International Comparison." *Kyklos* Vol. 42 (1989): 203-217.

Peterson, W. L., and Y. Hayami. "Technical Change in Agriculture." In *A Survey of Agricultural Economics Literature*, edited by L. R. Martin. Minneapolis: University of Minnesota Press, 1977.

Pezzy, J. *Economic Analysis of Sustainable Growth and Sustainable Development.* Environment Department Working Paper No. 15. Washington, D.C.: World Bank, 1989.

Piñeiro, M. *Agricultural Research in the Private Sector: Issues on Analytical Perspectives.* PROAGRO Paper No. 1. The Hague: ISNAR, 1985.

Piñeiro, M. *The Development of the Private Sector in Agricultural Research: Implications for Public Research Institutions.* PROAGRO Paper No. 10. The Hague: ISNAR, 1986.

Piñeiro, M., and E. Trigo. *Latin American Agricultural Research — The Public Sector: Problems and Perspectives.* ISNAR Working Paper No. 1. The Hague: ISNAR, January 1985.

Pingali, P., P. Moya, and L. Velasco. "The Post-Green Revolution Blues in Asian Rice Production." IRRI, Manila, 1990. Mimeo.

Pinstrup-Andersen, P. *Agricultural Research and Technology in Economic Development.* New York: Longman, 1982.

Pinstrup-Andersen, P. "Changing Patterns of Consumption underlying Changes in Trade and Agricultural Development." Paper presented at the International Trade Research Consortium, CIMMYT, El Batan, Mexico, 14-18 December 1986.

Pinstrup-Andersen, P. "Food Security and Structural Adjustment." In *Trade, Aid and Policy Reform*, edited by C. Roberts. Washington, D.C.: World Bank, 1988a.

Pinstrup-Andersen, P. "Macroeconomic Adjustment and Human Nutrition." *Food Policy* Vol. 13, No. 1 (February 1988b): 37-46.

Pinstrup-Andersen, P., and P. B. R. Hazell. "The Impact of the Green Revolution and Prospects for the Future." *Food Review International* Vol. 1, No. 1 (1985): 1-25

Pinstrup-Andersen, P., N. Ruiz de Londoño, and E. Hoover. "The Impact of Increasing Food Supply on Human Nutrition: Implications for Commodity Priorities in Agricultural Research and Policy." *American Journal of Agricultural Economics* Vol. 58, No. 1 (1976): 131-142.

Plucknett, D. L., and N. J. H. Smith. "Sustaining Agricultural Yields." *Bioscience* Vol. 36, No. 1 (1986): 40-45.

Plucknett, D. L., N. J. H. Smith, and S. Ozgediz. *Networking in International Agricultural Research.* Ithaca: Cornell University Press, 1990.

Porter, R. "The New Approach to Wilderness Preservation through Benefit-Cost Analysis." *Journal of Environmental Economics and Management* Vol. 9 (1982): 59-80.

Pray, C. E. *Agricultural Research and Technology Transfer by the Private Sector in the Philippines.* Economic Development Center Report No. 2. Minneapolis: University of Minnesota, 1986.

Pray, C. E. "Private Sector Research and Technology Transfer in Asia." In *Policy for Agricultural Research,* edited by V. W. Ruttan and C. E. Pray. Boulder: Westview Press, 1987.

Pray, C. E., and R. G. Echeverría. "Transferring Hybrid Maize Technology: The Role of the Private Sector." *Food Policy* Vol. 13, No. 4 (1988): 366-374.

Pray, C. E., and R. G. Echeverría. "Private Sector Agricultural Research and Technology Transfer Links in Developing Countries." In *Making the Link: Agricultural Research and Technology Transfer in Developing Countries,* edited by D. Kaimowitz. Boulder: Westview Press, 1989a.

Pray, C. E., and R. G. Echeverría. Unpublished interviews with private seed company officials. 1989b.

Pray, C. E., and C. F. Neumeyer. "Impact of Technology Policy on Research in the Agricultural Input Industries." Paper presented to the American Agricultural Economics Association Meetings, Baton Rouge, Louisiana, 1989.

Pray, C. E., and V. W. Ruttan. *Complete Report of the Asian Agricultural Research Project.* Economic Development Center Bulletin Number 85-2. St. Paul: University of Minnesota, April 1985.

Pray, C. E., S. Ribeiro, R. A. E. Mueller, and P. P. Rao. *Private Research and Public Benefit: The Private Seed Industry for Sorghum and Pearl Millet in India.* Resource Management Program, Economics Group Progress Report 89. India: ICRISAT, 1989.

President's Material Policy Commission. *Resources for Freedom.* Washington, D.C.: US Government Printing Office, 1952.

President's Water Resources Policy Commission. *A Water Policy for the American People.* Vol 1. Washington, D.C.: US Government Printing Office, 1950.

Prescott, E. C., and M. Visscher. "Organization Capital." *Journal of Political Economy,* Vol. 88 (1980): 446-461.

Pritchard, A. J. *Lending by the World Bank for Agricultural Research: A Review of the Years 1981 through 1987.* World Bank Technical Paper No. 118. Washington, D.C.: World Bank, 1990.

Pryor, F. L. "A Quasi-Test of Mancur Olson's Theory." In *The Political Economy of Growth,* edited by D. C. Muller. New Haven: Yale University Press, 1983.

Pryor, F. L. "Rent Seeking and the Growth and Fluctuations of Nations: Empirical Tests of Some Recent Hypotheses." In *Neoclassical Political Economy: The Analysis of Rent-Seeking and DUP Activities,* edited by D. Colander. Cambridge, Mass.: Ballinger Publishing Company, 1984.

Quirino, T. R. *Researcher's Profiles and Activities at Kenya's National Agricultural Research Center: An Experimental Application of ARIS.* ISNAR Staff Note 89-61. The Hague: ISNAR, 1989.

Ragoff, M. H., and S. L. Rawlins. "Food Security: A Technological Alternative." *BioScience* Vol. 37 (December 1987): 800-807.

Rausser, G. C., A. de Janvry, A. Schmitz, and D. Zilberman. "Principal Issues in the Evaluation of Public Research in Agriculture." In *Evaluation of Agricultural Research,* edited by G. W. Norton, W. L. Fishel, A. A. Paulsen, and W. B. Sundquist. Minneapolis: Minnesota Agricultural Experiment Station, 1981.

Rayner, S. "Nature Myths and Policy Design." In *Resource and Environmental Constraints on Sustainable Growth in Agricultural Production,* edited by V. W. Ruttan. Staff Paper Series P90-33. St .Paul: University of Minnesota, May 1990.

Reilly, J., and R. Bucklin. "Climate Change and Agriculture." In *World Agricultural Situation and Outlook Report*. Washington, D.C.: USDA, ERS, 1989.

Ribeiro, S. A. "Private Research, Social Benefits and Public Policy: The Case of Hybrid Sorghum and Pearl Millet in the Indian Seed Industry." M.S. thesis, Department of Agricultural Economics, Rutgers University, New Brunswick, 1989.

Roe, T. L., and T. Graham-Tomasi. *Competition among Rent Seeking Groups in General Equilibrium*. Economic Development Center Bulletin No. 90-2. St. Paul: Department of Agricultural and Applied Economics, University of Minnesota, September 1990.

Roe, T. L., and H. von Witzke. "Economic Growth and Distributive Justice: Questions of Market Failure and Collective Action in a Global Economy." Paper presented to the Center for International Food and Agricultural Policy, University of Minnesota, Minneapolis, November 1989.

Roe, T. L., and E. Yeldan. *An Open Economy Model of Political Influence and Competition among Rent Seeking Groups*. Economic Development Center Bulletin No. 88-1. St. Paul: Department of Agricultural and Applied Economics, University of Minnesota, February 1988.

Rose, R. "Supply Shifts and Research Benefits: Comment." *American Journal of Agricultural Economics* Vol. 62, No. 4 (1980): 834-837.

Russell, E. J. *A History of Agricultural Science in Great Britain 1620-1954*. London: George Allen & Unwin Ltd, 1966.

Ruttan, V. W. "Technology and the Environment." *American Journal of Agricultural Economics* Vol. 53 (December 1971): 707-717.

Ruttan, V. W. "Bureaucratic Productivity: The Case of Agricultural Research." *Public Choice* Vol. 35 (1980): 529-547.

Ruttan, V. W. *Agricultural Research Policy*. Minneapolis: University of Minnesota Press, 1982.

Ruttan, V. W. "Toward a Global Agricultural Research System." In *Policy for Agricultural Research*, edited by V. W. Ruttan and C. E. Pray. Boulder: Westview Press, 1987.

Ruttan, V. W. "Sustainability Is Not Enough." *American Journal of Alternative Agriculture Vol. 3 (1988): 128-130.*

Ruttan, V. W., ed. *Biological and Technical Constraints on Crop and Animal Productivity: Report on a Dialogue*. St. Paul: Department of Agricultural and Applied Economics, University of Minnesota, December 1989a.

Ruttan, V. W. "Why Foreign Assistance?" *Economic Development and Cultural Change* Vol. 37, No. 2 (January 1989b): 411-424

Ruttan, V. W. "Constraints on Agricultural Production in Asia: Into the 21st Century." In *CIMMYT 1989 Annual Report*. Mexico, D.F.: CIMMYT, 1990.

Ruttan, V. W., ed. *Resource and Environmental Constraints on Sustainable Growth in Agricultural Production*. St. Paul: Department of Agricultural and Applied Economics, University of Minnesota, forthcoming.

Ruttan, V. W., and C. E. Pray, eds. *Policy for Agricultural Research*. Boulder: Westview Press, 1987.

Ryan, J. G., and J. S. Davis. *A Decision Support System to Assist Agricultural Research Priority Setting: Experience at ACIAR and Possible Adaptations for the TAC/CGIAR*. ACIAR/ISNAR Project Paper No. 17. Canberra: ACIAR/ISNAR, March 1990.

Salmon, S. C., and A. A. Hanson. *The Principles and Practice of Agricultural Research*. London: Leonard Hill, 1964.

Samper, A. "National Systems of Agricultural Research in Latin America." In *Organizational Trends in National Research Management: A Global Overview*, edited by P. D. Pages and J. D. Drilon, Jr. (no publisher indicated), 1980.

Sandrey, R., and R. Reynolds, eds. *Farming without Subsidies: New Zealand's Recent Experience*. Wellington: GP Books, 1990.

Sardar, Z. *Science and Technology in the Middle East*. London: Longman, 1982.

Sarles, M. "USAID Experiments with the Private Sector in Agricultural Research in Latin America." In *Methods for Diagnosing Research Systems Constraints and for Assessing Research Impact — Volume I: Diagnosing Research Systems Constraints*, edited by R. G. Echeverría. The Hague: ISNAR, 1990.

Scandizzo, P. L. *Agricultural Growth and Factor Productivity in Developing Countries*. FAO Economic and Social Development Paper 42. Rome: FAO, 1984.

Scandizzo, P. L., P. B. R. Hazell, and J. R. Anderson. *Risky Agricultural Markets: Price Forecasting and the Need for Intervention Policies*. Boulder:

Methods for Diagnosing Research Systems Constraints and for Assessing Research Impact — Volume I: Diagnosing Research Systems Constraints, edited by R. G. Echeverría. The Hague: ISNAR, 1990.

Scandizzo, P. L. *Agricultural Growth and Factor Productivity in Developing Countries*. FAO Economic and Social Development Paper 42. Rome: FAO, 1984.

Scandizzo, P. L., P. B. R. Hazell, and J. R. Anderson. *Risky Agricultural Markets: Price Forecasting and the Need for Intervention Policies*. Boulder: Westview Press, 1984.

Scherer, F. M. *Industrial Market Structure and Economic Performance*. Boston: Houghton Mifflin Company, 1980.

Schiff, M. "A Structural View of Policy Issues in African Agricultural Development: Comment." *American Journal of Agricultural Economics* Vol. 69, No. 2 (May 1987): 389-391.

Schmitt, G. H. "What Do Agricultural Income and Productivity Measurements Really Mean." *Agricultural Economics* Vol. 2 (1988): 139-157.

Schmookler, J. *Invention and Economic Growth*. Cambridge, Mass.: Harvard University Press, 1966.

Schuh, G. E. "The Role of Markets and Governments in the World Food Economy." In *The Role of Markets in the World Food Economy*, edited by D. G. Johnson and G. E. Schuh. Boulder: Westview Press, 1983.

Schuh, G. E. "Accomplishments and Issues: Policy Environment Issues." Paper presented at the 20th Anniversary of CIMMYT, El Batan, Mexico, September 1986.

Schultz, T. W. *Transforming Traditional Agriculture*. Chicago: University of Chicago Press, 1964.

Schultz, T. W. "The Value of the Ability to Deal with Disequilibria." *Journal of Economic Literature* Vol. 13 (September 1975): 822-846.

Schumpeter, J. A. *Capitalism, Socialism, and Democracy*. Third edition. New York: Harper and Row, 1950.

Scobie, G. M. *Food Subsidies in Egypt: Their Impact on Foreign Exchange and Trade*. IFPRI Research Report 40. Washington, D.C.: IFPRI, 1983.

Scobie, G. M. *Partners in Research: The CGIAR in Latin America*. CGIAR Study Paper No. 24. Washington, D.C.: World Bank, December 1987.

Scobie, G.M., and W.M. Eveleens. *The Return to Investment in Agricultural Research in New Zealand: 1926-27 to 1983-84*. MAF Economics Division Research Report 1/87. Hamilton: Ruakura Agriculture Centre, October 1987.

Sehgal, S. M. "Private Sector International Agricultural Research: The Genetic Supply Industry." In *Resource Allocation and Productivity in National and International Agricultural Research*, edited by T. H. Arndt, D. G. Dalrymple, and V. W. Ruttan. Minneapolis: University of Minnesota Press, 1977.

Sen, A. "Isolation, Assurance, and Rate of Discount." *Quarterly Journal of Economics* Vol. 81 (1967): 112-124.

Senanayake, Y. D. A. *Overview of the Organization and Structure of National Agricultural Research Systems in Asia*. ISNAR Working Paper No. 32. The Hague: ISNAR, July 1990.

Simmonds, N. W. *The State of the Art of Farming Systems Research: A Review*. World Bank Technical Paper No. 43. Washington, D.C.: World Bank, 1985.

Spencer, D. S. C. "Agricultural Research: Lessons of the Past, Strategies for the Future." In *Strategies for African Development*, edited by R. J. Berg and J. S. Whitaker. Davis: University of California Press, 1986.

Sprague, G. F. "The Changing Role of the Private and Public Sectors in Corn Breeding." In *Proceedings of the 35th Annual Corn and Sorghum Industry Research Conference*. Washington, D.C.: American Seed Trade Association, 1980.

Sprow, F. B. "Evaluation of Research Expenditures Using Triangular Distribution Functions and Monte Carlo Methods." *Industrial and Engineering Chemistry* Vol. 59, No. 7 (1967): 35-38.

Srinivasan, T. N. "Neoclassical Political Economy, the State and Economic Development." *Asian Development Review* Vol. 3 (1985): 38-58.

Srinivasan, T. N. "International Trade and Factor Movements in Development Theory, Policy and Experience." In *Trade and Development: Proceedings of the Winter 1986 Meeting of the International Agricultural Trade Research Consortium*, edited by M. D. Shane. Washington, D.C.: USDA, ERS, Agricultural and Trade Analysis Division, 1986a.

Srinivasan, T. N. "The Costs and Benefits of Being a Small, Remote, Island, Landlocked, or Ministate

Stavis, B. "Agricultural Research and Extension in China." *World Development* Vol. 6 (1978): 631-645.

Stefanou, S. E. "Technical Change, Uncertainty, and Investment." *American Journal of Agricultural Economics* Vol. 69, No. 1 (1987): 158-165.

Stiglitz, J. "Markets, Market Failures and Development." *American Economic Review* Vol. 79, No. 1 (May 1989): 107-139.

Stoop, W. *Towards a Sustainable Agriculture: Some Implications for ISNAR's Activities with NARS and for Agricultural Research Management in General.* ISNAR Staff Notes No. 90-93. The Hague: ISNAR, September 1990.

Summers, R., and A. Heston. "Improved International Comparisons of Real Product and Its Composition, 1950-1980." *Review of Income and Wealth* Vol. 30 (June 1984): 207-262.

Summers, R., and A. Heston. "A New Set of International Comparisons of Real Product and Price Levels Estimates for 130 Countries, 1950-1985." *Review of Income and Wealth* Vol. 34, No. 1 (March 1988): 1-25.

Swallow, B. M., G. W. Norton, T. B. Brumback, and G. R. Buss. *Agricultural Research Depreciation and the Importance of Maintenance Research.* Agricultural Economics Report No. 56. Blacksburg: Virginia Polytechnic Institute and State University, November 1985.

Swindale, L. D. *The Impact of Agricultural Development on the Environment: An IARC Point of View.* Patancheru, India: ICRISAT, 1988.

TAC Secretariat. *Relations with Non-Associated Centers: A Preliminary Desk Analysis of Information Available in the TAC Secretariat.* Rome: TAC Secretariat, FAO, 1988.

TAC Secretariat. *Process and Criteria for Assessment of Non-Associated Centres with Respect to Their Possible Incorporation into the CGIAR.* Progress report by the Chair of TAC. Rome: TAC Secretariat, FAO, 1990a.

TAC Secretariat. *A Possible Expansion of the CGIAR.* Rome: TAC Secretariat, FAO, September 1990b.

TAC/CGIAR. CGIAR Priorities and Future Strategies. Rome: TAC Secretariat, FAO, 1987.

TAC/CGIAR. *Sustainable Agricultural Production: Implications for International Agricultural Research.* FAO Research and Technology Paper No. 4. Rome: TAC Secretariat, FAO, 1989.

Tadvalkar, R. "Funding International Agricultural Research — Trends and Prospects." CGIAR Secretariat, Washington, D.C., 1989. Mimeo.

Tanzi, V. "Fiscal Policy Responses to Exogenous Shocks." *American Economic Review* Vol. 72, No. 2 (May 1986): 88-91.

Tanzi, V. "Quantitative Characteristics of the Tax Systems of Developing Countries." In *The Theory of Taxation for Developing Countries*, edited by D. Newberry and N. Stern. Washington, D.C.: World Bank, 1987.

Terluin, I. J. *Comparison of Real Output, Productivity and Price Levels in Agriculture in the EC.* The Hague: Agricultural Economics Research Institute LEI, August 1990.

Thirtle, C. G., H. S. Beck, P. Palladino, M. Upton, and W. S. Wise. "Agriculture and Food." In *Science and Technology in the UK*, edited by R. Nicholson, C. Cunningham, and P. Gummett. London: Longmans, 1991.

Thorpe, P., and P. G. Pardey. "The Generation and Transfer of Agricultural Knowledge: A Bibliometric Study of a Research Network." *Journal of Information Science* Vol. 16, No. 3 (1990): 183-194.

Timothy, D. H., P. H. Harvey, and C. R. Dowswell. *Development and Spread of Improved Maize Varieties and Hybrids in Developing Countries.* Washington, D.C.: USAID, Bureau for Science and Technology, 1988.

Tisdell, C. *Transaction Costs/Property Rights and Market Failure in Relation to Science and Technology Policy.* Working Paper 86/32. Wellington: New Zealand Institute of Economic Research, 1986.

Tollens, E. "The Economic Returns to Agricultural Research in INEAC 1934-1959, Zaire." Katholieke Universiteit Leuven, Leuven, June 1987. Mimeo.

Treadgold, M. L. "Growth, Structural Change, and Distribution in a Very Small Economy: A Case Study of Norflok Island." *Journal of Developing Areas* Vol. 19, No. 1 (1984): 33-58.

Trigo, E. J. *Agricultural Research Organization in the Developing World: Diversity and Evolution.* ISNAR Working Paper No. 4. The Hague: ISNAR, 1986.

Trigo, E. J. "Private Sector Participation in Agricultural Research and Development: Notes on Issues and Concerns." In *The Changing Dynamics of*

Global Agriculture, edited by E. Javier and U. Renborg. The Hague: ISNAR, 1988.

Trigo, E. J., and M. E. Piñeiro. "Dynamics of Agricultural Research Organization in Latin America." *Food Policy* Vol. 6, No. 1 (1981): 2-10.

Trigo, E. J., and M. E. Piñeiro. "Funding Agricultural Research." In *Selected Issues in Agricultural Research in Latin America*, edited by B. Nestel and E. J. Trigo. The Hague: ISNAR, 1984.

Trigo, E. J., M. E. Piñeiro, and J. F. Sabato. "Technology as a Social Issue: Agricultural Research Organization in Latin America." In *Technical Change and Social Conflict in Agriculture — Latin American Perspectives*, edited by M. E. Piñeiro and E. J. Trigo. Boulder: Westview Press, 1983.

True, A. C. *A History of Agricultural Experimentation and Research in the United States 1607-1925*. USDA Miscellaneous Publication No. 251. Washington, D.C.: U.S. Government Printing Office, 1937.

True, A. C., and D. J. Crosby. *Agricultural Experiment Stations in Foreign Countries*. USDA Office of Experiment Stations Bulletin No. 112. Washington, D.C.: U.S. Government Printing Office, 1902.

Tsiang, S. C. "The Rationale of the Mean-Standard Deviation Analysis, Skewness Preference and the Demand for Money." *American Economic Review* Vol. 62, No. 3 (1972): 354-371.

Tsur, Y., M. Sternberg, and E. Hochman. "Dynamic Modelling of Innovation Process Adoption with Risk Aversion and Learning." *Oxford Economic Papers* Vol. 42 (1990): 336-355.

Ulrich, A., H. Furtan, and A. Schmitz. "Public and Private Returns from Joint Venture Research: An Example from Agriculture." *Quarterly Journal of Economics* Vol. 101 (1986): 103-129.

UN. *The World Food Problem: Proposals for National and International Action*. Provisional Agenda from the United Nations World Food Conference, Rome, 6-16 November 1974.

UN. *National Account Statistics: Main Aggregates and Detailed Tables, 1984*. New York: United Nations, 1986.

UNESCO. *International Standard Classification of Education*. Paris: UNESCO — Division of Statistics on Education, Office of Statistics, March 1976.

UNESCO. *Manual for Statistics on Scientific and Technology Activities*. Paris: UNESCO — Division of Statistics on Science and Technology, June 1984.

UNESCO. *UNESCO Statistical Yearbook*. Paris: UNESCO, multiple years (1975, 79, 83, and 87).

USDA. *Seeds for Planting: US Seed Exports*. Foreign Agricultural Circulars, 1957/87. Washington, D.C.: US Department of Agriculture, Foreign Agricultural Service, multiple years (1957 to 1987).

USDA. *Agricultural Statistics*. Washington, D.C.: US Department of Agriculture, multiple years (1960 to 1985).

Valverde, C. *Estructura y Organización de la Investigación Agrícola en Latino América y el Caribe: Una Visión Regional*. ISNAR Staff Notes 90-90-s. The Hague: ISNAR, 1990.

Venezian, E. *Chile and the CGIAR Centers: A Study of their Collaboration in Agricultural Research*. CGIAR Study Paper No. 20. Washington D.C.: World Bank, 1987.

Venezian, E. L., and W. K. Gamble. *The Agricultural Development of Mexico: Its Structure and Growth since 1950*. New York: Frederick A. Praeger, 1969.

Vocke, G. *Economic Growth, Agricultural Trade, and Development Assistance*. Agricultural Information Bulletin No. 509. Washington, D.C.: USDA, ERS, March 1987.

Vries, B. A. de. "The Plight of Small Countries." *Finance and Development* Vol. 10, No. 3 (September 1973): 6-8.

Wan, H. "Agricultural Research Organisations in Taiwan." In *National Agricultural Research Systems in Asia*, edited by A. H. Moseman. New York: Agricultural Development Council, 1971.

Watson, J. W. "Comparative Study of Agricultural Research Organization and Administration in the Near East Region." Paper prepared for the Workshop on Organization and Administration of Agricultural Services in the Arab States, Cairo, 2-15 March 1964.

Weaver, R. D. *Federal Research and Development and U.S. Agriculture: An Assessment of Role and Productivity Effects*. Department of Agricultural Economics and Rural Sociology Staff Paper 112. University Park: Pennsylvania State University, 1986.

Webster, B. N. *Index of Agricultural Research Institutions and Stations in Africa*. Rome: FAO, n.d..

Welch, F. "The Role of Investments in Human Capital in Agriculture." In *Distortions of Agricultural Incentives*, edited by T. W. Schultz. Bloomington: Indiana University Press, 1978.

Whitehead, A. N. *Science and the Modern World.* New York: Macmillan, 1925.

Wilson, L. A. *Institutional Perspectives on Regional Agricultural Research and Training in the Caribbean: An Alternative Framework for the 21st Century.* St Augustine, Trinidad: The University of the West Indies, June 1985.

Wilson, R. "Risk Measurement of Public Projects." In *Discounting for Time and Risk in Energy Policy*, R. C. Lind et al. Washington, D.C.: Resources for the Future, 1982.

Wise, W. S. "The Shift of Cost Curves and Agricultural Research Benefits." *Journal of Agricultural Economics* Vol. 35, No. 1 (1984): 15-30.

Wise, W. S. "The Calculation of Rates of Return on Agricultural Research from Production Functions." *Journal of Agricultural Economics* Vol. 37, No. 2 (1986): 151-161.

Wise, W. S., and E. Fell. "Supply Shifts and the Size of Research Benefits: Comment." *American Journal of Agricultural Economics* Vol. 62, No. 4 (1980): 837-840.

Wong, L. F., and V. W. Ruttan. "Sources of Difference in Agricultural Productivity Growth Among Socialist Countries." In *Application of Modern Production Theory: Efficiency and Productivity*, edited by A. Dogramaci and R. Fare. Boston: Kluwer Academic Publishers, 1988.

World Bank. *Agricultural Research — Sector Policy Paper.* Washington, D.C.: World Bank, 1981a.

World Bank. *China: Socialist Economic Development—Annex C.* Washington, D.C.: World Bank, 1981b.

World Bank. "Methodological Problems and Proposals Relating to the Estimation of Internationally Comparable per Capita GNP Figures." World Bank, Economic Analysis and Projections Department, Washington, D.C., November 1983. Mimeo.

World Bank. *Price Prospects for Major Primary Commodities — Volume III: Agricultural Raw Materials.* Report No. 814/86. Washington, D.C.: Commodities & Export Projections Division, World Bank, 1986.

World Bank. *World Tables 1988-89 Edition.* Diskette version. Washington, D.C.: World Bank, 1989.

World Bank. *World Development Report.* New York: Oxford University Press, multiple years (1978 to 1990).

World Commission on Environment and Development. *Our Common Future.* The Brundtland Report. Oxford: Oxford University Press, 1987.

Yadav, J. S. P. "Agricultural Education in India." In *Proceedings of Scientific Agricultural Manpower in Asia*, edited by H. P. M. Gunasena and H. M. G. Herath. Sri Lanka: National Agricultural Society, 1985.

Yamada, S., and V. W. Ruttan. "International Comparisons of Productivity in Agriculture." In *New Developments in Productivity Measurement and Analysis*, edited by J. W. Kendrick and B. N. Vaccara. Chicago: University of Chicago Press, 1980.

Yeganiantz, L., ed. *Brazilian Agriculture and Agricultural Research.* Brasília: EMBRAPA, 1984.

Yotopoulos, P. A. "Middle-Income Classes and Food Crises: The 'New' Food-Feed Competition." *Economic Development and Cultural Change* Vol. 33, No. 3 (1985): 461-483.

Young, L., and S. P. Magee. "Endogenous Protection, Factor Returns and Resource Allocation." *Review of Economic Studies* Vol. LIII (1986): 407-419.

Yudelman, M., G. Butler, and R. Banerji. *Technological Change in Agriculture and Employment in Developing Countries.* Development Centre Studies – Employment Series No. 4. Paris: OECD Development Centre, 1976.

Zuidema, L. *Report on Human Resource Management for the National Agricultural Research Board: The Gambia.* Ithaca, New York: Cornell University, 1990.

Author Index[1]

---

[1] Entries are to text citations, not reference items.

Subject Index[1]

---

[1] Region, sub-region and country entries in the appendix on national agricultural research expenditures and personnel estimates are not indexed here.